INTERNATIONAL COOPERATION
FOR THE
DEVELOPMENT
OF SPACE

INTERNATIONAL COOPERATION FOR THE DEVELOPMENT OF SPACE

LANGDON MORRIS

AND

KENNETH J. COX, PH.D.

EDITORS

FOREWORD BY
BRUCE MCCANDLESS II
CAPTAIN, US NAVY (RET.); 24-YEAR NASA ASTRONAUT

AN AEROSPACE TECHNOLOGY WORKING GROUP BOOK

IN PARTNERSHIP WITH

THE INTERNATIONAL SPACE UNIVERSITY

AND

THE INTERNATIONAL INSTITUTE OF SPACE COMMERCE

© **2012 BY EACH OF THE VARIOUS AUTHORS**
Except Chapter 16, also copyright 2010 by its authors.

Cover design by Langdon Morris

Photo Information

Front Cover:
Starfield: Hubble Image (NASA)

Front Cover Insets, Top to Bottom:
ISS Cupola, December 29, 2011 (NASA)
Lights of Earth from ISS (NASA)
Taikonaut Zhai Zhigang Space Walk, September 27, 2008 (CNSA)
Saturn, as photographed by Cassini spacecraft, March 27, 2004 (NASA)
Astronaut Robert L. Gibson (STS-71) shakes hands with cosmonaut Vladimir N. Dezhurov, Mir 18 Commander, June 1995 (NASA)
Atlantis (STS-71) Docks with Mir 18, June 1995 (NASA)

Back Cover Insets:
Center Image - ISS Truss Assembly (NASA)

Left Side Images, Top to Bottom:
Earth Image, Satellite Composite (NASA)
Artist Image of ISS (NASA)
Artist Image of Curiosity Spacecraft on Mars (NASA)

Right Side Images, Top to Bottom:
Mission Patch: ISS Expedition 27, 2011 (NASA)
Mission Patch: Soyuz-Apollo VII, 1975 (NASA)
Space Shuttle (NASA)

Copy Editor *Extraordinare*: Diane Castiglioni

All rights reserved.

ISBN 9 781478 186236

DEDICATION

THIS BOOK IS DEDICATED TO EVERYONE WHO ASPIRES TO A WORLD OF PEACE AMONG NATIONS, CULTURES, PEOPLES, FAMILIES AND WITHIN OURSELVES, BOTH ON EARTH AND IN OUR WONDERFUL ADVENTURES BEYOND.

•••

THE EDITORS OF THIS VOLUME AND THE MEMBERS OF THE AEROSPACE TECHNOLOGY WORKING GROUP WISH TO THANK EACH OF THE AUTHORS AND ADVISORS WHO GRACIOUSLY CONTRIBUTED THEIR EFFORTS TO THIS WORK.

IN THE SAME SERIES

Beyond Earth
The Future of Humans in Space
Krone, Cox, & Morris, Editors
2006

Living in Space
Cultural and Social Dynamics,
Opportunities, and Challenges
in Permanent Space Habitats
Bell & Morris, Editors
2009

Space Commerce
The Inside Story by the People
Who Are Making It Happen
Morris & Cox, Editors
2010

Our hope is that these books will contribute to humanity's continuing quest to explore and develop our universe *and* ourselves, and in so doing will deepen our appreciation and understanding of the opportunities that lie before us.

TABLE OF CONTENTS

THE AEROSPACE TECHNOLOGY WORKING GROUP	V
THE INTERNATIONAL SPACE UNIVERSITY	VI
THE INTERNATIONAL INSTITUTE OF SPACE COMMERCE	VII
PREFACE By Reinhold Ewald	VIII
PREFACE By Edgar Mitchell	X
FOREWORD By Bruce McCandless II	XI

1. INTRODUCTION: 3
   THE CHARACTER OF COOPERATION IN SPACE
   By Langdon Morris and Kenneth J. Cox, Ph.D.

PART I: COOPERATION – PAST, PRESENT, FUTURE

2. BROADENING THE BASE: 19
   COOPERATION AS A SPRINGBOARD FOR NEW PARTICIPANTS IN THE SPACE SECTOR
   By Michael Simpson, Ph.D.

3. SPACE AS A CATALYST FOR INTERNATIONAL   43
   POLITICAL COOPERATION
   By Walter Peeters, Ph.D.
   APPENDIX: A DRAFT MARS CONVENTION, BY OLIVER INGOLD   57

4. PUBLIC SPACE LAW, THE LEGAL PRACTITIONER,   67
   AND THE PRIVATE ENTREPRENEUR:
   DISTINGUISHING WHAT "OUGHT TO BE"
   FROM "WHAT IS"
   By George S. Robinson

5. THE SPACE DATA ASSOCIATION   81
   A NEW MODEL FOR INTERNATIONAL COOPERATION
   IN SPACE
   By Stewart Sanders and Christopher Stott

6. INTERNATIONAL SPACEPORT CHALLENGES: 89
   THE NECESSITY OF FORMULATING THE POLICY
   FOR DEVELOPING THE CRITICAL TECHNOLOGY
   By Thomas E. Diegelman

7. INTERNATIONAL SOCIAL COOPERATION IN 121
   SPACE AWARENESS:
   YURI GAGARIN, YURI'S NIGHT AND *FIRST ORBIT*
   By Christopher Riley, Ph.D. and Christopher Welch, Ph.D.

8. THE SCIENCE AND TECHNOLOGY OF 143
   SPACE EXPLORATION:
   LEVERAGING CYBERSPACE FOR GLOBAL
   COOPERATION IN THE DEVELOPMENT OF SPACE
   By Rita Lauria, Ph.D.

PART II: COOPERATION AMONG NATIONS AND REGIONS IN SPACE

9. AUSTRALIA'S PLACE IN SPACE 165
   HISORICAL CONSTRAINTS AND FUTURE OPPORTUNITIES
   By Brett Biddington

10. CURRENT STATUS AND FUTURE DEVELOPMENTS 209
    IN CHINA'S SPACE PROGRAM:
    INTERNATIONAL SPACE COOPERATION
    By Le Wang

11. THE LEADERSHIP COMPETITION 243
    BETWEEN JAPAN AND CHINA
    IN THE EAST ASIAN CONTEXT
    By Kazuto Suzuki, Ph.D.

12. THE EUROPEAN – AFRICAN PARTNERSHIP 261
    IN SPACE APPLICATIONS
    By Christina Giannopapa, Ph.D.

13. EUROPEAN SATCOM POLICY 287
    A TOOL OF INTERNATIONAL COOPERATION
    BETWEEN EUROPE AND AFRICA
    By Veronica La Regina

14. INTERNATIONAL COOPERATION IN SPACE, 313
    FROM USSR TO RUSSIA
    By Olga Zhdanovich, MSc

15. THE CAPABILITY CRITERION 345
    INTERNATIONAL COOPERATION AND NATIONAL
    PRIORITIES IN SPACE DEVELOPMENT
    By Captain Christopher M. Stone and Captain Brent D. Ziarnick

16.	Perspectives on Improving International Space Cooperation A Permanent Solution to Global Crisis By James D. Rendleman and J. Walter Faulconer	363
17.	Modeling the Nation-State in Space Commerce By Thomas E. Diegelman and Thomas C. Duncavage	393
18.	Overview of the Space Programs of Brazil, India, and Israel Compiled by the editors	413

Part III: ... and Beyond

19.	The 100 Year Starship Endeavor By Jeffrey Nosanov and Michael Potter	425
20.	An Incomplete Species The Continuing Evolution of Humanity, Society, Technology, and Law By George S. Robinson	437
21.	Conclusion: Prospects and Ambitions for Cooperation By Langdon Morris and Kenneth J. Cox, Ph.D.	461

Index 485

The Aerospace Technology Working Group

The Aerospace Technology Working Group, also known as ATWG, is an independent space policy research and innovation group led by seasoned professionals in aerospace and other fields who seek to further humanity's exploration of space while simultaneously benefiting people on Earth.

The ATWG was instituted by NASA Administrator Richard Truly in 1990 as an independent body to perform future planning for the nation's space efforts. Initially, the ATWG began identifying and seeking improvements in both existing and developing space systems through the planned application of emerging technologies and the development of new ways of doing business, including the application of distributed missions and innovative strategic concepts in operations.

Today, the ATWG is an independent entity, using semi-annual and regional Forums, technical and strategic dialogs, personal interactions, books, articles, and speeches to explore topics pertinent to developing a space faring people and prepare policy recommendations for national and global leaders.

Using the organization's substantial base of management, engineering and scientific expertise, the ATWG also provides strategic and technical advice, public speakers, and consulting teams to address specific aerospace tasks and broad conceptual and philosophical questions. The ATWG collaborates actively with other space-related national and international organizations.

In addition, the ATWG places special emphasis on promoting and stimulating education in the sciences, mathematics, the engineering disciplines, and other technical areas.

Participants in ATWG include experts from throughout NASA, aerospace contractors, systems suppliers, entrepreneurial businesses, professional societies, universities, and government agencies including the DOD, FAA, and DOE.

You can learn more about the ATWG at www.atwg.org.

The International Space University

The International Space University provides graduate-level training to the future leaders of the emerging global space community at its Central Campus in Strasbourg, France, and at locations around the world.

In its two-month Space Studies Program and one-year Masters program, ISU offers its students a unique core curriculum covering all disciplines related to space programs and enterprises – space science, space engineering, systems engineering, space policy and law, business and management, and space and society. Both programs also involve an intense student research team project providing international graduate students and young space professionals the opportunity to solve complex problems by working together in an intercultural environment.

In addition, the school offers an Executive MBA focusing on the space sector, a Technical Masters in Space Systems Engineering in cooperation with The Stevens Institute of Technology, several short courses including the Executive Space Course, and an innovative 5 week program focusing on space issues of particular concern to the Southern Hemisphere in Adelaide, Australia in cooperation with the University of South Australia.

Since its founding in 1987, ISU has graduated more than 3000 students from over 100 countries. Together with hundreds of ISU faculty and lecturers from around the world, ISU alumni comprise an extremely effective network of space professionals and leaders that actively facilitates individual career growth, professional activities and international space cooperation.

You can learn more about ISU at www.isunet.edu/.

The International Institute of Space Commerce

The International Institute of Space Commerce, or simply 'the Institute,' has been established on the Isle of Man through a partnership between the International Space University (ISU) and the Manx Government.

The Institute's mission is to become the leading think-tank in the study of the economics of space. It is intended to be the intellectual home for the industry and space academia around the world for which it shall perform studies, evaluations and provide services to all interested parties with the ultimate aim to promote and enhance world's space commerce to the general public. The vision for the Institute is to act as a resource for all, being an international and nonpartisan think-tank, drawing upon new ideas and solutions to existing and future problems the space industry faces by drawing together experts from academia, government, the media, business, international and non-governmental organizations, most notably those from the ISU and its extended network of people and resources. The aim of the Institute is to broaden the professional perspective and personal understanding of all those involved in the study, formulation, execution, and criticism of space commerce.

The Institute is a Not for Profit Foundation and has been located at the site of the International Business School (IBS) on the Isle of Man to capitalize on the Isle of Man's growing importance and position in the world's space industry.

You can learn more about the IISC at www.iisc.im/about.asp.

Preface

For a long period there was a strong push for autarky in space projects, which prevented people from seriously considering worldwide cooperation. Today, in an increasingly globalized world, the stereotypes of nationality do not make sense anymore. In fact, there are considerable advantages to be received by internationally sharing the tasks in cooperative projects, and employing the people available for a project according to the tasks in which they excel, no matter where they come from.

But to achieve cooperation one first needs to build up trust, and especially in technical projects, this requires never-tiring preparedness to settle the small and larger conflicts about the interfaces both between the hardware and the people involved in a constructive way.

Space has brought us to master this challenge and, after the Mir Station, the International Space Station is the visible sign of a peaceful exploration project uniting many cultures, nations, and individuals around the world in cooperation.

Assembling and operating the International Space Station means a daily effort to employ the best trained persons on board using the appropriate tools coming from a commonly shared toolbox, to maintain or repair elements and subsystems built by one of the Space Station partners, using spare parts delivered in one of the cargo spacecraft coming from four different launch sites around the world. Logistics and operations of ISS have truly united the World around the objectives of bringing forward our knowledge of the physical world around us, and improving our abilities to explore further.

ASE, the Association of Space Explorers, has in this context also performed a pioneering role. Joined by a similar drive and vision, astronauts and cosmonauts from all over the world, and now also taikonauts, have joined this organization irrespective of cultural and political boundary conditions. All of them have found a place to freely exchange their thoughts and ideas about the benefits of Space, and have initiated a forum of knowledge transfer.

In order to clearly illustrate ASE's intercultural character we have decided to put at the top of our ASE website[1] the wise words of our Arab astronaut colleague Sultan bin Salman bin Abdul-Aziz al Saud:

On the first day, we all pointed to our countries...
On the third day, we all pointed to our continents...
By the fifth day, we were aware of only one Earth...

Sultan bin Salman al-Saud, STS-51G Astronaut

May this global view become the view of all mankind, and may this book provide a meaningful contribution to achieve such a vision!

Dr. Reinhold Ewald
ESA astronaut, Soyuz TM-25/Mir
Founding member of ASE Europe
Member of ISU Adjunct Faculty
August 1, 2012
© Reinhold Ewald, 2012. All Rights Reserved.

---

[1] www.space-explorers.org

Preface

Throughout the twentieth century and continuing into the twenty first, every measure of human activity and productivity has exhibited exponential growth. New technologies abound in every field of endeavor. And although our technologies are magnificent and provide enhanced standards of living for most human families, the toll on the basic materials of production is overwhelming. Clearly, this is a condition that cannot continue unabated and unmanaged in a finite physical space, namely our own home planet. We are systematically overwhelming our finite local resources for continuous growth.

Going out from our planet to seek other possible sources and resources in our solar system may help sustain life here on Planet Earth for a significant period if we humans work together and cooperate in finding, exploring and sharing the potential.

However, even that will not be enough in the longer time frame. In due course our sun will burn out, and although that is in the far distant future it is destiny, as all stars, like every living system, have finite lifetimes.

Thus, we on planet Earth must become universal citizens in the long run. Going into space, exploring our Moon and other planets is just the beginning; we must eventually think in terms of other solar systems. Whereas our forebears of previous millennia did not need to think in such terms, in the twenty first century it is a necessity that we begin to consider travel and eventual colonization of far off places. Survival depends upon it.

We have just begun the journeys into space, and as we do so we must organize this as work not just involving nations, but rather in terms of sustaining and extending a planetary civilization far beyond Earth, and into a cosmos that we have just begun to understand.

Edgar Mitchell
Apollo 14 Astronaut
August 1, 2012
© Edgar Mitchell, 2012. All Rights Reserved.

Foreword

BRUCE MCCANDLESS II
CAPTAIN, US NAVY (RET.); 24-YEAR NASA ASTRONAUT

© Bruce McCandless II, 2012. All Rights Reserved.

"Cooperation in Space" has been glibly invoked as the solution to a myriad of difficulties in realizing major under-funded objectives. This work, by authors from the Aerospace Technology Working Group, the International Space University, and the International Institute of Space Commerce, comprehensively examines the prescription from a wide variety of perspectives.

In Chapter 15, Captains Christopher Stone and Brent Ziarnick present their "Capability Criterion" for cooperation, which proposes that "An actor should engage in international cooperation if and only if completion of the cooperative effort's objective will result in the actor achieving a higher level of enhanced space capability than it would be able to achieve by acting alone while exerting a reasonable but strenuous level of political will and resources." They then proceed to examine its three major elements in detail.

In Chapter 2, former ISU President Michael Simpson also takes a solidly analytical look at the legitimate reasons for cooperation, asserting that these include a reasonable prospect of economic return to all partners, political support and resolve rooted in more than just economic return, and support from partners with experience or niche expertise.

Note that the "economic return" may in fact be much broader than financial gains, and may include the acquisition of technology and/or development of a participant's business and industrial bases. Later in the volume Kazuto Suzuki, writing in Chapter 11 on Competition in an East

Asian Context, persuasively contends that *by denying China* (the People's Republic of China) access to our space technologies under the ITAR (International Traffic in Arms Regulations), we have stimulated that country to invest heavily in the independent development of these technologies, which are then freely used by China as elements of political and economic barter to "have-not" countries – countries from which the ITAR also bans receipt of U. S. space technologies.

Brett Biddington in Chapter 9 characterizes Australia's approach to space as intensely pragmatic and collaborative. "Grand plans" for space exploration and space industry development have been deferred in favor of more "down to Earth" (no pun intended) endeavors based upon Australia's unique and strategic geographical location, fostering and/or maintaining of alliance relationships, and obligations under the Outer Space Treaty. Geocircumferential distribution of certain US facilities (*e.g.*, Deep Space Network & Manned Spaceflight Flight Network antennas) has been necessary in order to achieve 100% coverage of the weak signals from lunar and interplanetary missions. Setback from coastlines, on the other hand, has been a consideration when siting facilities supporting satellites requiring discrete download of data without easy interception by "trawlers" or other marine assets.

Conversely, Russia's *lack of* low latitude human space launch facilities drove the inclination of the ISS up to 51.6 degrees, from a KSC-minimum-energy inclination of 28.5 deg, with a considerable performance penalty from the latter launch site.

Suzuki and others in several of the chapters discuss the constraints of ITAR in greater depth. Under ITAR, all satellite and space-related technologies (including "dual-use" technologies) have been placed on the US Munitions List (USML) – based upon their intended end-use (space). This means that export is prohibited, and in order to export same both a waiver from the USML and a license from the State Department must be obtained. This is a tedious and time-consuming process that casts a severe damper upon the sale of US space-related products, even things such as communications satellites, once regulated somewhat loosely by the Commerce Department.

In practice, ITAR hampers even *the domestic sale* of satellites by impeding the flow of technical data to foreign (*e.g.*, London) insurance underwriters. This discussion is continued by Capts. Stone and Ziarnick, who point out that this has led to the development in India, Europe, and China of "ITAR-free" satellites that may be sold freely without US involvement, and, consequently, to the dramatic erosion of the once-dominant position held by the US communications satellite vendors in the world market. In this context, it's clear and perhaps sadly ironic that the application of ITAR to these technologies has actually been counter to US interests.

In Chapter 10, Le Wang discusses interests and policies of the People's Republic of China. She notes that China's fundamental interests are in developing their economy and in realizing its modernization, and that the priority aims and principles of China's space program are to serve the country's national interests and to implement its development strategy, a view that contrasts somewhat with Suzuki's characterization of China's space activities and its decision making process as "opaque at best."

An emerging strong driver for international cooperation is the problem of orbital debris. Once discarded, pieces of interstage structures, entire upper stages, insulation fragments, and even paint flakes present hazards to the activities of *all* spacefaring nations. The decades to centuries-long persistence of these items in MEO (Medium-altitude Earth Orbit) and higher exacerbate the severity of the problem. Biddington points to the vigorous and swift Australian reaction to the 2007 destruction by the PRC of one of its own defunct satellites in-orbit, resulting in a very large in-space debris field. Others note that China, especially following this test, has been especially supportive of international debris alleviation efforts.

James Rendleman and Walt Faulconer in Chapter 16 posit that, from the US perspective, cooperation *actually costs more*, both financially and in terms of obstacles and other impediments to overcome. They then discuss the rationales for doing it anyway! They note that such collaborative efforts have come at the detriment of other highly productive US-only undertakings. It is suggested that NASA has an institutional belief in the promise of international cooperation, and Administrator Charles Bolden is cited as affirming that greater international cooperation is coming. Certainly the survival of the ISS at one point (October 21, 1993) rested on a single vote in the House of Representatives – and reportedly "squeaked through" on the basis of some Members wishing to honor international commitments.

Chapters on Space Law *vs.* Cyberlaw *vs.* "Post-human" Law as well as the "essence" of a species (*e.g.*, *homo sapiens*) and on "The 100-year Starship Endeavor" by DARPA (Defense Advanced Research Projects Agency) round out this robust work. Not a space trip, this latter effort looks principally at the organizational challenges to mounting a multi-generational effort to build and launch a human starship mission.

All totaled, the editors and authors of this work have done an outstanding job of pulling together a transcendent treatment of a very complex subject. But, this preface is neither an abstract nor a "Cliffs Notes;" it is rather an appetizer, and an invitation to enjoy the body of the book. The reader who takes the time to digest these insightful writings will be well nourished.

Bruce McCandless II
STS 31 and STS 41B Astronaut
August 1, 2012

BRUCE MCCANDLESS II

Bruce McCandless II is a graduate of the US Naval Academy and a Navy fighter pilot who was selected as a NASA astronaut in 1966. He served through the Apollo and SKYLAB Programs in a variety of supporting roles and flew on two Space Shuttle missions, STS 41-B and STS-31. During most of his career with NASA he was deeply involved in the development of astronaut maneuvering units, and was the first to fly the Manned Maneuvering Unit, untethered, in space, in 1984. In connection with the Hubble Space Telescope, Bruce had the Astronaut Office responsibility for insuring its on-orbit serviceability and flew on the Deployment Mission in 1990.

After retiring from the Navy he held a variety of middle management positions with Lockheed Martin Space Systems Company before transitioning to a "casual" employment status in 2005. In addition to his Annapolis education, Bruce earned an MSEE from Stanford University and an MBA from the University of Houston, Clear Lake Campus.

INTERNATIONAL COOPERATION FOR THE DEVELOPMENT OF SPACE

Chapter 1

Introduction
The Character of Cooperation in Space

Langdon Morris
PwC

And

Kenneth J. Cox, Ph.D.
ATWG Founder

© Langdon Morris and Kenneth J. Cox, Ph.D., 2012. All Rights Reserved.

What is Your Vision?

When you think of space, what comes to your mind?
  There are, of course, many possible perspectives. Do you imagine great voyages across incomprehensible distances? Perhaps space calls forth great scientific and engineering challenges in your mind, designing and constructing spacecraft that will travel billions of kilometers. Perhaps you think of commerce, and the enormous opportunities for successful business endeavors. In this regard you might be thinking of today's vast

satellite communications industry, or perhaps it is tomorrow's search for raw materials and energy throughout our solar system that interests you.

Is it physics that you think of first, the mysteries of the universe as evoked by the billions of galaxies, black holes, or the search for fundamental particles and plausible explanations for the very origins of the universe?

Or do you think of the great cinematic works of science fiction, the space epics such as *2001: A Space Odyssey, Star Wars*, or *Star Trek*?

You could also think about the role that space plays, and will in the future play, in the development of our home planet, in the sciences of the climate, agriculture, hydrology, soils, forestry, and oceanography.

And perhaps you also consider the role that space plays, and may yet play, in the political, social, and cultural development of our human civilization and of our planet.

Maybe you think of the role that space will play in human evolution, and the possibility or necessity that we will one day venture in vast numbers beyond our home planet as a matter of species survival, exactly as Edgar Mitchell suggests in his Preface.

Space is indeed all of these things, and more, an amazing canvas of possibilities upon which each of us imagines our own visions of compelling, thrilling, and rewarding futures.

And for some of us, of course, it is more than our imaginations that are engaged, for there are hundreds of thousands of people working all around our small globe whose daily endeavors have to do with making such visions into reality, in business, in science and technology, in engineering, in exploration, in literature and film.

In our previous books on the meaning and significance of space we have explored a broad range of these themes and topics, starting first in *Beyond Earth* with a broad examination of many different issues, and then in our second volume with a collection of writings about the challenges and opportunities that *Living in Space* will present. In the third volume in this series we examined the future of business in space, *Space Commerce*.

The underlying premise of each of these books is that space is not only an irresistible lure for our imaginations, but also an inevitable destination for our serious goals and ambitions. Collectively, the many authors who have contributed their works to this series are committed to the realities of life, science, culture, and business in space, and we all agree that extending civilization into space is both a valid destination that will ultimately be home to millions and billions of us, and so in a very real way it also will become humanity's destiny.

And yet it is, frankly, a strange thing that so many of us see so clearly what is necessary and what indeed will happen even as we also acknowledge that the tiny steps of progress that we've made so far are just that, tiny steps only.

So what stands in the way, what lies between our current reality and the far greater vision? One factor of great significance is the trying environment of scarcity and competition for resources that characterizes our current economic and political realities. We don't seem to have the resources, nor do we have, in many cases, the political will.

Indeed, if we think of the journey to space in pursuit of any purpose, scientific, commercial, or otherwise, as a matter of this country's space program or that nation's investment, in isolation from one another, we consign ourselves to the inevitability of a wait that could continue for centuries, if indeed we have that long.

However, we get a much different viewpoint when we consider the venture into space not as a group of projects of nations, but as one project of humanity. Hence this book, *International Cooperation for the Development of Space*, seeks to examine the tremendously greater benefits that come when we think beyond the boundaries of national interests and consider the venture and adventure into space as a matter of shared ambition, shared knowledge, shared risk, and shared benefit across all of humanity. This is the character of cooperation in space to which we aspire.

And indeed, we see examples everywhere we look. As you will discover in these pages, cooperation is now the norm among the 22 nations that have national civil space programs with budgets in excess of US$ 10 million,[1] 9 of them with budgets more than US$ 1 billion. The military budgets, of course, are not disclosed, and in many cases are much more.

Unlike most military efforts which tend to be nationalist and very secretive, cooperation among nations throughout the civil space sector is considered natural, desirable, and indeed essential to achieving the individual goals of each nation and the collective goals of humanity.

The most obvious example, The International Space Station, is a project of 16 nations, and so, too, are many of the major projects and programs among the current spacecraft and ventures, including the Saturn observation crafts Cassini-Huygens, numerous Earth-orbiting commercial and scientific satellite systems engaging nearly all nations, and of course all ESA projects, which are by definition cooperative among the 19 members.

---

[1] http://en.wikipedia.org/wiki/List_of_space_agencies. Accessed July 14, 2012.

ORGANIZATION OF THIS BOOK

The authors of this book are experts in their fields. They are university presidents and professors, chief executives and entrepreneurs, scholars, aerospace engineers, award winning artists, military officers, professional investors, and yes, astronauts. And they are also leaders, visionaries who have thought long and hard about the future, and who have committed a significant portion of their professional lives to the realization of these visions. We share a profound commitment to the significance of space for the present and the future of humanity.

In Part I, our authors offer seven perspectives on current and future of international cooperation, its prevalence, significance, and importance. We examine the patterns, methods, and degrees of integration reflected in the organization of today's cooperative activities, as well as the role of law in abetting (or impeding) such endeavors, an example of commercial cooperation, as well as challenges to commercial cooperation. We also learn about the cultural phenomenon and media diffusion of space-relate social events (Yuri's Night), and future possibilities for cooperation through virtual training.

In Part II, we explore cooperation from the perspective of nations and regions, with specific chapters on Australia, China, Japan, Russia, Europe (as noted above, the European Space Agency currently has 19 members), the United States, and across all of Africa.

> (We regret that we were unable to obtain chapters on India, Israel, or Brazil, but we will address the efforts of these nations in summary form in Chapter 18 at the conclusion to Part II.)

In Part III we summarize with three chapters that examine the longer term future by considering the evolutionary perspective, a 100 year goal, and finally a Conclusion that reiterates many of the key themes, topics, and insights of the previous 460+ pages, and offers its own perspectives about the next steps forward.

WHAT'S IN THE NEWS?

The last few years has seen a mixture of good and bad news affecting the space community.

The conclusion of the US Space Shuttle program was accompanied by cutbacks and layoffs at NASA, and contributes to a general sense that the American space program has lost its way.

Despite this setback, though, there is plenty of good news. To get a sense of how events are flowing, we have complied a selection of eighteen recent stories. We understand that this is by no means a complete set of every story that the space community may be interested in, and we also recognize that when you read this section, the stories here will no longer be "news." But we include them nevertheless because we found it interesting and revealing to see all of these stories together in one place, as it gives a sense of some overall trends in globalization of the space endeavor, which naturally also contributes to our theme of international cooperation.

1. April 2011: GPS in health care, helps treat asthma

"Spiroscout is an inhaler with a built-in Global Positioning System locator and (in advanced models) a wireless link to the internet. Whenever someone uses the inhaler, it broadcasts the location and time to a central computer. Asthmapolis plots and analyses the data, and sends weekly reports to participating patients and their doctors summarising the observations and making recommendations. That is useful for the individuals involved, since it may illuminate patterns of which they were unaware (the proximity of a particular kind of crop, for example). It could also help doctors identify those patients whose asthma is not under proper control. Use of the inhaler more than a couple of times a month suggests there is something wrong, and that the patient's medication may need to be changed. Patients do not, however, always report such problems, and so do not get the right drugs. The big public gain, though, will come from pooling all the data from the inhalers, once they have been suitably anonymised. That will open the way for a much more detailed analysis of what is going on, and may allow the triggers to be identified and ranked in order of importance."[2]

2. July 2011: The last Space Shuttle Mission

"Atlantis rocketed into orbit today at 11:29 a.m. EDT and is flying at 17,500 mph around the Earth. The mission, STS-135, will catch up with the International Space Station in two days. The Space Shuttle launch marks the last in NASA's history, closing out a government-funded space program that lasted 30 years. 'The shuttle's always going to be a reflection of what a great nation can do when it commits to be bold and follow through,' said astronaut Chris Ferguson, commander of the mission, from

---

[2] *The Economist* Magazine. "Inhaling information: How to collect data on asthma while, at the same time, treating it." April 7, 2011.

the cockpit of Atlantis just before pushing into space atop a billowing cloud of fumes. 'We're completing a chapter of a journey that will never end. Let's light this fire one more time, and witness this great nation at its best.' During their 12-day mission, Ferguson and his three crewmembers — veteran astronauts Doug Hurley, Sandra Magnus and Rex Walheim — plan to wrap up construction of the space station. They'll deliver a new room crammed with a year's worth of food, water and other supplies and perform a suite of experiments in orbit."[3]

3. NOVEMBER 2011: THE FIRST SCIENTIFIC PAPER SUBMITTED FROM OUTER SPACE

Soon after editor-in-chief of European Physical Society Journal (EPL) Michael Schreiber joked about publishing a paper from the ISS, in fact the first scientific paper submitted from space has been received and published by EPL. "On October 27, 2011 Russian cosmonaut Sergey Alexandrovish Volkov submitted a manuscript (published online 11 November) about complex plasmas in microgravity conditions from the ISS." Volkov was stationed on the ISS at the time of the submission.[4]

4. JANUARY 2012: SATELLITE DATA IN VIDEO GAMING

Data from the Aster Global Digital Elevation Map, a joint satellite project of Japan's Ministry of Economy, Trade, and Industry and NASA, has been used to provide a three-dimensional model of K2, the world's second tallest mountain on the border of Pakistan and China, for video game maker EA as the basis of the latest version of its SSX snow boarding video game. But not just K2. Mountain ranges around the world are rendered inside the video game. "Riders will now be able to helicopter-drop onto the icy heights of Kilimanjaro, grind on the Great Wall of China, descend Everest, deploy a wing suit, and race in real time against 100,000 other players around the world – all set against a backdrop that's as realistic as anything on Google Earth."[5]

5. FEBRUARY 2012: SCIENTISTS IN THE US STATE DEPARTMENT

"In the late 1990s, science and engineering leaders were deeply concerned that the US State Department lacked the scientific expertise that would be needed for the 21st century. Now, after a sustained effort, State

---

[3] Mosher, Dave. "The Last Space Shuttle Launches Safely Into Orbit." *Wired Science*, July 8, 2011
[4] *Science* Magazine, November 18, 2011, Vol 334, p 882.
[5] Goldstein, Melissa. "Shred Everest." *Wired* Magazine. January 2012. P. 60.

has built significant scientific strength and a promising capacity for science diplomacy. ... Starting in 2000 with the appointment of veteran scientist-diplomat Norman P. Neuriter, four senior scientists have been appointed to 3-year terms as science advisor to the Secretary of State. Fellowship programs now bring dozens of scientists every year to the State Department and the US Agency for International Development."[6]

6. FEBRUARY 2012: SWISS SPACECRAFT TO CLEAN UP SPACE DEBRIS

Space researchers in Switzerland are seeking funding to build a spacecraft, dubbed ClearSpaceOne, that would help reduce space debris in orbit around Earth. Researchers at the Swiss Space Center at the École Polytechnique Fédérale de Lausanne have been working on the necessary technology for 3 years." The satellite would grab an unneeded satellite and drag it down into the atmosphere to burn up upon reentry. The first target is expected to be a Swiss picosatellite. Center Director Volker Gass commented, "Switzerland is a country that likes to keep things clean, so we decided to first get our own satellite down."[7]

7. APRIL 2012: US MILITARY BUYING $1 BILLION OF COMMERCIAL SATELLITE BANDWIDTH

80% of the bandwidth used by the US Department of Defense is provided by commercial satellite operators. Further, the US military is expected to spend up to $1 billion to include its own equipment on commercial launch payloads over the next decade. "The requirement for bandwidth is insatiable," according to Jim Simpson, vice president of business development for Boeing Space and Intelligence Systems, largely driven by soaring demand by drone video and radio intelligence over Middle Eastern war zones. The $1 billion is calculated at 12 payloads at $82.5 million each.[8]

8. APRIL 2012: DISAGREEMENT OVER US FUNDING FOR R&D

"With the US budget under intense pressure, lawmakers increasingly are looking to save money by cutting federal spending on research and development. But at two Capitol Hill briefings, experts offered compelling

---

[6] "Experts See Progress, Challenges in Advancing Science Diplomacy." *Science* Magazine, Vol. 335. February 24, 2012. p 935.
[7] *Science* Magazine, February 24, 2012, Vol 335, p 896.
[8] McGarry, Brendan. "Military may spend $1 billion to get its payloads into space." *San Francisco Chronicle*, April 24, 2012. p D2.

evidence that federal investment can yield enormous dividends and warned that cuts to R&D could hinder future economic growth. 'We have very powerful evidence of the productivity' of R&D spending, said University of California-Davis sociologist Fred Block. 'When we're cutting back on R&D spending we are eating our seed corn, and upsetting the possibilities of future economic growth and development.' Under President Barak Obama's 2013 budget proposal, R&D investment would increase by 1.2% over 2012 funding levels." The House of Representatives proposed budget, however, would reduce R&D spending to 8% less than the president's proposal.[9]

9. May 2012: Space suit for Mars developed in Austria

The Austrian Space Forum has developed the Aouda.X Mars Space Suit Simulator to enable scientists to study the dynamics of suit design for an eventual voyage to the red planet. On April 28, 2012 an international team of researchers tested the suit in a mountain cave in Dachstein, Austria.[10]

10. May 2012: SpaceX Dragon docks at ISS

"High above northwestern Australia, a robotic arm on the International Space Station grabbed onto a cargo capsule floating 10 meters away. With that penultimate act, the Space Exploration Technologies Corporation of Hawthorne, Calif., or SpaceX, made history as the first private company to send a craft, the Dragon, to the station. The grab — which NASA refers to as a grapple — occurred at 9:56 a.m. Eastern time on Friday. 'It looks like we've got us a Dragon by the tail,' said Donald R. Pettit, the NASA astronaut on the station who was operating the robotic arm.... SpaceX launched the Dragon capsule on top of its Falcon 9 rocket."[11]

11. June 2012: Cassini spacecraft finds an ocean on Saturn's Titan

The Cassini space probe orbiting Saturn's moon Titan has provided gravitational data that scientists on Earth have used to determine the

---

[9] Lane, Earl and Becky Hamm. "The Payoff of Federal R&D: iPod, Google, and Human Genome Project." *Science* Magazine, Vol 336. April 27, 2012. P. 433.
[10] *Science* Magazine. "That Age-Old Question: What to Wear on Mars." May 11, 2012, Vol 336, p 656.
[11] Chang, Kenneth. "First Private Craft Docks With Space Station." *New York Times*, May 25, 2012.

existence of an ocean surrounding the moon beneath 100 kilometers of ice. "It's an enormous technical achievement," according to planetary physicist Francis Nimmo of University of California at Santa Cruz.[12]

12. JUNE 2012: CHINESE TAIKONAUTS DOCK 2 SPACECRAFT IN ORBIT; FIRST CHINESE WOMAN IN SPACE

"Chinese astronauts carried out a manned docking with an experimental space module on Monday, the latest milestone in China's ambitious campaign to build a space station. The Shenzhou 9 and its three-person crew, which includes China's first woman in space Liu Yang, linked with the Tiangong (Heavenly Palace) 1 module, with state television showing the pictures live. Almost three hours later, the blue jumpsuit-wearing mission commander, Jing Haipeng, entered the module followed by colleague Liu Wang and Liu Yang, the first time China has been able to transfer astronauts between two orbiting craft. Rendezvous and docking exercises between the two vessels are an important hurdle in China's efforts to acquire the technological and logistical skills to run a full space lab that can house astronauts for long periods."[13]

13. JUNE 2012: VOYAGER PASSES HELIOSPHERE BOUNDARY AT 18 BILLION KILOMETERS FROM EARTH

The Voyager spacecraft, now in its 35th year of flight, is estimated to be about 18 billion kilometers from Earth. "An increase in cosmic rays hitting the spacecraft could mean that it's leaving the heliosphere and entering interstellar space."[14]

14. JULY 2012: VIRGIN GALACTIC SET TO FLY BRANSON'S FAMILY

"The first space flight of Richard Branson's Virgin Galactic venture will be a family affair: The billionaire adventurer confirmed that he will be joined by his two adult children." Virgin says it has booked 529 paying passengers, which happens to be one more than the total number of people who have traveled to space since Yuri Gagarin's 1961 flight. The cost of the two hour trip on Virgin Galactic is $200,000, and includes a one week

---

[12] Kerr, Richard A. "Cassini Spies an Ocean Inside Saturn's Icy, Gassy Moon Titan." *Science* Magazine, Vol 336, June 29, 2012. P 1629.
[13] Martina, Michael and Ben Blanchard, "China astronauts complete successful space docking." Reuters, June 18, 2012.
[14] *Science* Magazine. "18 billion kilometers." June 29, 2012. Vol 336, p 1626.

training program prior to the flight at Virgin's New Mexico Spaceport location.[15]

15. JULY 2012: EXOPLANET SCIENCE TEACHES ABOUT EARTH'S PHYSICS

"Discovered in 2005 and orbiting a star 63 light-years from us, exoplanet HD 189733b doesn't generally get much attention from climate scientists on Earth. However, thermal observations of the gas giant have suggested that the planet's hot spot has shifted about 30 degrees aware from the 'high noon' position, suggesting strong winds." A study of HD 189733b led planetary scientists Adam Showman and Lorenzo Polvani ... "to the discovery that the models currently used to study terrestrial climate had left out some key physics concerning the transfer of momentum. 'We're learning about Earth by studying exoplanets and about exoplanets by studying Earth' commented Yohai Kaspi of the Weizmann Institute of Science in Israel.'"[16]

16. AUGUST 2012: INDIA ANNOUNCES MISSION TO MARS FOR 2013

"Top Indian space department officials revealed on Thursday that India would launch a Mars Mission in 2013. The country intends to undertake an orbital probe mission around Mars in attempts to grasp deeper understanding of its geology and climate. The 2013 mission will be a milestone in India's greater plans to launch its first ever manned mission in 2016. This is a representation of the increased aggression in India's space programs. According to reports, an Indian Polar Satellite Launch Vehicle rocket will be used to transport the orbiter spaceship. This 320 ton rocket will blast from the state run Indian Space Research Organization launch site in Sriharikota, Andhra Prandesh."[17]

17. AUGUST 2012: CURIOSITY LANDS ON MARS

"In a flawless, triumphant technological tour de force, a plutonium-powered rover the size of a small car was lowered at the end of 25-foot-long cables from a hovering rocket stage onto Mars early on Monday

---

[15] Associated Press. "Bransons will ride on company's debut flight." *San Francisco Chronicle*, July 12, 2012. P D6.
[16] Klintisch, Eli. "Earth and Planetary Scientists Search for Common Ground." *Science* Magazine, Vol 337. July 13, 2012. P 145.
[17] Yieke, Lennox. "India To Step On Mars Next Year As Space Program Expands." *ValueWalk*. August 2, 2012.

morning. The rover, called Curiosity, ushers in a new era of exploration that could turn up evidence that the Red Planet once had the necessary ingredients for life — or might even still harbor life today. NASA and administration officials were also quick to point to the success to counter criticism that the space agency had turned into a creaky bureaucracy incapable of matching its past glory."[18]

18. August 2012: Asteroid mining venture

"Planetary Resources Inc., an asteroid-mining venture backed by Google Inc. executives, said it added more billionaire investors and is nearing a partnership agreement with a 'top-10' mining company. The company wants to be the first to harness potentially trillions of dollars of minerals including platinum group metals by using robotic technology to mine asteroids."[19]

...

In this small and somewhat random sampling of news highlights we read about many different spacecraft and projects pursuing many different destinations, Saturn, the Moon, Low Earth Orbit, Mars, asteroids, the edge of our Solar System, and even the debris field surrounding Earth. We mentioned efforts originating in Austria, Switzerland, the US, China, India, Spain, and the ISS, and public and private sector ventures and adventures. We discussed tourism, physics, astronomy, communications, exploration, commerce, health care, diplomacy, and other themes as well.

Space, as it turns out, is all of these things, and many more. It is vast, of course, as it encompasses the entire universe that surrounds our small, blue home planet, and it is filled with promise, potential, and peril.

And one further thing that "space" is, if you wish to believe the authors of this book; it is our future. The purpose of this book, then, is to help us understand the future that we may create and are in fact we are

---

[18] Chang, Kenneth. "Curiosity Rover Lands Safely on Mars." *The New York Times*, August 6, 2012.

[19] Riseborough, Jesse and Thomas Bieseuvel. "Asteroid-mining company attracts new billionaires." *San Francisco Chronicle*, August 8, 2012, p D4.
This theme is explored in detail in a previous book in this series, in Taylor, Thomas and Haym Benaroya, "Developing a Space Colony from a Commercial Comet Mining Company Town." in *Living in Space: Cultural and Social Dynamics, Opportunities, and Challenges in Permanent Space Habitats*, Bell and Morris, editors. Aerospace Technology Working Group, 2009.

creating for ourselves, for humanity, beyond the bounds of our home planet.

A Cooperative Effort

It is worth noting that each of the authors of this work has volunteered his or her effort in exactly the sort of cooperative spirit that the future development of space will require. We've put this volume together not for money (since none of us are getting paid), but out of our individual and shared commitment to our shared vision. On behalf of the co-sponsors of this volume, ATWG, ISU, and IISC, we would like to extend our sincere gratitude to each author.

This is a rather large volume, but of course there is a great deal more to say even on the topics that are covered explicitly here. (With hundreds of footnotes, you'll be directed to plenty of other materials to explore on every topic we address.)

We will conclude this Introduction with the we hope that our efforts are rewarding for you, the reader. We also hope that you come away from thse writings with a deeper appreciation for the essential role that cooperation has played in the successes of the past and present, and a profound commitment to cooperation in pursuit of vastly grander dreams and aspirations for the future of humanity in space.

•••

LANGDON MORRIS

Langdon Morris is a Director of PwC, and a leader in its Growth & Innovation practice. Formerly a partner of InnovationLabs (www.innovationlabs.com), he is recognized globally as a leader in the field of innovation. His recent clients include organizations such as NASA, GE, Gemalto, Total Oil, the Federal Reserve Bank of the US, Johnson & Johnson, Tata Group, France Telecom, Stanford University, Wipro, L'Oréal, Accor Hotels, and many others.

He is a member of the leadership team of ATWG, and is formerly Senior Practice Scholar of the Ackoff Center at the University of Pennsylvania. He is a Senior Fellow of the Economic Opportunities Program of the Aspen Institute, Associate Editor of the International Journal of Innovation Science, and a member of the Scientific Committee of Business Digest, Paris. He has taught MBA courses in strategy at the Ecole Nationale des Ponts et Chaussées in Paris and Universidad de Belgrano in Buenos Aires.

He is the author and co-author of numerous white papers and five highly acclaimed business books, with editions in Japanese, Chinese, Korean, and French. He is highly sought after as a speaker and workshop leader who participates frequently at conferences and workshops worldwide.

KENNETH J. COX, PH.D.

Dr. Kenneth J. Cox earned his bachelor's degree in 1953 and his master's degree in 1956 in electrical engineering from the University of Texas/Austin. He earned his PhD at Rice University in 1966.

From 1960-1962 Ken led the first evaluation of digital flight system applications for the Martin Company. In 1963 he joined NASA to develop the flight control system for the Little Joe II Booster Vehicle. Later, Dr. Cox became the Technical Manager for the Apollo Digital Control Systems, which included the Lunar Module, the Command Module and the Command/Service Module, the first spacecraft to fly with a digital flight control system. He later was Chief Technologist, a member of the NASA Institute of Advanced Concepts, Technical Manager of the Apollo-Soyuz Androgynous Docking Demonstration, Space Shuttle Technical Manager for Guidance, Navigation, and Control, and Division Chief of the Avionics System Division.

His awards include the AIAA Mechanics and Control Flight Award in 1971, the NASA Medal for Exceptional Engineering and Achievement in 1981, the AIAA Digital Avionics Award in 1986, and the NSS Space Pioneer Award in 2007.

Ken is coeditor of this book series, and coauthor of Chapter 21. He has been the leader of ATWG since it was established in 1990.

Part I

International Cooperation Past, Present, and Future

CHAPTER 2

BROADENING THE BASE
COOPERATION AS A SPRINGBOARD FOR NEW PARTICIPANTS IN THE SPACE SECTOR

MICHAEL SIMPSON, PH.D.
THE SECURE WORLD FOUNDATION

© Michael Simpson, Ph.D., 2012. All Rights Reserved.

INTRODUCTION

Numerous examples of international cooperation among well-established spacefaring states have emerged over the last several decades. Although the prime example may be the International Space Station, other missions are similarly worthy of mention: Cassini-Huygens, Chandrayan 1, Hayabusa, and literally every mission launched by the European Space Agency. As a result of such cooperation our knowledge of our solar system, including our home planet, has been increased dramatically by missions whose success exceeded the capacity of single states, even large and wealthy ones, to organize and fund alone.

Less attention has been paid to the role that cooperation within and between national states has played in helping boost countries with less space experience into a position where they could benefit economically from participation in space activity and also contribute materially to the success of space missions.

Among such national states several patterns of cooperation are evident.

First, there is the case of economically well-developed countries that have consciously adopted a policy of niche specialization designed to facilitate inclusion of strong national competencies into space missions organized by more powerful spacefaring countries.

Second, there is the situation of emerging space states that have benefited from the express policy of spacefaring countries to encourage their development as participants in the space sector.

Third, and often related to one or both of the above cases, there is the pattern of national pursuit of space participation as part of an overt strategy of national economic development and capacity building.

Fourth, there is the example of sub-national efforts at regional economic development that specifically emphasize space activity as a means for enhancing the diversity of the local economy and enhancing the market for high technology products and services.

Lastly, there may be another pattern emerging, as private entities collaborate to bring order to potentially chaotic situations where political authorities have failed to act either by design or neglect.

A number of examples will help make these patterns clearer.

PATTERN 1: ECONOMICALLY DEVELOPED COUNTRIES AND NICHE SPECIALIZATION:
LOOKING AT CANADA AND AUSTRALIA

The best example of Pattern 1 is Canada. Active in the very early days of the space age with the development of the Alouette satellites first lofted into space on September 29, 1962, it chose the cooperative path from the beginning, electing to launch aboard an American Thor-Agena vehicle. Speaking about this program thirty years after its launch, C. A. Franklin noted that the cooperation was far from accidental, stating, "It was to be a cooperative undertaking between Canada and the US, with each country paying its own costs in the project."[1] With the subsequent addition of the United Kingdom to the endeavor, it would become even more international.

In the years since the launching of Alouette 1, Canada has shown continued openness to cooperation, integrating the iconic Canadarm robotic assembly into both the American Space Shuttle program and the International Space Station and following these successes with the addition of the Dextre fine motor control capable robot, now part of the ISS advanced tool kit. In addition to a resumé of the Canadarm's early history, the Canadian Space Agency's website highlights the cooperative nature of the project: "The Canadarm project remains a sterling example of successful international space cooperation."[2]

Another physical example of the success of Canada's cooperative strategy is the presence of a Canadian LIDAR aboard the US Phoenix mission. This instrument played a central role in the detection of snow in the Martian atmosphere, confirming the role water continues to play in the ecology of that otherwise dry planet. Other technological examples include synthetic aperture radar applications, biomedical technologies, and specialized clothing for human spacefarers.

Canada's cooperative strategy also includes a political dimension, which has allowed it to be a privileged partner in the US space program while also being an active associate member of the European Space Agency. In an evaluation of Canada's relationship with ESA prepared for the Canadian Space Agency by management consultants Goss Gilroy Inc., the expected dividends of a cooperative strategy are expressed clearly and early:

"Generally speaking, the Agreement contributes to maintaining Canada's world leadership in its traditional niches (e.g., civilian radar technology for Earth observation, and advanced satellite communications services) and enhancing the international competitiveness of the Canadian manufacturing industry through the development of space technologies, innovative advanced systems, and terrestrial applications."[3]

One manifestation of this strategy has been the substantial investment

---

[1] Franklin, C. A., "Alouette/ISIS: How it all Began," a speech to the IEEE International Milestone in Engineering Ceremony, Shirley Bay, Ottawa, Canada, May 13, 1993, as viewed on May 30, 2011 accessible on-line at http://ewh.ieee.org/reg/7/millennium/alouette/alouette_franklin.html

[2] "Canadarm," web page included in the Canadian Space Agency site as viewed on May 20, 2011, accessible online at http://www.asc-csa.gc.ca/eng/canadarm/default.asp

[3] "Summative evaluation of the 2000-2009 Canada/ESA Cooperation Agreement, Final Report," February 22, 2010, Goss Gilroy Inc., Management Consultants, Suite 900, 150 Metcalfe Street, Ottawa, ON K2P 1P1.

in the development of human capital capable of supporting the space sector, and especially those elements that contribute to meeting terrestrial needs. In the nearly 25 year history of the International Space University, the number of Canadian graduates, numbering 410 as of early 2012, is second only to those of the United States (466) among that school's more than 3300 alumni from over 100 countries.

Focusing on what it does well, investing in innovation where new opportunities present themselves, and concentrating resources within centers of excellence, Canada has demonstrated the utility of the cooperative model not only for the development of space activity, but for its own development as well.

Australia has pursued a similar strategy, but with less consistency and breadth. From the earliest days of the space age, it recognized that it occupied a privileged geographical position as regards monitoring both natural radio frequency signals from space, and those produced by human-made spacecraft. This led it to play a central role in monitoring communication from the early days of the US human space flight program, and to establish critical components of the deep space network on its territory. The cooperative advantage of the latter has been important in the monitoring of the continuing faint signals of the two Voyager spacecraft, and was critical to the success of those missions during their encounters with the outer planets. This emphasis continues today, as it supports an active bid in cooperation with New Zealand to host the Square Kilometer Array, intended to dramatically enhance the ability of radio astronomy to increase our knowledge of deep space phenomena, continuing its substantial interest in space-based telecommunications technologies and the ground infrastructure necessary to support them.

Additionally, Australia has invested in launch infrastructure, recognizing that its favorable latitude and sparsely inhabited interior provide excellent conditions especially for sounding rocket launch and recovery operations. Although efforts to establish a launch facility on Christmas Island in support of orbital missions have not borne fruit, the central Australian facility at Woomera from which Australia successfully launched its only orbital flight in 1967, continues to be an attractive site for sounding rocket missions, and the wide open territory in its range proved an ideal landing site that Japan has used to bring its Hayabusa spacecraft safely back to Earth.

Unlike Canada, however, which has maintained a focused and consistent emphasis on space activity, Australia's interest has ebbed and flowed substantially over the years. With the tasking of the Australian

Space Research Program to overcome the sense of drift described in a 2008 report to the Australian Senate entitled "Lost in Space?," the country has embarked on an important initiative to build both intellectual and physical capacity.

As an example of cooperation, the joint effort between the University of South Australia and the International Space University to offer a space studies program focusing on the needs of states in the Southern Hemisphere is particularly interesting. The mandate for this program has been intentionally international from the beginning and its inaugural session at Mawson Lakes, Australia in 2011 attracted students from 10 countries and 5 continents. Tasked with identifying space activities and technologies that were of particular value and interest to the states of the southern hemisphere, the program reflected Australia's willingness to advance along a cooperative path and its readiness to play a leadership role.

This was particularly evident with the focus of the 2012 program. Working to produce a white paper entitled, "REACH2020: Tele-reach for the Global South," its mission was "to develop a sustainable framework under which states can collaborate on economic and social needs, and maximize Information and Communication Technology (ICT) to provide space and terrestrial tele-reach applications".[4]

The "Lost in Space?" report also made it clear that there is a big difference between cooperation and dependency. Noting the extent to which the country relied on non-Australian sources for critical satellite-based services, the report summarized the arguments in favor of increasing numbers of satellites being built and operated by Australia with the phrase, "while the risk of being denied access to satellite data is not necessarily large, it would have severe consequences if it eventuated."[5] The report also noted that Canada harbored enough of its own concerns to also be engaged in the development of independent capabilities.[6]

These concerns were balanced, however, by awareness of the advantages of collaboration. Cooperative opportunities with Japan, the United States, China, Korea, and Canada were specifically highlighted.[7]

---

[4] "Southern Hemisphere Summer Space Program," University of South Australia, Division of Information Technology, Engineering & the Environment, http://www.unisa.edu.au/itee/spaceprogram/whitepaper.asp

[5] The Commonwealth of Australia, "Lost in Space? Setting a New Direction for Australia's Space Science and Industry Sector," The Senate, Standing Committee on Economics, PO Box 6100, Parliament House, Canberra, ACT 2600, November 2008, p. 16, paragraph 2.40. Online at asri.org/system/files/private/report.pdf

[6] *Ibid.* p. 15, paragraph 2.36.

[7] *Ibid.* pp. 16-17, paragraphs 2.41-2.44.

Special emphasis was placed on Australia's ability to provide mission critical support through its ground station network, thus emphasizing that Australia, like Canada, was conscious of its strengths, and intended to develop them in an environment where important technical capabilities became both centers of excellence and centers of economic return.

PATTERN 2: EMERGING SPACE STATES: NIGERIA AND THE PECS COUNTRIES

Good examples of pattern 2 include Nigeria and the countries participating in ESA's Plan for European Cooperating States (PECS). Although possessed of a good technological infrastructure of its own, Nigeria has benefitted substantially from technical assistance, support, and resource development provided by several international partners. For their part, the PECS countries have benefitted from a close mentoring relationship with ESA that has enabled them to identify where they can contribute to space activity and to build capacity to be more involved in the future.

Like many countries new to space activity, Nigeria entered the space arena with high hopes of building capacity and encouraging technological development.[8] It has worked with China, the United Kingdom, Russia, and the international Disaster Monitoring Constellation (DMC) to increase its technical ability, develop independent satellite control capability, and advance toward a goal of acquiring indigenous technology in Earth observation and telecommunications.[9]

Nigeria's 2011 plans called for the launch of three satellites. NIGCOMSAT-1R, developed in cooperation with China Great Wall Corporation as a replacement for a satellite that failed in 2007, was launched in December 2011. NigeriaSat2, developed with Surrey Satellite Technology Ltd. (SSTL) and NigeriaSat-X were successfully launched in August 2011 from Kourou, French Guiana. Although the first two of these projects involve contracts with non-Nigerian firms, they were designed to increase technology transfer to Nigeria in preparation for increasing the country's domestic share of future projects. The NigeriaSat-X project, by contrast, was completed in country, and represents an important step

---

[8] "National Space Policy and Programs," Federal Ministry of Science and Technology, 2001, Abuja, Nigeria, p. 3.

[9] "Paths to Progress: Space and the Southern Hemisphere," a white paper prepared by the students of the Southern Hemisphere Space Program (a joint academic venture of the University of South Australia and the International Space University) 2011, Mawson Lakes, Australia, p. 19.

forward in Nigeria's effort to build technical capacity in space applications and systems.

Nigeria's commitment to international cooperation has been rooted in its National Space Policy document of 2001. Chapter 9 emphasizes the importance of multilateralism in Nigeria's space planning, and also underscores that economic development is the goal more than space development.[10]

The initial role in space development for many smaller countries will often come as a side effect of their pursuit of solutions to terrestrial problems, but as Nigeria's participation in the DMC shows, a major space asset can also be assembled through the participation of several like-minded and technologically advancing states.

Similar conclusions can be drawn about the PECS states, but with an interesting twist. All current members of this group were members of the Soviet Bloc during the formative days of the space exploration. While working in Poland and the Baltic States as part of a European Union initiative to increase the participation of these countries in the economic activity of the space sector, I was struck by the number of stories related by veteran engineers who worked on projects in the 1960s and 1970s, and whose end purpose they only understood years later. In all cases these had been projects related to Soviet space missions, but the information about their eventual use was so compartmentalized that no one working on them in the client states knew the ultimate applications. Even in this information vacuum, however, both capacity and interest grew. In general, both the veterans and their younger colleagues showed considerable interest in the developmental challenges and opportunities of space itself, even while never forgetting the objective of economic development as well.

They and their non-technical colleagues also showed considerable enthusiasm for the role ESA was willing to play in their capacity-building efforts through the PECS system. And they had good reason for their enthusiasm. Two former PECS countries, the Czech Republic and Romania, are now full members of ESA, and a third, Poland, has just initiated the process to accede to membership. For Hungary, Estonia, and Slovenia, there is ample evidence that close cooperation can lead to opportunities to participate not only in locally focused applications, but also in scientific and exploratory missions as well. Equally as important, countries including Latvia and Lithuania are showing a growing interest in the PECS, and may soon apply to participate.

Cooperation with ESA can mean contracts and cash flow of course,

---

[10] "National Space Policy and Programs," section 9.2 (b), p. 33.

but the greatest boost to space activity often comes from the role the European agency plays in facilitating performance and capability audits. Enterprise Estonia, a national agency dedicated to supporting the country's entrepreneurial growth, devotes an entire page in its space business prospectus to the results of ESA's capability audit. With growing capabilities such as ground systems support, robotics, software, and advanced materials among many others, Estonian suppliers could now move with some confidence to market these capabilities in conjunction with documented advantages in turnaround time and workmanship.

The PECS countries also benefit from another source of mentorship as they seek to enter the space sector. Under the 2009 Treaty of Lisbon, the European Union acquired a so-called "shared competency" in space policy with its member states. This has led it to a more active pursuit of space issues not only as part of its foreign policy, but also in support of economic development objectives.

This conjunction of interests played a role in the establishment of the NordicBaltSat project, aimed at accelerating the entry of European Member States from the Baltic region into active participation in the space economy. In so doing it has placed a high priority on building human capital through training sessions, briefings, and workshops. During the NordicBaltSat sessions, presenters have often heard how participants had only recently discovered that a technology in which they had considerable expertise was of interest to the space sector. During one series of sessions in February 2011, representatives of several laser technology laboratories or enterprises expressed both surprise and delight that there were so many diverse applications for lasers in space missions, both on the ground and in flight. Although a commercial value cannot immediately be put on such an epiphany, it is evident that the team devoting some of their creative energies to the utility of laser technology in the space sector has increased in size. With that increase, the potential for innovations and even breakthroughs capable of advancing the development of space has also grown.

Although each country participating in the space sector has its own economic motives, even the newest arrivals are capable of advancing the broader and shared overall goal of pushing forward the frontiers of humanity's knowledge of the cosmos. What is especially true of the smaller, newer participants, is that if along the way the development of space provides insights to solving practical problems on Earth as well, those implications will be very welcome indeed.

PATTERN 3: SPACE COOPERATION AND ECONOMIC RETURN

All countries in the space sector are well aware of the potential for economic return in the form of technology sales and/or economic stimulus. Many also place high priority on the use of space technology in support of domestic economic sectors that are critical to them politically or commercially. If there is a difference between the largest and most powerful spacefarers and the newcomers, it is in the absence of exploration or cosmic rhetoric in their expressions of space objectives. The objectives of the newcomers always include capacity building and economic benefit; they rarely include reference to a human imperative to explore.

A good example of space policy by a country with some historical experience of space activity but only just now rediscovering the sector is the Lithuanian statement of June 7, 2010. It gets to the economic goals without delay in paragraph 1.1, declaring an unambiguous objective:

"To create and develop competitive business and science sector of Lithuania acting in the field of space (hereinafter referred to as the space sector);"[11]

Of special significance here is that the statement was issued by the Ministry of Economy (MoE), which is given the specific mandate of coordinating space policy. Like many new or returning entrants to the sector, Lithuania is strongly attracted to the potential for both direct and indirect economic benefits, although its initial expression of interest in the arena was advanced under the aegis of the Ministry of Education and Science (MoES). The mandate for creating and executing space policy shifted in early 2010 to the Ministry of Economy. MoES continues to have an important role to play in research issues, but the focus of the Lithuanian government's interest in space clearly centers on economic development.

Even the young Lithuanian Space Association, a group of universities and research centers with a common interest in the usual broad range of academic subjects included in the field of space studies, chose to focus on the economics of the sector at its second International Space Conference at Vilnius in November 2011. Exploring the topic of "Space Economy in the Multipolar World," the association shows that it, too, has its eye on the

---

[11] "National Programme on Development of Research, Technologies and Innovation in the Space Sector for 2010-2015," Government of the Republic of Lithuania, Ministry of Economy, Dainius Kreivys, Minister, Vilnius, Lithuania, Order No 4-436, June 7, 2010, p. 2.

sector's economic prospects.[12]

With Lithuania's emerging interest in the space sector and Estonia's participation in the PECS already mentioned, it should come as no surprise that neighboring Latvia is also showing substantial interest. Like its neighbors, it has placed considerable emphasis on existing technology and expertise that can be adapted to meet the needs of space missions, identifying materials, components, equipment, infrastructure and R&D capabilities that could be "spun-in" to space projects undertaken by other countries.[13] As with other countries in the former Soviet Bloc, Latvia does have experience dating from the early days of space flight, and it is eager not only to resume active participation in the sector but also to participate at as advanced a technological level as possible. A recent informational brochure published by the Latvian Space Technology Cluster makes it clear that the country is aware that it had a significant role in the development of early liquid fuel rocket technology through the work of Friedrich Zander, and that awareness leads to a strong belief that it can be a key contributor to future developments as well.[14] Evidence of the political support enjoyed by this perspective was the participation of Dana Reizniece, a member of the Latvian Parliament, in the International Space University's nine week long Space Studies Program held during the summer of 2011 in Graz, Austria.

Emphasizing its historical competencies in materials sciences, engineering, and astronomical observation, Latvia has highlighted recent accomplishments in scientific instrumentation, cryogenic insulators, radio astronomy, and printed circuitry as evidence that it has a current role to play in the space sector. Significantly, the country's first satellite Venta-1, is being developed as part of a cooperative project between the University of Latvia, Riga Technical University, and the University of Bremen, Germany, in cooperation with the German satellite manufacturer, OHB Systems. Of the four stated objectives for the Venta-1 project, acquiring practical skills in satellite design, studying theory of satellite design, developing satellite engineering in Latvia, and promoting the proficiency of

---

[12] "Call for Speakers: Space Economy in a Multipolar World," Lithuanian Space Association, accessible on line at http://www.space-lt.eu/pranesejams.htm?lid=4

[13] Cf. Simpson, Michael K. "Spin-out and Spin-in in the Newest Space Age," Chapter 5 in *Space Commerce: The Inside Story*," Langdon Morris, ed., Aerospace Technology Working Group, International Space University, and the International Institute of Space Commerce, September 2010 for a discussion of the economic potential of "spin-in."

[14] "Latvia and Space Technologies: History and Nowadays," Latvian Space Technologies Cluster supported by the Ministry of Economics of the Republic of Latvia.

high-technologies in Latvia, three are unabashedly economic in nature.

In addition to the numerous relationships Baltic countries are building with partners outside the region, the broad base of interest in the space sector has created a fertile field for intraregional cooperation as well. In the case of the NordicBaltSat program already mentioned, this kind of cooperation has been specifically sponsored and nurtured by outside partners such as The European Union, ESA, and ISU. Another such initiative, Baltic Sea Region Stars, although sponsored in part by the EU, is more homegrown.

Aimed broadly at stimulating innovation and not just at the space sector, it has nonetheless target strengths and attributes of considerable importance to the space sector: health sciences, energy, sustainable transportation systems and digital services.[15] Within this project, so-called StarDust clusters not only invoke the space mystique as an innovation stimulus, but also look toward transnational linkages in the region that can support breakthroughs useful to the space sector. Interestingly, many participants at NordicBaltSat workshops saw immediate parallels between their growing interest in the space sector and the collaborative, pan-regional approach inherent in the BSR Stars model. This model also holds out the possibility of pursuing collaborative projects with experienced ESA member states such as Sweden, Norway, Denmark, Germany, and Finland, as well as with Poland, whose application to join ESA has now been submitted.

As 2011 drew to a close, Estonia demonstrated its determination to continue on the cooperative path in space activity by formally signing the accession agreement to become a full member of the European Organization for the Exploitation of Meteorological Satellites (EUMETSAT). After five years as a cooperating state, Estonia's action will now enable it "to be fully involved in the strategic decisions of [EUMETSAT's] ruling Council."[16]

Elsewhere in Europe, Ireland provides another excellent example of how countries can contribute niche expertise to the space sector while aggressively pursuing economic growth and improved technical capacity. In a history of Irish space activity published by ESA in 2008, author Paul Clancy devotes an entire chapter to a chronicle of the role of international

---

[15] "BSR Stars Executive Summary," Lead Partner: Vinnova, Stockholm, Sweden, October 2010.

[16] "Estonia signs accession agreement with EUMETSAT," EUMETSAT website, December 14, 2011, http://www.eumetsat.int/Home/Main/News/Press_Releases/814189

cooperation in Irish space activity.[17] Clancy identifies 44 separate partnerships, only two of which are governmental; both governmental partnerships are led by institutions focused on economic growth, Enterprise Ireland and the Department of Enterprise, Trade, and Employment.

Among the partnerships singled out for mention, EUMETSAT is a particularly interesting example of how a small, developed state like Ireland can benefit from cooperation while also contributing to an extremely successful space activity. One of twenty members of the organization, and contributing only a bit over 1% of its budget, Ireland nonetheless is able to help sustain one of the world's preeminent providers of space-based weather data.[18]

Ireland can cite numerous examples of participation in European space initiatives and technological development, and one particularly interesting example was cited in ESA's 2001 report on "everyday uses for European space technology."[19] Through research sponsored by ESA, the Irish firm Parthus Technologies, with substantial expertise in mobile device technologies, developed a suite of space applications for Bluetooth Technology. These applications permitted reliable application of wireless technology both for data transfer among spacecraft components and for crew communications. Through this cooperation, ESA gained technology that permitted significant reduction in launch weight, and Ireland gained technology with substantial terrestrial value under the brand name BlueStream. Along the way BlueStream also generated 35% of the revenue that ultimately enabled Parthus to employ 400 people, and eventually become acquired by CEVA, Inc.

Ireland's pathway to such a result was a bumpy one. Starting with a firm resolve in December 1976 that the benefits that Ireland stood to gain from joining ESA were "potentially, very considerable," Ireland eventually ratified its membership in the agency in March 1979. Just over a decade later, however, the Irish Ministry of Finance recommended withdrawal as a means of saving £3.2 Million in annual fees. In the ensuing debate, the utility of space cooperation as an economic engine was the critical issue.[20]

For several years the Ministry of Finance and the Ministry for

---

[17] Clancy, Paul, *A Short History of Irish Space Activities*, European Space Agency Publication HSR-40, September 2008, distributed by ESA Communications Production Office, ESTEC, Nordwijk, The Netherlands, pp. 47-50.

[18] *Ibid.*, p. 50.

[19] Brisson, P. and J. Rootes, "Look No Wires, *Down to Earth: Everyday Uses for European Space Technology*," European Space Agency Publication BR-175, June 2001, pp. 100-101.

[20] Clancy, *op. cit.*, p. 37-40.

Enterprise and Employment argued opposite sides of the membership issue, until finally the question of withdrawal was abandoned in 1993. By 2001, the economic benefit arguments fully mated with the sense of participation in a major international undertaking worthy of Ireland had gained the upper hand as Mr. Noel Treacy, Minister for Science, Technology and Commerce made clear in a news release following an ESA ministerial meeting:

"[Ireland's] expenditure [for ESA membership] will not only assist Europe to build on its considerable strengths in Space, but will ensure that high technology Irish companies, and researchers, will continue to play a key role in European and global space activities, and enhance the growing capacity of Irish companies to exploit leading edge technologies in global aerospace and telecommunications markets."[21]

Two elements of that debate are important to us here. First, Ireland ultimately recognized that its involvement in space activity was not only part of its strategy of economic development, but also part of its partnership in a cooperative program of European dimensions in which its part, though small, was important. In short it was prepared to be a mission partner and not just a vendor state. Second, economic serendipity in the form of the advent of prosperous times and the era of the "Celtic Tiger" provided the political lubricant to get beyond the voices of austerity claiming that expenditures close to home were more important than investments in space. It didn't hurt that the Irish Government's resolve to stay within ESA in spite of the economic difficulties of the 1990s had bought the country the time it needed to discover the "down to Earth" qualities that the Parthus story and its many parallels were revealing by the beginning of the new millennium. Sustaining cooperative partnerships will likely always require a combination of political commitment and economic return, and both should be cultivated and managed carefully.

Ironically, Mr. Treacy's statement also shows that part of the political commitment necessary to sustain space activity in the emerging countries seems to be rooted in a desire of medium powers to improve their competitiveness with the major space actors. Ultimately this will lend notes of both cooperative participation and competitive zeal to the environment of commercial space, not only for Ireland but for many other states as well. To see the breadth of this phenomenon, it is worth looking at Futron Corporation's recently released fourth edition of its Space Competitiveness Index. Announcing the publication in a press release on August 15, 2011, Futron's Space and Telecommunications Director, Jay Gullish, is quoted as saying, "The 2011 results show that even as countries

---

[21] *Ibid.*, p. 38-39.

collaborate in space, competition has intensified."[22] As the Irish Minister's remarks make clear, this is not a new phenomenon, but is accelerating as more and more countries discover their capacity to participate. At the very least we will need to keep in mind that no matter how much they may be facilitated by cooperation, space-based endeavors of the future are as likely to see intense competition as have those of the past. One difference, certainly, is that the number of potential collaborators and competitors is increasing rapidly.

PATTERN 4: SPACE COOPERATION AND SUB-NATIONAL ENTITIES

The next pattern of interest in our review of cooperation opportunities, beyond those between the marquee actors among Earth's spacefaring states, addresses cooperation by political entities and associations that are not at the national level. The world's countries are composed of a dizzying array of sub-units, including states, provinces, regions, departments, territories, and the list goes on. Many of these entities have developed their own offices or agencies with mandates to pursue space opportunities for the benefit of more local populations and interests. Others have simply added a space portfolio or two to already existing bodies that were pursuing research, technology development, or economic growth.

The United States, with its traditions of decentralization and constitutional empowerment of states, has given rise to a large number of such activities. Recent observations at American space conferences have revealed organizations including Space Florida, the Virginia Commercial Spaceflight Authority, Spaceport America (and the New Mexico Spaceport Authority), and the Colorado Space Coalition.

All of them seek to maximize local benefit through partnerships with the national space efforts of the United States federal government, and thus NASA, the Department of Defense (DOD), the National Oceanographic and Atmospheric Administration (NOAA), and the Federal Aviation Administration (FAA) are on all their priority contact lists. All of them also cultivate important business partnerships, although the size of their target audiences differs with local priorities and objectives.

Space Florida, with a massive space infrastructure economically

---

[22] "U.S. Space Lead Erodes for Fourth Straight Year As Middle-Tier Nations Ascend," Press Release, Futron Corporation, Bethesda, MD, August 15, 2011. http://www.futron.com/1254.xml?id=1045

threatened by the end of the Space Shuttle Program, has devoted particular attention to job creation possibilities surrounding the reprogramming of former shuttle support assets at Kennedy Space Center. To this end, it has brought non-Federal assets to the table, especially with partners that are able to reciprocate with investments of their own, and/or with job creation. Further north, as the operating authority for the Mid-Atlantic Regional Spaceport and with a smaller and more specialized facility to promote, the Virginia Commercial Spaceflight Authority has signaled its interest in partnering with other states in pursuing space missions to be launched out of its Wallops Island site. As the Authority's website makes clear, neighboring Maryland is well represented.[23]

Spaceport America, and the New Mexico Spaceport Authority which oversees its financial bonding capabilities, has an unusually international focus. Its prime tenant, Virgin Galactic, is rooted in the British corporate constellation of Sir Richard Branson, and the flight paths of planned sub-orbital flights may extend into nearby Mexican airspace, either as part of an eventual agreement, or as the result of an emergency scenario. Like the Mid-Atlantic Regional Spaceport, Spaceport America intends to signal with its very name an openness to participation with other states. At the annual International Symposium on Personal and Commercial Spaceflight held in Las Cruces, New Mexico it has even suggested that the term "America" be interpreted it is broadest context to cover potential cooperation with other countries in the Americas.

A particularly interesting aspect of Spaceport America's cooperative model is the integration of two New Mexico counties, Doña Ana and Sierra, into the project. Citizens of both municipalities voted sales tax increases in support of the spaceport project, thus demonstrating that the cooperative model can be manifested by countries that are new to the space sector, and is applicable to sub-national units as well. As various space sector participants look for funds to underwrite their expensive start-up costs, they should not overlook possibilities within countries as well as between them.

The Colorado Space Coalition uses the structure of a not-for-profit organization to assemble a wide-ranging collection of space companies that are working across the full spectrum of sector markets, including military, civil, and commercial space into a coherent voice on behalf of Colorado's space potential. Without a launch site to promote, its focus is industrial, research, and technological capacity. While acknowledging on its web site

---

[23] Home page, Mid-Atlantic Regional Spaceport Web Site. http://www.marsspaceport.com/about-mid-atlantic-spaceport, June 2011

that Colorado is the fourth largest beneficiary of NASA contract awards, it also highlights the concern that the state's capabilities in the space field are as "top secret" as some of the highly classified projects being pursued in the state for government clients.[24]

While the not-for-profit association model seems to be working well for Colorado, an even bigger space participant among US states has had less success with it. On June 10, the California Space Authority, which in spite of its official sounding name is in fact a private association of space sector participants, announced that it intended to dissolve and was ceasing operations immediately. Recently focused on a major project to create a 500,000 sq. ft. (44,000 sq. m.) California Space Park connected to Vandenburg Air Force Base, the major DOD launch site on the US West Coast, the Authority stated in its announcement that it believed that it faced a prolonged environmental approval process with the local government holding jurisdiction over the proposed site.[25] Given the panoply of motivations and priorities affecting political units and human organizations, it obviously pays to pick your potential partners carefully before entering arrangements that can only succeed through cooperation.

Looking beyond the United States, there are several more interesting examples of space cooperation organized by and around sub-national entities including the Catalonia region of eastern Spain, several sites in northern Scandinavia, and the Association of Ariane Cities.

Two Catalan organizations, CTAE (Centre de Tecnologia Aeroespacial) and BAIE (Barcelona Aeronàutica i de l'Espai [Barcelona Aeronautics and Space Association]), have labored to transform the region's centennial aeronautical history into a growing tradition in space activity as well. While CTAE has been particularly focused on developing regional technology and capacity in cooperation with Spain's ESA partners, and in fostering international cooperation among universities and research centers,[26] BAIE has asserted its goal of building partnerships throughout the space community, even while being attracted to the special opportunities presented to the city of Barcelona by its relative proximity to

---

[24] Home page, Colorado Space Coalition Web Site, http://www.spacecolorado.org/ June 2011.

[25] "Space group disbands after failed Vandenberg project ," *Pacific Coast Business Times* online edition. June 10, 2011.
http://pacbiztimes.com/index.php?option=com_content&task=view&id=2402&Itemid=1

[26] "Mission and Objectives," Centre de Technologia Aerospatial Web Site, http://www.ctae.org/mission-objectives, June 2011

France's considerable aerospace infrastructure in Toulouse.[27] Understanding the Catalan focus on space activity provides some clues as to why some local communities are prepared to make significant organizational investments in space development and participation.

As BAIE's website reveals, the space sector in Catalonia has a definite sense of its historical roots. With clear reference to its cooperative history, the site describes Barcelona's role as the first city in Western Europe to publicly capture Sputnik 1's signal, and as a contributor to the soil-sample analysis instrumentation on NASA's Viking Mars Lander. It also notes with pride that it was a Catalan astronomer, Josep Comas I Solà, who was the first to describe the atmosphere of the Saturnian moon Titan. In addition to historical roots, the sector also is aware of its technical capacity, reporting that Catalonia handles 9% of Spain's ESA contracts, and that its ratio of aeronautic to space activity is 76% / 24%, compared to the European mean ratio of 91% / 9%. One means of generating the energy needed to sustain institutional commitment to space activity at the local level is thus to combine a sense of historical roots with the hard realities of an engaged economic infrastructure. Supporting a capacity already in place can thus be as important as reaching out to generate new business through attracting outside firms.

Northern Scandinavia presents us with another model where local interests have sought to take advantage of high latitudes suitable for polar research and data collection from polar orbiting satellites. Above the Arctic Circle in the city of Kiruna, Spaceport Sweden has assembled a coalition of local business, public infrastructure, and the Swedish Space Corporation to develop and promote its far northern location for commercial space flight. Although the city has been a center for sounding rocket launches for over 40 years, it attracted broader media attention when it announced an agreement with Virgin Galactic to provide an eventual European launch site for personal spaceflight trips aboard Spaceship 2. Here the local investment is driven by an evolution away from the singular mining focus of the city's creation, with the objective of optimizing the area's long but modest participation in the launch business. Given that Arctic tourism has been a big factor in the area's economic diversification, the conjunction of launch experience with the emerging business of personal spaceflight has been a strong factor in stimulating local interest in the spaceport.

Not far away across the spine of Norway, Kiruna faces launch

---

[27] "About BAIE," Barcelona Aeronautics and Space Association Web Site, http://www.bcnaerospace.org/public/who.php, June 2011.

competition from the Andoya Rocket Range. The institutional model there is quite different, however, with the major partner being the Norwegian Ministry of Trade and Industry with the participation of the Kongsberg Defense and Aerospace Company. The business model is also quite different, with less emphasis on tourism and more on research and education. The model provides for substantial cooperative opportunities both within ESA and beyond, and fits quite comfortably into Pattern 3 as described above.

Before ending our discussion of Pattern 4, a few words about the Community of Ariane Cities (CVA) are in order. This organization assembles its 36 members, consisting of cities in France, Germany, Spain, Switzerland, Belgium, and French Guiana, in a not-for-profit association with a specific mandate to encourage collaboration and cooperation not only among the members, but also with the Caribbean and South American countries whose geographical proximity to the Ariane launch site in Kourou, French Guiana creates the foundation for a natural partnership. With a structure that includes both cities and institutional partners, the CVA provides a forum for space policy discussions at a sub-national level while also providing a forum for industrial and political actors to consult and cooperate as they seek to optimize the favorable impact of space activity on their operations.

Of particular note in the activity of the CVA has been the organization's heavy emphasis on outreach, public communication and education. Recognizing the importance of an informed public and a supportive constituency, the CVA has sponsored programs for such diverse groups as school children, teachers, and young engineers as it seeks to increase the mix and diversity of talent available to the future of the launcher industry.

PATTERN 5: PRIVATE COOPERATION FOR SPACE SUSTAINABILITY

With only one clear example, it may be too early to elevate this activity to the level of a pattern, but the Space Data Association (SDA) may well provide a model for structured cooperation in space where political regulation is lacking.[28] The purpose of this cooperative endeavor is to ensure safe satellite operations through the exchange of space situational awareness data. Formed as a not for profit corporation on the Isle of Man

---

[28] For more information, refer to the SDA's website: http://www.space-data.org/sda/

in 2009, the SDA is a response to the growing concern that orbits are becoming increasingly crowded with operational satellites, derelicts, and loose debris.

Using automated systems operated by Analytical Graphics Corporation, the Association benefits from the support of the Isle of Man Government, which while unable to legislate for space due to its relationship with Great Britain has at least been able to provide a safe harbor in which the new association could be legally established. In this cooperative environment, the SDA seeks to reduce the likelihood of on-orbit collisions or broadcast interference, in an environment where international public regulation has not been established. Here, then, the principle that cooperation permits a favorable result which was otherwise beyond the means of the cooperating parties is extended to the private sector. Whether this model will survive the competitive impulses of the members themselves or the anti-trust fears of potential regulators remains to be seen, but in the growing number of areas where countries have been unable to create agreed upon rules of order, it seems likely that private operators, like pioneers on the old frontiers, will have a strong incentive to organize their environment as best they may, at least until the law arrives.

Some Lessons from the Examples

Cooperation is too common in the space sector to be a random occurrence, but our examples also show that it needs several pre-conditions to flourish. Ideally these conditions include as many of the following elements as possible:

A reasonable prospect of economic return to all partners

Even the most established spacefaring countries justify their space missions in terms of jobs created and economic development expected. All the new entrants are unabashed in their emphasis on expected economic returns. Increasingly commercial participants also anticipate profit and return on investment.

Political support and resolve rooted in more than just economic return

When economies go into down cycles there must be something more than short term returns to sustain them. The European Idea pulled Ireland out of its withdrawal crisis in the 1990s, and a sense that Australia should not be left behind helped revive interest in space activity down under in the new millennium. Frequently, the impact of space activity as an inspiration to the young, or in maintaining a technical infrastructure, are advanced where economic returns fall short or face protracted delays. Similar non-economic arguments reinforce space initiatives almost everywhere that they persist. That part of this resolve may originate in the desire of previously less active space players to compete with the well-established spacefaring states, may ultimately prove to be both an incentive to cooperation and an inducement to commercial competition. In either case the opportunities for synergy should increase substantially, and the need for creative procedures to provide structure to cooperation will become more urgent.

SUPPORT FROM PARTNERS WITH EXPERIENCE OR NICHE EXPERTISE

The hardest part of this element to deliver is the awareness that the assistance of others can be useful. Slowly, even the well-established spacefaring countries are coming to accept that building a market for space-based applications and extended space operations is beyond the resources of even the most powerful and wealthy players. The principle of "spin-in" is moving beyond national frontiers and increasingly will lead to acquisition of mission critical skills, equipment, and materials wherever they can most easily and economically be found.[29] Interestingly, even in the emerging pattern represented by the SDA, the niche expertise of Analytical Graphics and the opportunity to share its cost across several partners is a key feature of the association's recruitment message to potential members.

A SENSE OF HISTORIC CONNECTION TO THE SPACE SECTOR OR ITS ANTECEDENTS.

Just as countries of the former Soviet bloc are energized by the knowledge that they played important roles in the earliest days of the space age, so, too, will the beneficiaries of today's mentorship efforts come to view the missions that they contribute to as part of their space traditions in the future. As that tradition grows in more and more countries, the prospect for a well-rooted expansion of space commerce also grows. Thus efforts such as ESA's PECS system, China's outreach to the states of the

---

[29] Please see "Spin-Out and Spin-In in the Newest Space Age" by Michael Simpson in the previous volume in this series, *Space Commerce: The Inside Story by the People Who Are Making it Happen*. Aerospace Technology Working Group, 2010.

emerging space sector, and the CVA's effort to include Caribbean and South American representatives under the growing Ariane tent, all hold the promise of enhancing the size and value of tomorrow's commercial activity for the space sector.

LEADERSHIP WITH VISION

Cooperation seems to generate many enemies among those who prefer the ease of protected markets or the convenience of familiar commercial patterns. Leadership, political and commercial, will need to see beyond the "threats" of including participants from "outside" their national or economic networks. They will need to see that rapidly expanding markets can create opportunities enough to permit spreading benefits widely. The visionary Space Data Association is a solid example of what can be accomplished when leaders cooperate to address a problem beyond the capacity of any one of their institutions to resolve.

Like the vision shown by the SDA, those of governments will need to be focused as well as ambitious. Promises of economic benefit will need to be tempered with an understanding of the potentially long time scales required for infrastructure development and to achieve mission milestones. Ultimately vision will be measured by the skill that leaders show in seeing what is achievable, not in the rhetoric they pronounce about potentially exciting destinations. President Kennedy's challenge to go to the Moon and back was based after all on a hard-headed study about what objectives could reasonably be obtained in the context of a space race that had caught the USA by surprise.[30]

EFFECTIVE PUBLIC COMMUNICATION

The public is hungry for space-based information, but is also skeptical about the beneficial results of space activity. The hunger and the skepticism both need to be addressed. And in both cases the cooperation theme itself is broadly appealing and is emphasized increasingly in communications from the major space agencies.

For those of us who have followed the websites of major space participants over that last couple of decades, there is an understanding that we have come a very long way in our ability to communicate with the public. In the process, to the chagrin of some, space information, especially images, have become something of an entertainment industry.

---

[30] Logsdon, John, *John F. Kennedy and the Race to the Moon*, New York, Palgrave MacMillan, 2010. [This is by far the most thoughtful treatment of John Kennedy's historical challenge to the American People and his ultimate willingness to have made the voyage one of international cooperation.]

Given the cash flows in entertainment this may not be altogether such a bad thing. It may even be one more incentive for us to cooperate as we seek to keep the relationship between science, applications, exploration, and storyline in balance.

Conclusion

Looking at five distinct but combinable patterns, we have seen just how deeply rooted and widely sown cooperation has become in the space sector. Whether crossing national lines or enriching development within the diverse regions of a single country, cooperation is releasing synergy and enabling collaboration on a scale that permits even those coming late to space activity to hope reasonably for political, economic, and social rewards.

From the examples defining these patterns we have also been able to infer important lessons about what makes the soil fertile for cooperation, and how to keep it that way. In a sector increasingly interested in the challenges and importance of sustainable activity in space, it will be useful to keep a close eye on the role cooperation will play not only in developing creative new solutions to such problems as debris, traffic management, RF interference, access, and shared participation in economic benefits, but also in increasing the number of people, institutions, and countries with a reason to care.

Ultimately by broadening the base we will have provided the entire space sector, from the first participants to the most recent, with the foundation we need to make the future of space activity exceed even the wildest dreams of its greatest visionaries. This should more than adequately compensate us for the hard work ahead.

•••

DR. MICHAEL K. SIMPSON

Dr. Michael K. Simpson joined the staff of the Secure World Foundation in September 2011 following seven and one half years as President of the International Space University (ISU). He became the Foundation's Executive Director on January 1, 2012

Simpson's academic career extends over 36 years and five continents. In addition to his tenure at ISU, he has been president of Utica College and the American University of Paris with a combined total of twenty-two years of experience as an academic chief executive officer.

Simpson has taught courses in political science, international relations, business management, leadership and economics at Universities in both the United States and France and holds a post as Professor of Space Policy and International Law at ISU. He received his Bachelors Degree magna cum laude from Fordham University in 1970 where he was elected to Phi Beta Kappa. He has also been elected to academic honor societies in the fields of political science and business management. He is a corresponding member of the International Academy of Astronautics.

After graduating from Fordham University, Simpson accepted a commission as an officer in the US Navy, retiring from the Naval Reserve in 1993 with the rank of Commander. He completed his Ph.D. at Tufts University, The Fletcher School of Law and Diplomacy, holds the Master of Business Administration from Syracuse University; and two Master of Arts degrees from The Fletcher School. He has also completed two prestigious one year courses in Europe: the French advanced defense institute (Institut des Hautes Études de Défense Nationale) and the General Course of the London School of Economics.

He is the author of numerous scholarly papers, presentations, articles and book contributions and his practical experience includes service as a Political Military Action Officer, observer representative to the UN Committee on the Peaceful Uses of Outer Space, and member of the Association of Space Explorers International Panel on Asteroid Threat Mitigation. He currently serves on the Commercial Spaceflight Safety Committee of the IAF, the Executive Committee and Board of Directors of the World Space Week Association, and the Board of Governors of the National Space Society in the United States. He is a founding Trustee of Singularity University and an emeritus director of the Utica College Foundation.

Past board service includes over ten years as a trustee of a large mutual savings bank and eight years as a director of a billion dollar retirement fund.

Chapter 3

Space as a Catalyst for International Political Cooperation

Walter Peeters, Ph.D.
Dean and Vice President for Academic Affairs, The International Space University, and
Director, The International Institute of Space Commerce

© Walter Peeters, Ph.D., 2012. All Rights Reserved.

Introduction

At a recent symposium held at the International Space University (ISU) in Strasbourg, France, the topic of the International Space Station (ISS) was discussed in a broad context. One of the presenters, highlighting the international character of this project, made the statement that if the ISS had been operational by the beginning of the twentieth century, we probably would not have had two World Wars, by which he meant that working together in space brings people closer to each other and promotes mutual respect. His statement was of course purely conjecture, but his quite valid point was that if international projects such as the ISS had existed earlier in the 20th century, there would have been less likelihood of international world conflicts.

This statement is no doubt very bold and would be impossible to prove, but it is provocative and there seems to be a core of truth in it.

A prime driver for international space cooperation is no doubt the desire to receive economic benefit from the knowledge of the respective partners, and in this respect, international space cooperation makes sound economic sense since people from various cultures have developed different skills using different technologies, and by pooling these skills the individual countries do not have to acquire them from scratch at high cost, but instead can benefit from the expertise readily available within the group as a whole.

We need to make distinctions between the differing motivations of various participants in space projects. International cooperation driven by the public sector may have an economic dimension, but carries a more paramount geopolitical motive. Private companies generally have economic goals, so if the products derived from a satellite service cover different areas of the world, as is certainly the case with Low or Medium Earth Orbit satellites, companies will attempt to sell their services to all entities within the footprint of the service, irrespective of national borders. Publicly financed satellites, in particular if designed for dual-use (i.e., civil and military), may prefer to have limited footprints and restrict their offers to a single country, their own.

A third category includes projects such as the International Space Station, which stem from a political will and are used as vehicles for broader cooperation, but they may also result in economic savings, as well as in valuable cross-cultural exchanges.

Besides the tangible rationale, we shall not ignore the symbolic, non-tangible rationale driving international cooperation in the space sector. Probably due to the high visibility of space activities, policy makers have on several occasions in the past used this symbolism as a precursor for a broader political 'glasnost,' and we may expect them to do so still more in the future.

In this chapter we will examine these different types of cooperation, and the evolution of space efforts from a competitive environment to a much more cooperative one.

Present International activities in the space sector: from competition to cooperation

The Space Age really was born in a competitive environment reflecting the Cold War conflict between the US and Russia. Competition characterized nearly all activities in space for the first few decades, until gradually the benefits of cooperation became more apparent to all participants.

Today we recognize that international cooperation between space agencies often has an economic motivation. Similar to strategic alliances in industry, national space agencies have increasingly accepted the concept that know-how in certain areas is more effectively and cheaply obtained by cooperation than by developing it with one's own (limited) resources.

Indeed, in a study on international cooperation in space,[1] A. M. Schaffer concludes that from the US perspective, the benefits for international cooperation are primarily that:

It increases the total resource level available,
It utilizes knowledge that already exists,
It eliminates duplication of effort, and
It improves international relationships.

Interestingly, R. Launius, a well-known NASA historian, points out that the ESA defines its objectives rather differently from those of the US:[2]

Is this field of science worth pursuing?
Is the technology that may be acquired interesting to us?
Are there savings possible in resources? and
Is there a political advantage?

Broadly sketched, then, whereas the US perspective is generally utilitarian and focused on economic considerations and "what's in it for us?," the European assessment carries in addition to this utilitarian element a more learning-oriented perspective on the end goals, such as "How can science benefit from it?"

An example of NASA's approach is its interest in participating in a series of Mir flights in the early 1990's as preparation for ISS operations,

---

[1] Schaffer, A.M., "What do nations want from international collaboration for space exploration?" *Space Policy* 24 (2008) 95-103.
[2] Launius, R.D., "The historical dimension of space exploration: reflections and possibilities." *Space Policy* 16 (2000) 23-28.

which was certainly intended to leverage knowledge that the Russians already had. Indeed, many years of Mir station operations by the Russians as a 'closed-loop' system led to the accumulation of considerable experience in the field of logistics, and also in the handling of anomalies while in orbit, such as repairs due to onboard fire, meteorite impact and regular maintenance necessitated by wear and tear over many years of operations. In contrast, the STS system logistics and decontamination issues were resolved on the ground after it had landed after each flight; as the Mir system stayed permanently in space, all issues had to be resolved in the complex and constrained environment of uninterrupted flight mode.

A very specific example of this learning is the problem of leaks in the Mir cooling system, which occurred due to an unforeseen form of material fatigue. This knowledge led to a redesign of the International Space Station cooling system. Previous to the ISS, this type of information had not generally been exchanged.

Similarly, the longer stays of cosmonauts in space during the Mir era enabled the Russians to develop extensive knowledge of medical issues and countermeasures, which was subsequently used to prepare astronauts for longer duration ISS flights.

For all of these reasons and more, the International Space Station is the best current example of extensive multi-nation cooperation. Complete 'building blocks' have been delivered by the project's partners, each based upon their respective experiences and expertise. Some of the obvious examples include:

- The first core modules, provided by Russia, are based upon similar modules which were proven on Mir.
- Approach maneuvers are adapted from US GPS-based technology.
- The Russian Soyuz capsules, used as rescue vehicles on board ISS, also have a record of proven reliability from the Mir era.
- The US Shuttle provided flexible upload and especially download capacity, which was lacking in the Mir implementation.
- The Canadian ISS robotic arm is a further development of the Canadian-supplied robotic arms on board the Space Shuttle.
- Columbus, the European module, is substantially based on the proven and successful Spacelab concept.

In terms of scientific research accomplishments on ISS, many authors have pointed out that a significant portion of the results achieved have been possible due to the collaboration of scientists with different backgrounds, having access to methods and equipment which previously were not mutually available.[3]

Interestingly, the ISS developed from a purely American project to an international project over a period of ten years, from 1984 to 1994, as we see in Table 1:[4]

Date	Event	Partners
25.1.1984	President Reagan directs NASA to develop the Space Station Freedom	USA
29.9.1988	Multilateral intergovernmental agreements are signed involving new partners	USA, ESA, Japan and Canada
25.1.1994	President Clinton directs NASA to involve Russia in the development of the space station.	USA, ESA, Japan, Canada and Russia

Table 1: ISS evolution in terms of international cooperation

The various steps towards international cooperation can be described as follows. Initially the space station was mainly linked to the interplanetary exploration goals of the USA, and designed solely as a NASA project (hence the full critical path control...). Later, US politicians saw the project as a chance to link their 'allies' into the program, and in this way promoting international space cooperation.

Cooperation became a symbol of the end of the Cold War era, and was combined with the political objective of stopping proliferation of ballistic missile technology from Russia to rogue nations, while at the same time compensating Russia and providing economic support.

Although the sharing of these resources clearly had an economic objective, it is impossible to ignore the political, and even philosophical, dimension in this type of public international cooperation.

A precursor to the international cooperation evident in the ISS was the Apollo 18 - Soyuz coupling in 1975, an early example of a similar intent regarding international cooperation and, at that time, an important triumph for political détente. Officially called ASTP (Apollo-Soyuz Test Project), there is no doubt that the technical merits associated with this

---

[3] Taverna, M.A. and Morring, F., Space cooperation enhanced by space station, Spacelab collaboration. *Aviation Week and Space technology*, 173 (20) (2011) p.62.

[4] Peeters, W., *Space Marketing*, Kluwer, Dordrecht, 2000.

project were important, but certainly not as important as the political ones that occurred at a point in time when the Cold War period was in a transition phase.

The political and cultural effects of the program were significant. It will be no surprise, given the lingering Cold War climate at that time, that the commanding astronaut and cosmonaut selected were both vetted politically; both the Russian Alexei Leonov and the American Thomas Stafford were trained military pilots, both were considered nationalists, and both reached high military ranks as generals.

Figure 1: Apollo-Soyuz ASTP crew in 1975 (photo: NASA)
It is interesting to note that the crews wore uniforms of different colors and designs, which certainly served to emphasize their differences.

Twenty-six years later I had the opportunity to meet Leonov and Stafford in Moscow at the occasion of the 50th anniversary of Yuri Gagarin's flight (April 2011), and it was evident that they were two old friends. Both told many stories and anecdotes from this period, and joked willingly, which was not always fully appreciated by the political officers from both sides who accompanied them.

A similar story was repeated in the 1990s. To learn more about long-duration flights, a series of NASA astronauts participated in Russian Mir expeditions, and the initial hesitancies fast disappeared when mutual respect prevailed. This was not only the case for the selected astronauts, but also for the support crews consisting of hundreds of engineers on both sides.

Figure 2: A. Leonov and T. Stafford discussing with the author, Moscow, 2011. (Photo: L. Peeters)

Having personally witnessed this process, which took place in Star City, near Moscow, it was evident that the initial feelings of mistrust rapidly changed to trust and respect once they began working together. My observation is that two factors explain this difference between the flight crews and the support crews:

- Astronauts and cosmonauts, irrespective of their nationality, pass similar psychological filters in the selection process, and therefore tend to think according to similar patterns, typically oriented to the fulfillment of their missions.
- Astronauts are fully aware that they will have to fully trust each other in space, not only in emergency situations but in everyday operations as well, and that their safety depends on the other crew members, irrespective of their nationality.

It is therefore no surprise that long before geopolitical systems changed, astronauts already exchanged thoughts amongst themselves in their own organization, ASE, the Association of Space Explorers.[5]

---

[5] It is no coincidence that ASE originated and was politically tolerated as early as 1985, as a meeting of astronauts exchanging ideas in an informal and non-political forum. See also www.space-explorers.org.

Nevertheless, collaboration was definitely hindered on both sides by the unavoidable political 'observers' who were ubiquitous. But this could not deter a growing respect between astronauts and engineers on the technical skills and capacities of their counterparts. Although they had completely different educations, the common languages of physics and mathematics considerably facilitated mutual understanding that led eventually to mutual respect.

Probably one of the most explicit examples of such political signals is illustrated by a comment written in 1962 by Yuri Gagarin:

When I circled the Earth in my space capsule, I saw how beautiful our planet in reality is. People from the planet Earth, I hope together we will manage to preserve or augment this indescribable beauty, but in any case never to destroy it.

We shall not forget the year this was written, the year of the Cuban Missile Crisis, the height of the Cold War and the time when we may have been the closest the world has come to nuclear holocaust, and when every public statement was carefully considered and had to be endorsed and finally approved by the Russian political leadership. This text is a clear example how space activities were used to send political and cultural messages.

Future developments in international peaceful space cooperation:
From Cooperation to Inspiration?

In the first decade of the 21st century, some authors have speculated about how future cooperation or competition would develop. Looking at the various scenarios, Dupas and Logsdon[6] identified the following possibilities:

- Russia to focus on lunar exploration as a stepping stone to Mars,
- Europe to focus on Martian robotic missions,
- China to seek collaboration with Russia to reach the moon before 2020.

---

[6] Dupas, A., Logsdon, J.M., "Creating a productive international partnership in the Vision for Space Exploration," *Space Policy 23* (2007) 24-28.

The theme underlying this work assumes a new competitive scenario, similar to the Space Race to put the first man on the moon in the 1960s.

The financial crash of 2008, however, has changed this perspective considerably. Both financing possibilities as well as financial priorities have conspired to redistribute space budgets over the coming years, which will accelerate not only the shift to an economic motivation for space endeavors, but also a shift of the geopolitical balance.

Indeed, if we extrapolate space budgets in accordance with expected GDP growth, it is evident that by 2050, China and India will be equal partners with the USA simply based on the size of their economies. This effect will already be notable around 2030, and hence will be an integral factor in space exploration plans as we approach mid century.[7]

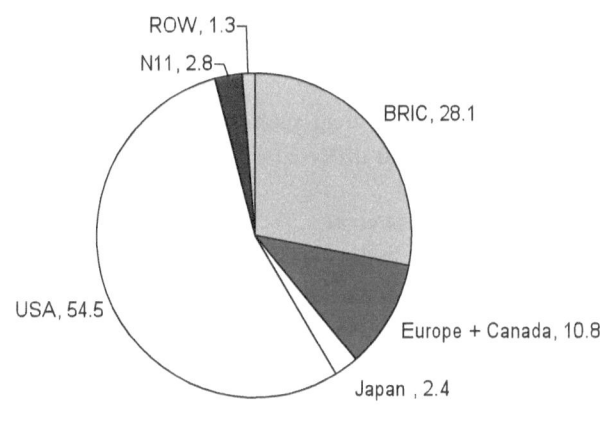

Projected space budget distribution in 2030

Perhaps recognizing this reality, US Secretary of State Hillary Clinton has noted,[8]

---

[7] Peeters, W., "Forecasting the consequences of the 'Crash of 2008' on space activities." In: *ESPI Yearbook on Space Policy*, Springer (2010) 164-178.

[8] Logsdon, J.M., "Change and continuity in US space policy," *Space Policy* 27 (2011) 1-2.

We cannot go back to Cold War containment or to unilateralism... We will lead by inducing greater cooperation among a greater number of actors and reducing competition, tilting the balance away from a multi-polar world and toward a multi-partner world.

Indeed, the next step in space exploration will be a huge and a very expensive one, such as permanent settlements on Mars. Several authors, including Ehrenfreund, Sadeh, Sadeh, and Lester[9,10] have suggested that only a multinational effort can lead to success, and that the new space powers including China and India cannot be ignored.

However, the cooperation format we see in the ISS under the leadership of the USA seems very unlikely. The partners will have to accept that others are contributing equally, although some of the elements will be on the critical path and others will not. In this context authors have in general pointed out the influence of cultural differences in future space exploration,[11] but probably of more paramount importance will be – again – the political motivation.

Exactly this level of critical path involvement is the major factor which Zelnio develops in four different cooperation levels.[12]

LEVEL 1: COORDINATION

Each country participates relatively independently, but there is technical and scientific coordination. The countries exchange results and data resulting from this form of cooperation.

LEVEL 2: AUGMENTATION

Countries participate independently in major parts of the project. This is the type of cooperation we presently see in the International Space Station, whereby participating states are responsible for well-defined parts or modules of the project, under the overall system level coordination of the USA.

---

[9] Ehrenfreund, P., Peter, N., "Towards a paradigm shift in managing future global space exploration endeavors," *Space Policy 25(4)* (2009) 244-256.

[10] Sadeh, E., Lester, J., Sadeh, W., "Modeling international cooperation in human space exploration for the twenty first century," *Acta Astronautica 43* (1998) 427-435.

[11] Ehrenfreund, P., Peter, N., Schrogl, K.U., Logsdon, J.M., "Cross-cultural management supporting global space exploration," *Acta Astronautica* 66 (2010) 245-256.

[12] Zelnio, R. "A model for the international development of the moon," *The Space Review*, Dec. 5, 2005.

LEVEL 3: INTERDEPENDENCE

Cooperation takes place also at the level of critical path elements. In this case each participating country is co-responsible for the overall progress and the technical achievements of the overall project.

LEVEL 4: INTEGRATION

A full combination of resources occurs. All funds are pooled, and management is made by fully integrated teams. Coordination takes place at higher level in accordance with predefined rules. ESA projects are an example of this type of coordination.

Going from level 2 to level 3 in this scheme for a post-ISS exploration program will be a major step, in particular for the US. It will require a courageous and inspired political leadership to take this step, but under the constraints imposed by the current financial situation it is hard to imagine an alternative approach for the coming decades.

It is equally hard to imagine, in view of the improving political contacts between the major space powers, that we will return to a competitive scenario.

There may still be a lot of barriers to be overcome before we reach higher cooperation levels, and a stepping stone approach is therefore a very interesting gradual implementation scenario as suggested by COSPAR[13] and detailed in these pragmatic programmatic steps:[14]

1. Establish an International Earth-based field research program
2. Support the "Science exploitation of the ISS enabling exploration"
3. Promote an "International CubeSat program in support of exploration" for developed and developing countries.

These could then be the start for further activities such as:

- The "Global Robotic Village"
- "International human bases" (Moon, Mars) using research activities in Antarctica as a model
- "Synergies between space exploration and Earth science"
- "Protecting the lunar and Martian environments for scientific research"
- "Environmental stewardship" to protect the Earth–Moon–Mars space.

---

[13] COSPAR, *Toward A Global Exploration Program. A Stepping Stone Approach* (Paris, France, June 2010).

[14] Ansdell, M, Ehrenfreund, P. and McKay, C., "Stepping stones toward global space exploration." *Acta Astronautica* 68(2011), 2098-2113.

However, there is a second, more fundamental reason why we will need an inspired approach for future space exploration: the reality of physiological change that extended duration in different gravity conditions will cause to those who venture to Mars and other off-Earth destinations. The impact on humans residing off Earth shall not be underestimated from a medical as well as a consequent societal point of view.

In previous centuries, people emigrated to other continents with the intention to build up a new life for themselves and their children, fully aware that they would never return to the place of origin. This logic will be the same for settlements on Mars or other planets, and it will be compounded by the very different environmental conditions on Mars compared to Earth. When emigrants went from Europe to Australia, there were surely considerable differences in climate, and the journey took a long time, but the gravity levels and most other essential physiological parameters were the same.

We know from experiences on board of Mir and ISS that even long stays in microgravity can be managed by proper countermeasures, and that the medical effects are reversible, but longer stays in lower gravity (if we take the example of Mars) will lead to modifications in inter alia the skeletal, muscular and cardiovascular system. After a few generations, people born on Mars will most likely not be able to return and live under 1g conditions on Earth without considerable artificial aid, as their bodies will have adapted to the lower gravity conditions.

If we add elements such as vast distances and significant time differences, there are solid reasons to assume that we will not have a form of classical colonialization, but that we will most probably be faced with the emergence of a different civilization. In such case, we should start to reflect upon the unique opportunity for humanity to build up a new, hopefully better, society and political system.[15] Indeed, on the basis of the Outer Space Treaty, all prerequisites are present to establish societal rules from scratch, based upon the many years of – often negative – experience of political systems on Earth.

As an illustration of this, in the framework of a Mars terraforming project executed at ISU,[16] and in the spirit of considering interdisciplinary

---

[15] Dator, J., *Space settlements and new forms of Governance* (1995) http://www.futures.hawaii.edu/publications/Space/GoverningSpaceSettlements1995.pdf (accessed: 1 August 2011).

[16] ISU, *Visysphere Mars, Terraforming meets life adaptation*, ISU Team Project report, International Space University publications, April 2005, available under http://www.isunet.edu/index.php/sturep-masters (ISU, 2005) (accessed: 1 August 2011).

and intercultural aspects, the need for a Convention was analyzed and a draft developed by one of the co-authors of the report O. Ingold. This draft, which is a personal reflection from the author, is presented as an appendix to this chapter.

Conclusion

A long process has taken place over the last few decades, from competition in space towards space cooperation. Space activities have strongly assisted in creating a more cooperative spirit between the major powers in a broader, geopolitical context. Therefore, we can without hesitation state that space cooperation has been a strong catalyst in ending this Cold War period. The International Space Station shows us this process, having evolved from a purely national (US) project to a project involving many, ideologically different partners.

As a next level of cooperation, additional partners including China and India will participate in the joint space activities. This will in particular be the case for space exploration towards other celestial bodies, which will be too expensive to be undertaken by the ISS partners. Let us hope that after such a broader basis of cooperation is achieved that humanity will have the courage to look further and work in the direction of … new inspiration.

Will it inspire people to try to create a new societal structure on such other celestial bodies, learning from all the mistakes we have made? The Outer Space Treaty provides the formal baseline for this; we can only hope that visionary politicians will grasp this unique opportunity to support the creation of a better society.

...

WALTER PEETERS PH.D.

Walter Peeters is President at International Space University (ISU) and Director of The International Institute of Space Commerce (IISC).

Following initial management positions in the construction and petrochemical industries, he joined the European Space Agency (ESA) in 1983 in a number of project control and management functions, among others in the HERMES project in Toulouse, France. Since 1980, he has been involved in astronaut activities as Head of the Coordination Office of the European Astronaut Center in Cologne, with strong involvement in the EUROMIR missions. He joined ISU in 2000, after serving as visiting professor (Non-profit Marketing) at the University of Louvain, Belgium. He was nominated as Dean of ISU in 2005.

He is the author of articles on incentive contracting, project management, space commercialization and organization in the space sector, and wrote the book *Space Marketing* (Kluwer, 2000). Recent research has focused on space commercialization, financing models of space projects and space tourism. He is a contributor to various organizations in the area of space commercialization (Working Group 2000, JAXA, OECD, IAC, IBA, EC).

Recent consultancy assignments include Space Policy (Luxemburg, Estonia), and Space Tourism (Singapore spaceport, Gallactic Suite, Excalibur)

He earned bachelor's degrees in engineering and applied economics at the Catholic University of Louvain, Master of Business Administration at Louvain, and Cornell University, and acquired a Ph.D. degree in Industrial Organization at TU Delft, the Netherlands.

Appendix

A DRAFT MARS CONVENTION

Author: Oliver Ingold[1]

Preamble

The States Parties in this Convention,

Prompted by the desire to settle, in a spirit of mutual understanding and cooperation, all issues relating to the exploration and settlement by humans of the planet Mars and aware of the historic significance of this Convention as an important contribution to the future of all the people and of Humankind as a whole,

Noting that the developments since the United Nations Outer Space Treaty of 1967 have accentuated the need for a new and generally acceptable Convention on Space law,

Conscious of the problem of the depletion of natural resources, the overpopulation and pollution on planet Earth,

Recognizing the desirability to establish through this Convention a legal order which will facilitate the peaceful human habitation of the planet Mars, and will promote the equitable and efficient utilization of its resources, the study, the protection and the conservation of their possible living resources,

Bearing in Mind that the achievement of these goals will contribute to the realization of a just and equitable interplanetary economic order which takes into account the interest and needs of mankind as a whole,

Believing that the codification of a legal framework for human activities on Mars achieved in this Convention will contribute to the

---

[1] Author's present affiliation : European Space Agency, Directorate of Human Spaceflight and Operations. This work was performed by the author during his prior studies leading to an MSc degree, at ISU.

strengthening of peace, security, cooperation and friendly relations among all nations in conformity with the principles of justice and equal rights,

Desiring by this Convention to facilitate the principle of a permanent human presence in the solar system and on Mars in particular, which implies the modification of Martian environment and the adaptation of the Human Species to the consequently engineered Martian environment,

Reaffirming that it is in the interest of all mankind that Mars shall continue to be used for peaceful purposes and shall not become the scene or object of international discord,

Affirming that matters not regulated by this Convention continue to be governed by the rules and principles of general international space law,

Have agreed as follows:

Part One: Use of Terms and Scope

Article 1: Use of terms

For the purpose of this Convention:

a. "Area" means Mars ground for the purpose of this Convention

b. "Areosphere" means the solid (lithosphere) and liquid (hydrosphere) portions of Mars (Ares).

c. "Concession" means the right to use land or other property for a specified purpose or to conduct specified operation in a particular area, granted by a controlling body.

d. "Genetic engineering" means modifying an organism by manipulating its genetic material

e. "Martian original form of life" is life resulting from and evolutionary process specific to Mars and without any influence from organisms coming from Earth.

f. "Resources" means all solid, liquid or gaseous mineral resources in situ on the planet.

g. "Terraforming" means a process of planetary engineering specifically directed at enhancing the capacity of an extra-terrestrial planetary environment to support life.

The ultimate in terraforming would be to create an unconstrained planetary biosphere emulating all the functions of the biosphere of the Earth.

Article 2: Scope

The provisions of this Convention apply to all the activities conducted on planet Mars and in the Solar System in the perspective of the terraforming of Mars. This process shall be developed in parallel with the adaptation of life to this newly engineered planet. It does not entail activities conducted on the planet Earth.
All activities shall be carried out in accordance with International Law in particular the Charter of the United Nations.

Part Two: Status of Mars

Article 3: Freedom of Mars

The planet Mars is open to all States and other entities. All willing States or organization shall be entitled the right to participate in this endeavor. Freedom of Mars is exercised under the conditions laid down by this Convention and by other rules of international law. It comprises:

(a) Freedom of access
(b) Freedom of movement
(c) Freedom of over flight
(d) Freedom of scientific research
(e) Freedom of use

These freedoms shall be exercised by all Parties with due regard for the interest of other Parties and with due regard for the overall project of Mars terraforming.

Article 4: Scientific Research

Freedom of scientific investigation on Mars and cooperation toward that end shall continue subject to the provision of this present Convention.
In order to promote international cooperation in scientific investigation, the Contracting Parties agree that, to the greatest extent possible and practicable:

1. Information regarding scientific plans for scientific programs on

Mars shall be exchanged to permit maximum economy of and efficiency of terraforming operations.

2. Scientific observations and results shall be exchanged and made freely available.

Article 5: Reservation of Mars for peaceful purposes

Mars shall be used for peaceful purposes only. There shall be prohibited, inter alia, any measure of military nature.

The present Convention shall not prevent the use of military personnel or equipment for scientific research. However, any nuclear explosion on Mars shall be prohibited

Article 6: Legal status of the Area

The Area and its resources are the heritage of mankind and shall be used in the perspective of a permanent Human establishment on Mars.

No acts or activities taking place while the present Convention is in force shall constitute a basis for asserting or supporting appropriation.

No State shall claim or exercise sovereignty or sovereign rights over any part of the Area or its resources, nor shall any State or natural or juridical person appropriate any part thereof.

The exploitation of the resources is possible against the payment of fees to the Authority. The fees shall take the form of a percentage of the benefits whose amount shall be fixed by the Council.

All rights on the Area or in the resources of the Area are vested in mankind on whose behalf the Authority shall act.

Part Four: Duties and Obligations of States

Article 7: Terraforming process

With respect to other provisions of this Convention, including the role and prerogatives of the Authority, Parties shall conduct all policies and use all techniques necessary to the terraforming of the planet Mars. The modification of the Martian environment shall be authorized and recognized as a necessity.

The transformation of the Martian environment must be understood as a whole. It concerns the areosphere, and the atmosphere of the planet. Introduction of Earth type life on the planet shall not be considered as a harmful contamination.

The Parties shall accept a priori control of their long-term projects by the Authority.

Article 8: Adaptation of life

To establish a permanent human presence on Mars, Parties are free to conduct scientific research in the domain of gene therapy and genetic engineering.

Research on how to enhance human adaptability to the living conditions on Mars shall be authorized under the control of the Authority.

Genetic modification of living organisms shall be encouraged to accelerate the transformation of Mars environment during the terraforming process.

Article 9: Measures for conservation of the living resources of Mars

Necessary measures shall be taken by all Parties in accordance with this Convention with respect to activities in the Area to ensure effective protection of any Martian original form of life from harmful effects, which may arise from such activities.

These measures shall be taken under the control of the overall Martian Authority and shall not compromise the terraforming process.

If any form of original Martian life should be discovered, international environmental law shall be applied.

Article 10 Rights of the people on Mars

The people living on Mars shall benefit from all Human Rights recognized by international law.

Intellectual property rights of inventions made on Mars shall also be respected. Each discovery shall be registered under the national law of the person discovering it.

Article 11: Jurisdiction

Every State shall effectively exercise its jurisdiction and control in administrative, technical or legal matters over their nationals including personnel, materials or installations.

Article 12: Rescue duty

Parties shall adopt all practicable measures to safe guard the life and health of persons on Mars. For this purpose they shall offer shelter in their installations and other facilities to persons in distress on the planet.

Article 13: Responsibility and liability

States parties shall bear responsibility to ensure that national activities conducted in the area shall be carried out in conformity with this Convention.

International organizations shall bear responsibility to ensure that the activities conducted by their representatives in the area shall be carried out in conformity with this Convention.

Damages caused by the failure of a State Party or of another entity to carry out its responsibilities under this Convention shall entail liability. The Outer Space treaty of 1967 and the Liability Convention of 1972 shall govern any other damages.

Part Five: The Authority

Article 14: Establishment of the Authority

The Parties shall put in place an international body (called the Authority). The activities conducted on Mars shall be organized and controlled by the Authority on behalf of mankind. All States Parties are ipso-facto members of the Authority.

The Authority is composed of two organs:

1. The Assembly meets every 2 years during the Conference. The Assembly has the power to modify the present Convention and to give

recommendations to the Council. Extraordinary Conferences can take place according to the request of all members. States are members of full right and other organizations participate to the Assembly with the status of Observer.

2. The Council is the executive organ of the Authority and ensures the application of the Convention. It is composed of fifteen members of full right of the Assembly. Ten permanent seats are given to the most contributing States and the Assembly shall elect five other members for a period of 6 years by way of majority secret ballot. The Council also acts as the arbitrator in case of unsolvable conflict between two Parties and shall propose to the Assembly to sanction Parties refusing to comply with the Convention dispositions.

3. The Council shall consult the Advisory Board whose five members represent the interests of private entities participating to the terraforming project.

Article 15: Role of the authority

The sole purpose of the Authority is to support and help States and other organizations to coordinate their efforts in the perspective of the colonization of the planet Mars.

The Authority shall exert its overall control over the different techniques and activities used for the terraforming of the planet.

In order to ensure the observance of the Convention, all areas of Mars, including all stations, installations and equipment within those areas, shall be open at all times to inspection by observers designated by the authority.

The Authority shall take measures in accordance with this convention to acquire technology and scientific knowledge relating to terraforming activities or to bioengineering activities.

To encourage activity, the Authority shall have the power to attribute concessions to any State, International Organization or Private Person to conduct any activities necessary to the establishment of a permanent human colony using terraforming techniques and assisted life adaptation.

The terms of the concession agreement shall be established on a case-by-case basis. The agreement shall include a termination clause and shall respect the provisions of this Convention, noting that a concession shall not last more than 100 years.

Article 16: Financial principles

The Authority shall be financed by the contributions of the State Parties to the Convention. The amount of the contribution shall be determined between the parties on the basis of each State's growth domestic product.
Each Party contributing to the Budget requirements of the Authority shall receive capital repayment and compensation for use of this capital.

Part Six: Final Provisions

Article 17: Settlement of disputes

The States, concerned in any case of dispute with regard to activities conducted on Mars, have the duty to consult together and under the coordination of the Authority with a view to reaching a mutually acceptable solution. If negotiation shall fail, States shall solve their conflict by arbitration with the Authority Council acting as arbitrator.

Article 18: Modification of the Convention

The present Convention may be modified or amended at any time by unanimous agreement of the Contracting Parties.
Every 15 years from the entry into force of the Convention, the Parties shall undertake a general and systematic overview of the manner in which the international regime of the Area established in this convention has operated in practice. In the light of this review, the Parties shall meet in a Conference and could decide to amend the Convention by a vote at the majority of two thirds.

Article 19: Opening to signature, ratification and accession to the Convention

The Convention shall be subject to ratification by the signatory States. It shall be open for signature and accession by any State, which is a Member of the United Nations and is willing to participate and invest into the colonization of Mars by the Human Species.

Article 20: Entry into force

Instruments of ratification shall be deposited with the Secretary General of the United Nations hereby designated as the depository organization.

The Convention shall enter into force one month after deposition by the twentieth State, being recognized that these twenty States shall include at least five States with autonomous access to space.

The Convention is undividable and shall not include exclusion clauses or exemption clauses.

Article 21: Withdrawal

The right to withdraw shall be granted to all parties. In the process, the Authority shall promote the transfer of technology and scientific knowledge amongst the Parties so that the Enterprise may not suffer there from.

Notice of withdrawal from the Convention shall be given to the depository organization. Withdrawal shall take effect one year from the date of receipt of the notification.

Article 22: Deposition

The original of the present Convention, of which Arabic, Chinese, English, French, Russian, and Spanish texts are equally authentic, shall be deposited in the archives of the United Nations, which shall transmit duly certified copies thereof to the Governments of the acceding States.

•••

CHAPTER 4

PUBLIC SPACE LAW, THE LEGAL PRACTITIONER, AND THE PRIVATE ENTREPRENEUR:
Distinguishing What "Ought To Be" From "What Is"

Dr. George S. Robinson

© George S. Robinson, 2012. All Rights Reserved.

For much of the lay public, and perhaps even for many in the legal profession, when the question of "space law" arises, thoughts turn to "special treaties" and "international conventions," and to noble acts of diplomacy involving concepts like "adventurous and peaceful" activities in space, magnanimously conducted for the "benefit of all mankind." The documents containing these pronouncements have been prepared by people from numerous professions and disciplines, including politicians, legislators, and their staffs and analysts, with the support of diplomats, scientists, engineers, budget analysts, and the like. Once drafted, they are then studied by lawyers who attempt to determine where the proposed activities of their clients may fit with the often amorphous, wishful, and imprecise provisions, phrases, terms, and words that appear in those very same documents. This shows us that law is a process intended to provide clarity and direction, but which instead too often results in ambiguity and uncertainty. Such is the nature of "space law" as it was written in 1967, and as it is practiced today. How we may overcome these shortcomings in the current manifestations of space law, and what laws we will need to see and, indeed, enable the opening up of space for full commercialization is the focus of this chapter.

I. Thoughtful Anticipation Or Wishful Thinking?

In the everyday practice of space law, the tenuous nature of such issues as the governance and commercial use of space resources, humankind migration, and off-Earth settlement, are particularly reflected in two provisions set forth in every space-related treaty. First is the normal penultimate article setting forth the procedures for amending the document, and second is the provision setting forth the protocol whereby a signatory nation can withdraw from that treaty.

In other words, if a treaty is not consistent with a given nation's political, economic, or defense requirements of the moment, then the necessary amendments can be executed. If amendment is unlikely to occur, or is too time-consuming, then the relevant treaty provisions either will be ignored, or the treaty withdrawal provision will be implemented.

If the issue is sufficiently important, and time is not of the essence, the parties can turn to the International Court of Justice, i.e., the World Court, but if the issue is pressing then the International Arbitration Commission might be preferred.

So while the well-intentioned 1967 treaty provisions may be commendable, and may help define the noble spirit and intent for national and global collaboration with respect to space activities, in the practical world they are apt to be dysfunctional and, perhaps even worse, intentionally designed to be useless.

The ambiguous terminology of the Outer Space Treaty makes it anything but functionally self-executing, leaving many of the (ambiguous) terms subject to disparate and even contradictory interpretations as applied by the signatory nations as they formulate and then implement the relevant domestic legislation in their countries.[1, 2, 3]

II. LAYING THE FOUNDATION FOR WHAT MIGHT HAVE BEEN

The Outer Space Treaty of 1967 was and remains the controlling document defining how the international community of nations intended ... and superficially continues to intend ... for space to be explored, occupied, settled, and used, most notably peaceably and for the "benefit of all mankind." The Outer Space Treaty was followed by several other space-related treaties, and also bilateral/multilateral agreements, negotiated and brought to fruition under the aegis of the United Nations, all addressing international agreement regarding how we must go into space and use its resources; not just that an increasing number of nations were developing the technology to accomplish these objectives.

However, there is serious doubt that the Outer Space Treaty will

---

[1] On January 11, 2007 China used a ground-based missile to hit and destroy one of its older weather satellites orbiting more than 500 miles in space. For a discussion of the immediate reaction of the US, Australia, and Canada to the use of a ground-based missile in the satellite "shootdown," *see* "US Condemns China Satellite-Killer Test" at [http://www.spacewar.com/reports/US_Condemns_China_Satellite_Killer_Test_9 99.html]. *See* also, by G.S. Robinson, "The 2010 United States National Space Policy: New Emphasis on 'Trust' but Verify Will be Required of All Participants," in *German Journal of Air and Space Law*, ZLW 59. Jahrg. Hedft 4 S. 534-550, Dec. 2010.

[2] For a more detailed commentary regarding the 2010 US National Space Policy, see by G.S. Robinson "The US National Space Policy: Pushing the Limits of Space Treaties?" in the *German Journal of Air and Space Law*, ZLW 56, Jg. 1/2007.

[3] In 2009, Russia claimed sovereign rights over certain areas of the Arctic seabed adjacent to and a part of Russia's continental shelf. The area apparently is rich in oil deposits and other minerals. International law currently asserts that the countries ringing the Arctic, i.e., Russia, Canada, the United States, Norway, and Denmark (which "owns" Greenland), are limited to control over a 200-mile economic zone around their coastlines. Russia is pursuing the establishment of military bases in its respective areas of Arctic seabed control and management. For an interesting discussion of the practical and legal issues relating to this claim and subsequent assertions of a similar nature by other nations, *see*, generally, "Putin's Arctic Invasion: Russia Lays Claim to the North Pole…and All Its Gas, Oil, and Diamonds," online at [http://en.wikipedia.org/wiki/Territorial_claims_in_the_Arctic]. For a general discussion of the law relating to this claim, *see* "Territorial Claims in the Arctic," online at [http://en.wikipedia.org/wiki/Territorial_claims_in_the_Arctic].

continue to serve as a realistic beacon of international cooperation in conducting space activities; doubt based in part on the seemingly naive foundation upon which the major superpowers constructed the Treaty in the early 1960s. Neither the United States nor the former Soviet Union, nor their respective allies, appeared to know what the true space capabilities of the others were at the time. Nor did they have a clear picture of what any nation's capabilities would be twenty or fifty years hence.

Already by the early 1970s, the former USSR tested the spirit inherent in the phrases "for the benefit of all mankind" and "no weapons of mass destruction will be placed in space" by constructing the Fractional Orbital Bombardment System (FOBS), designed to carry nuclear warheads from space to targets on Earth. This was a rude wake-up call suggesting how space would in fact be used as soon as the necessary technology evolved, regardless of how the treaty suggested it "ought to be" used.

The question now for the pragmatist policy-maker and space lawyer is whether the evolution of space-related technology has caught up with and passed the politically "transcendent" motivations reflected in the 1967 Treaty. Current space-related technology available for the application of military strategies is, of course, far beyond FOBS, and now that the military and private sector firms are fully intertwined as they jointly pursue their respective goals, programs, and projects, military objectives and profit motives are fully entangled in creating new challenges to the spirit and letter of the 1967 Treaty.[4]

---

[4] For purposes of this discussion, it should be noted that public-private partnerships are common in the European community as they relate primarily to launch capabilities, space telecommunications, and remote sensing. In Germany, there is a strong cooperative relationship between military and private industry space activities. Until the present, these types of partnerships have been very mixed in terms of resulting successes and failures. In large part, fiscal considerations and accountability have left the efforts in the hands of NASA, involved members of the Congress with significantly participating constituencies, and the Department of Defense. The reluctance of NASA personnel to yield any of their respective authorities, not to mention employment opportunities and other incentives for short-term planning disfranchises necessary evolution of mission formulation and management authority to the private sector.

III. Does "Majesty Of The Law" Flow From Transcendent Motivations Into The Realities Of Emerging Commercial Space Activity Opportunities And Customary International Law?

If there is any "majesty" in space law, it will be seen in the spirit of the text of the treaties, conventions, and even in the emerging customary international law[5] for space. Not only is the "solemn intent" of the treaty negotiators to be seen in the text of these laws, but in practice this must be followed by an unwavering, routine, and manifest willingness of the governmental authorities and private industry, and a persuasive lay public to recognize and enforce the collective spirit and intent of these space treaties and related documents. Without enforcement, in other words, the treaties will mean little, or nothing. The signatory nations and organizations must fully support and defend not only the letter of the law, but the spirit and intent as well. Unfortunately, this is not happening aggressively, consistently, realistically, and effectively to the extent required for success; and there are very practical reasons for these circumstances.

The United Nations and many of its public international organizations appear to ignore distinct violations of the spirit, intent, and often the clear letter of the law embraced by the Outer Space Treaty; at least at those points where clarity does in fact exist in the document. For the most part, and particularly as many governments transfer portions of their civilian and military space responsibilities to the private sector, decisions are being made in a self-serving and practical context regarding what is permissible. Those decisions are likely to be supported by the "best" legal rationalizations that can be mustered, and once the international community is faced with a de facto reality, the consequences will result either in a direct series of confrontational reactions, or it will be considered a fait accompli and pass into the emerging body of "customary international space law," regardless of how little it may conform to the letter and/or spirit of the 1967 Treaty and its collateral documents.

---

[5] Customary international law is generally considered to consist of rules of law derived from the consistent conduct of States acting out of the belief that the law required them to act in that fashion, i.e., a widespread repetition by States of acts over a period of time, and which are likely to be incorporated formally in international treaties and conventions. For an excellent discussion of customary international law, its origins and practices, *see* online "Customary International Law" at [http://en.wikipedia.org/wiki/Customary_international_law.]

IV. National Security Interests v. Globally-Shared Transcendent Principles

While it is common sense that every nation will protect its space-related defense interests, it also is necessary to recognize that many of the current space-related policy and legal issues are still very debatable.[6] Consequently, it is critical that nations not allow the increasing militarization activities in space to become an unchallenged end unto themselves. Lawful military use of space has a role, an essential role, in national, regional, and indeed in global security. But it is not an unbridled role. Space must not become solely the high ground for securing military and other defense-related assets in space and on Earth, dragging civilian and commercial space activities around as budgetary coattails on efforts organized primarily for military interests. Instead, it must be premised primarily on promoting civilian migration through an evolving private commercial presence, based on traditional concepts of international cooperation as well as proven principles of marketplace competition.

What really shapes the uses of near and deep space are:

(1) how fast the evolving technology becomes available for military and other national security applications;
(2) the effectiveness of the diplomacy exercised between and among governments;
(3) the existing and rapidly evolving military space capabilities of nations and alliances; and
(4) the competitive dictates of the private sector in a global economy.

While lawyers must advise their clients and negotiate on behalf of the parochial interests of their clients, they must also provide sound legal advice as demanded by space commerce, public or private, with awareness of the shifting commitments of spacefaring governments pertaining to the peaceful uses of outer space for the benefit of all humankind. This is certainly not an easy task, particularly in view of the contempt in which the international public legal profession is sometimes held by industries, governments, and, sadly, by large segments of the general populace.

---

[6] *See*, therefore, by G.S. Robinson, "United States National Security Space Strategy (Unclassified Summary of January 2011)," in the *German Journal of Air and Space Law* (ZLW 60.Jg. 2/2011, pp 274 – 279).

V. TREATIES, LIKE ALL LAWS, ARE DESIGNED TO BE BROKEN ... AND TO BECOME OBSOLETE

Significant amendments to the Outer Space Treaty of 1967 and related agreements are needed to support and enable the emerging era of space commerce. Perhaps, however, a completely new controlling treaty for space commercialization is a better solution, i.e., one that is based on the merger of commercial interest and the pursuit of human species or specieskind survival through migration to, and settlement of, near and deep space. Can the lay public and private investors understand and agree to this need? Is the global community ready for this?

These new agreements must take into consideration the mutually shared ignorance of the state of space technology in the early to mid-1960s.[7] In the long run, the private sector, and certainly the commercial entrepreneur (large and small) must keep the humankind species there.

In the interim, what the practicing international space lawyer must focus on even more intensely to support commercial activities and a durable presence of humankind in near and deep space is a thorough familiarity with the growing body of international conflicts of law as they bring spacefaring nations together to explore, develop, migrate, and settle near and deep space.

Looking back to lessons of the early English charters and American corporations as models for commercial "exploitation" of in situ space resources, it's evident that establishing a habitat society with private and quasi-private corporate governance is not a new concept.

Two key factors must be kept in mind. First, no sovereign ownership of space and its resources can be permitted, at least at present, which would otherwise provide the authority behind private ownership of space resources, including interstitial space. Yet such ownership is a critical requirement for private commercial investment in basic research, space exploration, resource capture, development, marketing and sales. Second, there are no established space-indigent cultures with which to interact.

---

[7] In this respect, see, for example, by G.S. Robinson, "Space Law for Humankind, Transhumans, and Post Humans: Need for a Unique Theory of Natural Law Principles?" in *Annals of Air and Space Law*, McGill University (2008). Also by that author, "Space Law: Addressing the Legal Status of Evolving 'Envoys of Mankind'" in *Annals of Air and Space Law*, McGill University (2011); and with co-author C. Smith, "Quantum Physics and the Biology of Space Law: The Interstitial Glue for Global Support for Space Migration and a Proposed Commercial Management Infrastructure," in *Annals of Air and Space Law*, McGill University (2010).

The historic precedent of the United Kingdom shows the example of the very effective charter issued to "The Governor and Company of Merchants of London, Trading into the East Indies" (i.e., the East India Company). Between 1603 and 1606, another charter was issued to the Virginia Company, and in 1670, yet another to the Hudson's Bay Company. While their culturally imperial policies were abhorrent to our modern viewpoint, the ensuing societies and nations nevertheless were primarily based upon and secured by growth and expansion through commercial exploitation (in the meanest sense of that word). Control of physical assets was at the core of the phenomenon.[8]

VI. THE PRAGMATISM OF EVOLVING INTERNATIONAL SPACE LAW

The law of various countries consists of some provisions which are shared by many or all nations, and some which are unique to various nations. For example, antitrust laws in the United States and laws of other countries prohibit some activities which clearly are designed to restrain trade. In addition, there are laws relating to a great many factors (what follows is a rather long and not altogether complete list): the promotion and securing of free trade and economic globalization, and others pertaining to the funding of the (expensive) commercial development and exploitation of space activities; technology export and re-export laws supporting national defense and national security interests (such as the International Traffic in Arms Regulations of the Arms Export Control Act in the United States, i.e., ITAR); a variety of securities laws, fiscal laws, contract laws, corporate and tax laws; international trade agreements and attendant national implementing legislation and regulation; private and public placement laws; insurance laws and other evolving risk management principles and policies; health care laws and quarantine protocols; transportation laws pertaining to (and this is just a partial list) land, water, air, the moon, and space tourism; domestic, foreign, and off-Earth employment laws; human, humankind, and post-human rights, policies, and laws; domestic, international, space-based, and interplanetary communications laws; law

---

[8] *See*, therefore, by G.S. Robinson, "No Space Colonies: Creating a Space Civilization and the Need for a Defining Constitution," in 30 *J. Space Law*. No. 1, pp.169-179 (2004); and also by that author "Transcending to a Space Civilization : The Next Three Steps Toward a Defining Constitution," in *J. Space Law* Vol. 30, No.1 (Fall 2006).

pertaining to the rights and obligations of space-transiting vessels and their crews (analogous to the law of the sea and oceangoing vessels); and all aspects of intellectual property rights, domestic as well as international and global (including effective enforcement of non-disclosure agreements undertaken in internationally collaborative pursuits).

The full list will occupy the studies of a good portion of Earth's law school students for many decades to come, and of course it will also occupy the legislators who draft the laws, and the regulators who will define the practical measures and means, and the authorities who will enforce them. It may or may not be a good living, but there certainly is a living to be made in these endeavors.

Some key areas of space law that will have to be addressed in this century, and which include many specifically related to collaborative private commercial space activities, are the following.

1. Securing and protecting international venture capital or other start-up funding and resources for space ventures.

2. Changes in legal education and curriculum that will allow for the formulation of the employment laws that will be necessary to meet radically changing characteristics of human capital in space, both technically and in the management of the business affairs of global, transnational, and perhaps even transglobal corporations ("transglobal" referring to corporate entities housed in/on and doing business in/on multiple celestial bodies, possibly none of which are Earth).

3. Indeed, the "transglobal" phenomenon will arrive upon the scene imminently, but as yet we have no legal or regulatory frameworks capable of addressing this new situation. There is also a significant jurisdictional issue as well.

4. Policies and laws must recognize the requirement for drastic changes in educational objectives and methodologies pertaining to research conducted at universities and non-profit research organizations, which also may be located off-Earth, such as on the International Space Station.

5. Communications law will have to keep pace with rapidly changing technical improvements, some of which address the

importance of "communities of knowledge," intellectual property rights unique to exploring, commercially developing space resources, and the settling of near and deep space.

6. Laws that address and protect "inclusiveness" where necessary in decision-making, domestic and global, concerning space policies and activities, rather than those based only or primarily on an "either-or" decision-making methodology. In this context, space jurisprudence will most likely reflect the awareness that a viable space society and civilization may be critical to the very survival of the human species, and/or its nature and essence(s), and that this is likely to take the form of descendant transhumans and, very likely, post-humans.

7. Legal expertise will reflect the need to protect against favoritism of governments toward major corporations, which might detract from the competitiveness of a start-up or mid-level entrepreneur's products or services.

8. Protection of intellectual property rights through effective international enforcement procedures, particularly as they may relate to small companies or individual entrepreneurs who are subject to various forms of theft by major corporations and, indeed, by governments as well.

9. Laws relating to the risks of forward, cross-, and back contamination of planets and other celestial bodies, both natural and fabricated.

10. Existing and evolving regimes of law relating to establishment of space commerce research and development carried out by private transglobal entities with quasi-sovereign authority, fiscally based upon open global investment by the general public.[9]

---

[9] Incipient efforts are being made at present to establish centers of research excellence relating to the role(s) of private commercial space activities in the forthcoming decades. These centers may involve partnerships with federal, state, and local entities/interests, including universities. They will also rely on the successful but aging generation of space pioneers and participants in the first manned lunar landing and subsequent governmental space exploration and commercially oriented activities, and the necessary learning interactivities of the present and future generations.

11. Identification of the best forums, national or international, for providing authoritative decisions regarding private ownership of space loci and resources.

12. "Customary space law" established through the application of the rebus sic stantibus principle, i.e., as a result of changed circumstances.

13. Metalaw relating to human biotechnologically integrated transhumans and post-humans.

14. The laws relating to telepresence, teleportation, advanced intelligent and autonomous biorobotics in extremis, and the related economic transactions.

15. And, to conclude this particular but not all-inclusive list, laws relating to nation building efforts in space and in cyberspace.[10]

VII. Conclusion

Regardless of how one may feel about the 1967 Outer Space Treaty, it is inescapably clear that legal regimes must evolve in the twenty-first century so that private and governmental activities in space reflect both sound principles of economics as experienced on Earth, and sound economic principles yet to be discovered. They also must steadily evolve in our impending space development transglobalism. This will occur, that is, unless recidivistic characteristics of ethnic, religious, economic, political/military, and cultural parochialisms are the survivors of our current social/political/cultural tumults. If the latter occurs, an entirely different scenario will evolve relating to space exploration, migration, and commercial uses of space resources, one that will most likely have strong militaristic overtones.

Whatever the direction, the applicable regimes of law will evolve hand-in-glove to reflect those policies, that is, they will reflect "what is" as opposed to "what ought to be." And that is what space lawyers must deal

---

[10] In this context, *see* by G.S. Robinson and R. Lauria "Legal Rights and Accountability of Cyberpresence: A Void in Space Law/Astrolaw Jurisprudence," in *Annals of Air and Space Law*, McGill University (2004).

with – not the parochially-interpreted ambiguous but transcendent objectives embraced in the suspect drafting of the Outer Space Treaty. Space law in the twenty-first century must and will embrace and respond to the requirements of, and for change in, the private and public sectors – their work forces, military applications, private commercial ventures, transglobal trade agreements, and the like, all while recognizing the independent trading nature of long-duration and permanent space communities.[11]

Above all, it will require that the legal experts work closely in a totally interdisciplinary fashion with all individuals and sections of domestic, international, and global societies and civilizations, who and which are affiliated with the aerospace industries and commercial users, or who represent start-up efforts. Regardless of where they come from, all of them are likely to be convinced that they know not only the technical and business worlds, but all of the relevant laws as well. This despite the fact that their "knowledge" may have come right out of a "How To" manual freshly procured over the Internet.

Establishing working relationships with clients under these circumstances demands the greatest of legal statesmanship, and it will also require of the space lawyers a significantly advanced form of knowledge about the empirical methods and data of secular sciences if they are to help their clients focus on "what is" as opposed to an unfounded leap to "what ought to be," thus supporting the evolution of the pragmatic, essential, and privately underwritten commercial exploration and settlement of near and deep space.

•••

---

[11] It should be noted that the US national policy dealing with space exploration and related activities has embraced the concept of encouraging and, indeed, facilitating commercial space development since the 1980s. Policy and law have not reflected that objective, at least until the present as observed by NASA Administrator Charles Bolden who noted at the 14th Annual Federal Aviation Administration Commercial Space Transportation Conference in Washington, D.C. (Feb. 9, 2011), that "NASA has always thrived on innovation... Industry has always been our partner. We have never built a big rocket. It has always been a NASA/industry team.... When I retire the space shuttles, that's it. For NASA access to low-Earth orbit – we need you... We can't survive without you." See, therefore, online at [http://www.space.com/10811-commercial-space-nasa-survival.html] .

DR. GEORGE S. ROBINSON

Dr. Robinson has been in the private and public practice of law since 1963, and has taught or lectured in space law at numerous universities around the world. After serving at NASA and the Smithsonian Institution in Washington, DC, for thirty years, he retired into private law practice and concentrates primarily on space matters. Dr. Robinson has served on hospital boards of directors, as well as various science and space-related boards of trustees and governmental and private advisory committees. He earned his AB degree from Bowdoin College, an LL.B. degree from the University of Virginia School of Law, an LL.M. degree from the McGill University Graduate Law Faculty, Institute of Air and Space Law, and the first Doctor of Civil Laws degree in space law awarded by that Institute. He presently is in private practice with his two sons and daughter-in-law. He is also the author of Chapter 20 in this volume.

CHAPTER 5

THE SPACE DATA ASSOCIATION:
A NEW MODEL FOR INTERNATIONAL COOPERATION IN SPACE

STEWART SANDERS
SVP, PLANNING & PROCUREMENT, SES
CHAIRMAN, SPACE DATA ASSOCIATION

AND

CHRISTOPHER STOTT
CHAIRMAN AND CEO, MANSAT
INTERNATIONAL INSTITUTE OF SPACE COMMERCE (IISC)

© Stewart Sanders and Christopher Stott, 2012. All Rights Reserved.

A NEW MODEL FOR COOPERATION

The formation of the Space Data Association (SDA) represents an interesting model for international cooperation in space activities. The SDA is a not for profit organization incorporated in the Isle of Man, part of the British Isles, and was formed by the world's three largest satellite operators, SES, Intelsat, and Inmarsat. What makes the SDA unique is that unlike other international organizations that are largely focused on the pursuit of either academic or scientific goals, the SDA is focused on directly improving the safety of the orbital operational environment by focusing on both radio frequency interference mitigation and collision avoidance. The SDA is also of interest in that it has been formed voluntarily by commercial interests to act as an industry self-regulating body whose beneficiaries are the customers, shareholders, and ultimately the general public which the satellite communications industry serves.

Although founded by the industry's three largest players, membership is open to every satellite and spacecraft operator in the world and at the time of the writing of this paper, the SDA has attracted a number of additional members. The SDA is not established as an exclusive organization, but rather as an inclusive one.

The Association was created as an organization that can embrace the various aspects of the commercial and governmental space markets. In this way it can cross international boundaries to better and practically encourage global cooperation for improvement in the use of the space environment crossing the industry/government divide. In essence the SDA has been established to provide the necessary legal framework for the sharing of specific data essential to the safe and efficient operation of spacecraft and for the identification of point sources of common interference to satellites. It meets its objectives by creating and maintaining a series of secure databases around the world that house critical operational data relating to two primary areas of space awareness: collision avoidance monitoring, and radio frequency interference reduction, but also supports points of contact data sharing and anticipates other types of data sharing such as Space Weather data.

The Isle of Man was chosen as the home jurisdiction because of its internationally recognized corporate legal system, high standing in the arena of international financial regulation with the OECD, IMF, US Government and others, the pro-space stance of its government, the advanced state of its satellite finance industry and associated space-based financial services markets, and the absence of an official space agency, allowing a neutral and level playing field to attract the greatest number of potential members. It also follows closely the UK Generally Accepted Accounting Practice, guaranteeing transparency for all members.

EVOLUTION OF THE INDUSTRY

The advent of the SDA marks an evolution in the satellite communications industry with the industry voluntarily seeking to regulate itself and to provide a logical, and more importantly, formal solution to an issue that has existed since the satellite communications industry began. Prior to the SDA, the question of spatial awareness, as it pertains to collision avoidance in the geostationary arc, and radio frequency interference was handled on an ad hoc basis by the individual operators. Any sharing of the data necessary to avoid collisions or to address issues of interference had been handled by convention via a gentlemen's agreement. Every satellite operator would keep such data at significant cost to itself with no way of knowing for certain the accuracy of the data, aside from that in respect of its own satellites. No formal process existed. Interaction between operators was on a best efforts, manual basis. Existing regulations do not properly address the practicalities of the interactions required. The SDA is voluntarily codifying this prior convention to give greater clarity of process in an increasingly mature space environment which in turn aids new entrants into the industry and ensures the efficient use of resources. In this regard, the SDA embodies the growth and maturity of the industry in its approach to good governance, logical self-regulation and a new level of global corporate citizenship.

The first set of data submitted by members to the SDA databases relates to situational space awareness in order to prevent potential collisions in space. Where exactly are your own satellites in relation to those of your neighbors? What if you are moving a satellite? Can you do so safely? In the past answers to such questions, which are crucial for the safe operation of space assets, could come only via a series of meetings and calls with your competitors, probably facilitated through little more than personal relationships and informal networks within the industry. In the days of INGOs this might have been acceptable, but in today's commercial environment, given the nature of the assets employed, the many commercial players, new uses of the geostationary arc and the associated potential risk of lost business, this is simply unacceptable. Reliance on third parties such as the USSTRATCOM to perform a 'neighborhood watch' on behalf of the satellite industry has been an ill-fitting compromise which does not properly address the space industry's needs. Hence the logic behind the creation of the SDA.

The member operators voluntarily make their data accessible to all other members to enable what is essentially air traffic control in geostationary orbit. Having such accurate and up-to-date data on file from the operators of the spacecraft themselves greatly increases the utility of the data. It reduces the risk of false or untimely data being used inadvertently, as was potentially the case with the former informal system. The need for maneuvers of existing craft is thus minimized thereby reducing their use of fuel and allowing costly assets to remain safely on orbit. There is also an incentive upon the members to ensure the data they provide is accurate. Any errors in the system can only adversely affect the operators themselves and hence a strong feedback loop is in place to ensure proper use and oversight by all members.

As well as the operators, the insurance companies are beneficiaries since they have a greater surety of reducing the risk of collisions. Similarly governments that, along with the operators, would ultimately be financially responsible for such accidents under the UN Outer Space Treaty and Conventions.

The establishment of the SDA database brings other advantages. A single standard is applied to the data meaning it can be read accurately at any time by any user of the system thus removing errors inherent in using data provided in differing formats. Member data is more current and accurate than that held by, and required by, national governments and international governmental organizations whose operations are not intended and, therefore, not properly suited to serve the needs which the SDA is seeking to fulfill. The SDA member data is standardized and supports all users. With the big three establishing the SDA as a non-commercial not for profit organization and fronting the costs, the industry's largest players have, at their own expense, voluntarily subsidized the creation of this vital facility for all players in the industry. A remarkable example of co-opetition: competitors acting together for the common good.

The SDA database will support any commonly used orbit regime: GEO, MEO, LEO and is currently supporting both GEO and LEO members' Conjunction Assessment processing

US Government data on orbital positioning, via ground-based radar from NORAD for example, and others are still crucial and still of use, most especially in the tracking of dead or drifting satellites and debris. In fact, the SDA database complements the existing military information sources. A logical step is to utilize the best available data sources, including redundant sources where available, to ensure the highest confidence in data accuracy. Accordingly, the SDA intends to utilize multiple data sources and expends considerable efforts, through its technical adviser, Analytical Graphics, Inc (AGI) and the Center for Space Standards & Innovation

(CSSI), in evaluating the accuracy of different data sources and the data of its members data. Once accepted as fit-for-purpose, the SDA's Space Data Center (SDC) system automatically validates data supplied by its members.

With regard to the tracking of point source interference to the radio frequencies operated by the worlds' satellite fleets in geostationary orbit, the SDA also plays a key role.

Radio frequency interference is a constant issue for those operating satellites. Radio interference can prevent the use of specific or, in some cases, all of the transponders of satellites operating from geostationary orbit and hence impact adversely on the quality and availability of services to customers with the consequent effect on revenues. It is estimated that the approximate cost of such interference impacts the global satellite industry by millions of dollars annually. Individual operators and companies can experience the same source of interference. Each would often find themselves spending many man-hours and revenues solving the same interference issue leading to orders of magnitude of needlessly duplicated efforts and costs. Often they would also find themselves in different stages of solving such interference issues: what one operator would be solving today another might have solved a month or more prior, more duplicated effort. Also, each operator would have to keep their own databases of past events and their solutions. More duplication and worse, duplicated cost that in turn causes prices to rise and profits to decline to shareholders.

Measurement techniques for identifying the sources of interference signals require accurate, up-to-date data on the satellites in the neighborhood of a satellite being interfered with; and any delay in obtaining the data will extend the service degradation or outage caused by the interference. The SDC provides a repository for this data to be available immediately when a member has need. Accurate information on radio frequency interference requires many sets of differing data, which in turn meant that many operators would manage their own data to different data protocols and data standards. The SDA now provides one standard for al operators. In the past such data would be shared on an ad hoc basis by the operator, if at all. There has also never been a standard agreement to protect commercially sensitive data, all of this adding to cost and risk.

However, with the SDA database the operators now have one source for all such data, drawing from the now combined historical databases of the members and up-to-date data provided in near-real-time. The data is now continuously available in a timely and uninterrupted fashion along with solutions. All members of the SDA sign up to data management controls to ensure that this data is used appropriately and not for commercial gain. The SDA membership legal framework resolves issues of risk and liability, protecting its members from undue liability but also providing significant penalties for misuse of the data.

At present, membership of the SDA is open to all commercial satellite operators. Additionally, relationships with commercial entities will be entered into when it is in the best interests of the members; members will choose which activities they are involved in and to which they provide their data. The authors can envisage a time when membership will also be open to governments and government agencies working in space, specifically in the case of the civilian agencies such as NASA, ESA, NOAA and others who also utilize the geostationary orbit. This is equally true for those agencies and statutory bodies who regulate the use of radio frequencies in Earth Orbit such as the ITU, Ofcom, and the FCC.

In regards to other government users of the geostationary orbit, most specifically the military (USAF and NATO) the author can envisage a special membership where they could have access to the commercial data available, but without having to list the position or status of their space assets. Using the analogy of a very crowded harbor, the SDA is in essence placing navigation lights on all of the commercial traffic in that harbor so the others users of the harbor can note their positions and avoid them.

There could be a time when launch vehicle service providers and insurance companies are also invited to join, perhaps as observing or associated members. This is logical given the need for insurers to calculate risk in terms of orbit insertion and collision avoidance thus seeing membership of the SDA as a means to reduce insurance premiums, for example with members getting better rates than those who are not members. In the case of launch service providers, understanding the geostationary asset and debris environment is crucial to ensure clean windows for orbital insertion either directly, with the case of the Proton launch vehicles, or GTO with others.

The SDA databases specifically in regards to interference issues could also be opened up to the users of satellites, for example those who purchase and utilize VSAT and other networks from satellites, as they will have a vested interest in identifying and preventing radio frequency interference to the satellites they depend upon for their businesses.

CONCLUSION

Ultimately the entire space sector is benefiting from the creation of the

Space Data Association and this proactive voluntary move by the founding satellite operators, not only in terms of the enhanced and now formalized orbital situational awareness, point source interference identification and resolution, reducing costs and increasing safety, and efficient use of resources throughout the space sector, but also in providing a viable model for international cooperation. The SDA's formation and work also stands as proof of good corporate governance in space: public good from private resources ensuring the most efficient use of the most limited natural resource known to humanity, radio frequencies. Its founders are to be commended.

...

Stewart Sanders

Stewart Sanders is responsible for the SES fleet planning activities, including procurement of new satellites and associated launch vehicles; this role also has responsibility for SES's technical innovation activities. Prior to this role, Stewart was SES's SVP for Customer Service Delivery, responsible for the monitoring and control of all real-time customer operations on the SES fleet. Having studied electronics and telecommunications in the UK while working for British Telecom International (BTI), Stewart joined Intelsat in 1987, initially working in Japan. Since then he has travelled extensively during his more than 30 years' experience in the satellite communication industry, having fulfilled various satellite and payload operations roles for BTI, Intelsat, ICO, NEW SKIES SATELLITES and SES. Stewart was actively involved in setting up the Space Data Association, working with industry colleagues from Intelsat and Inmarsat in particular, and is a Director and the Chairman of the data sharing organisation. He also worked closely with other industry bodies such as sIRG, WBU-ISOG, GVF, EUI and other satellite operators in a number of initiatives targeting RF Interference. Stewart has an MBA from the University of Liverpool and lives in Luxembourg with his wife and three well-travelled children. In his spare time he enjoys music, reading and sports, in particular rugby.

CHRISTOPHER STOTT

Mr. Christopher Stott is Chairman and Chief Executive Officer of ManSat, the global orbital frequencies and regulatory services company headquartered on the Isle of Man with offices in London, Houston, and Cape Canaveral. Founder of the Space Industry on the Isle of Man, Chris has also served as the Manx Government's Honorary Representative to the Space Community, and in 2010 was named Celton Manx Isle of Man Business Person of the Year.

In addition to his work with ManSat, Chris is passionate about space education and serves on the Main Boards of the Society of Satellite Professionals International (SSPI) where he is Vice President of Education, the International Institute of Space Commerce (IISC), and the International Space University (ISU) where he is also on Faculty and is Co-Chair of the School of Management and Business, the United Space School Foundation, the Conrad Foundation, and the Challenger Centers.

Chris is also presently the Chairman of the Manna Energy Foundation and serves on the Board of the Amar Appeal, working in clean water, clean energy, clean telecoms, and human rights in Africa and the Middle East. Working with Google's TIDES Foundation, the Manna Energy Foundation's Geeks Without Frontiers program recently completed the world's first open source 80211s aimed at bringing over one billion people more to the Internet.

Chris is one of the Founders of Odyssey Moon, the first entrant to the Google Lunar X Prize, and serves on its Main Board. Formerly of Lockheed Martin Space Operations and the Boeing Company, Chris has also worked extensively in British and American politics. In London working in the British House of Commons and House of Lords, and Washington DC working in the US Senate and on the Chairman's Staff of two US Presidential Campaigns.

Prior to his work in space, Chris was Special Projects Director with Life Education International, a children's health education and drug prevention program, a United Nations Non Governmental Organization (N.G.O.), working with fifty million children in over 27 nations around the world.

Educated at Millfield School in Somerset, England, Chris attended the University of Kent, Canterbury where he obtained a Bachelor of Arts Degree, with Honours, in American Studies Politics and Government. Chris also received a Diploma from the University of California, San Diego in International Relations and Marine Policy (Scripps Institute of Oceanography). Chris holds his Masters Degree in Space Sciences (Msc) from the ISU.

A published Fellow of the Royal Astronomical Society, the International Institute of Space Commerce, and a member of the International Institute of Space Law, Chris was the co-author of Europe's first work on space privatization and commercialization, "A Space For Enterprise; the aerospace industries after government monopoly", Stott & Watson, Adam Smith Institute, London, 1994. Chris has also contributed to a number of other publications, most recently *Space Commerce*, the previous volume in this series.

Chapter 6

International Spaceport Challenges:
The Necessity Of Formulating The Policy For Developing The Critical Technology

Thomas E. Diegelman
NASA

© Thomas E. Diegelman, 2012. All Rights Reserved.

Introduction

Commerce is the lifeblood of every nation, and of the world as a community of nations. Cut the supply of commerce and a nation will die; compromise the commerce, and it will atrophy. And to be sure, commerce and its continuing success in our globally competitive markets is predicated on the continuing evolution of technology and the essential competition for goods that are made faster, cheaper, and better. This has been true through thousands of years of recorded history, and it will remain true in the coming millennia as well as we transition from Earth-bound commerce to ventures that span the solar system.

In this chapter we will examine issues, challenges, technologies and other factors that feed the quest for the international cooperation required to initiate and sustain viable Earth-to-space and space-to-Earth commerce. The approach taken here is based on the proposition that the commercialization of space will follow the same path as terrestrial history, but with greater distance, cost, risk, and countless factors that are literally "unknown unknowns."

In addition, the colossal humankind endeavor into space will be enabled by a rate of technology change that is unprecedented and unparalleled in recorded history. That, too, is the challenge for international cooperation – each contributing nation will likely possess different levels of expertise, and yet despite the variation, or perhaps because of it, the nations of the world will find an inescapable need to participate in space commerce, as it will become central to their long term economic sustenance and success.

FORCES AFFECT INTERNATIONAL TERRESTRIAL COMMERCE SYSTEM TODAY AND THE INFLUENCES IN INTERNATIONAL EARTH-TO-SPACE COMMERCE

Commerce and the development of commercial prowess have been among the key drivers of human civilization. Ocean ports, for example, have always been not only the hubs of commerce, but usually of finance, culture, art, and higher learning. In commercial centers, acceleration of change is self-propelled through the interaction of people, ideas, ambitions, and challenges.

And acceleration also occurs as trade itself accelerates. With improving technology from oar to sail to steam, and then exponentially with air travel, commerce grows. The advent of integrated, intermodal transportation of cargo has led to yet another advance in the equation,[1] as it adds significant speed to the pace with which goods are moved. With commerce now moving into space, the "oceans" become regions of even greater challenge, near absolute zero in temperature, a near perfect devoid

---

[1] See Chapter 16, *Space Commerce: The Inside Story By the people Making It Happen*, Langdon Morris and Kenneth Cox, editors, Aerospace Technology Group, ISBN 978-0-578-06578-6, 2010.

vacuum, and unlike the watery ocean, a domain that cannot support life as we know it.[2, 3, 4]

But speed costs money, both to develop and to utilize, whether in the air, or on the computer chip, it is a question of economic investment and subsequent payback against the assumptions made in the business case.[5]

Today, for example, it may be counterintuitive to some, but container ships often take longer to cross the oceans than the Cutty Sark in the 1870s, as owners adopt a 'super-slow steaming' approach to reduce fuel consumption.[6]

Figure 2: Shipping 1869 and 2009: The Cutty Sark and The Emma Maersk

Hence, the decisions that are made in any given situation may not coincide with what one would have assumed, as narrow assumption-driven approaches may not work in the development of space commerce.

LINKING HISTORY AND THE WORLD TODAY: TECHNOLOGY AND ITS EFFECTS ON INTERNATIONAL COMMERCE

---

[2] http://en.wikipedia.org/wiki/1948_Tucker_Sedan; http://en.wikipedia.org/wiki/Preston_Tucker.

[3] "Fuel Efficiency of Commercial Aircraft: An Overview of Historical and Future Trends", Peeters, P.M., Middel, J., Hooolhorst, A., Nationaal Lucht-en Ruimtevaartlaboratorium, National Aerospace Laboratory, NLR, NLR-CR-2005-669, November 2005.

[4] *National Review*, October 3, 2011 Volume LXIII, No. 18. "Swift Blind Horseman?", Peter Theil, p 30.

[5] *Ibid*, p 31.

[6] "Modern cargo ships slow to the speed of the Sailing Clippers", by John Vidal, *The Observer*, July 24, 2010.

Space transportation systems designs, and consequently their cost, has not benefited significantly from innovation in fuel system design, celestial fuel depot storage and replenishment, or propulsion system technology since Wehrner Von Braun first speculated in the 1930s about how to get to space.[7] This same vision continues to be reflected in how NASA sees the next 50 years of access to space exploration.[8]

Figure 3: 75 Years Of Unchanging Fuel And Rocket Power Technology: The Model For The Future?

There is just not a priority and the commensurate funding being applied for the breakthrough that will be required to economically sustain the commercialization of space. Therefore, the pace of space conquest, including space commercialization, has slowed dramatically. How fast humanity achieves commercially viable space access is directly dependent upon the funding, both private and public, that is allocated for integrated transportation / fuel systems.

In this regard, the evolution of these systems doesn't just "happen," it's forced. And often this occurs when an immature technology is suddenly required in a more mature form, and all of a sudden new options appear to address the need.

In the case of space technology, while there is certainly no going back to the "glory days" of NASA, what will be the spark that ignites large-scale space commerce?

---

[7] *Colliers* Magazine June, 1953, feature article.
[8] NASA proposed Constellation Aries V Heavy Lift vehicle.

THE FUTURE POINTS TOWARD EFFICIENT SPEED TO ACCESS SPACE

As shown in Figure 3, the prevalent space lift system is vertical ascent and vertical (ballistic) re-entry or down. The space shuttle was the first step in the journey toward truly affordable and productive space access: horizontal (guided, controlled, and targeted) re-entry. The next step will be the horizontal ascent, and horizontal re-entry, or return. (That fundamental tenet of the development of space access was summarily ignored by the Constellation Exploration Program, which was proposed, initiated and then canceled by Congressional budget action.) This next step work is being pursued in England as a follow-on to the HOTOL (Horizontal Take-off and Landing) aircraft of the 1980's. The key technology is the engine that combines the jet with the rocket seamlessly and collects the rocket oxidizer (atmospheric oxygen) on the leg of the journey from take-off to the edge of the atmosphere while on jet power, as shown in Figure 4.[9]

This approach addresses the mass fraction-to-payload issue discussed in Chapter 12 in *Space Commerce*, Heavy Lift Boosters.[10] This is but one of several approaches to the continuation of the concept to the Orient Express or National Aerospace Plane (NASP) as proposed during the Reagan administration.

Figure 4: The National Aerospace Plane (NASP)

The technology required to build either of these crafts may still

---

[9] Reaction Engines Ltd., www.reactionengines.co.uk, Building D5, Culham Science Centre, Abingdon, Onon, OX14 3DB.

[10] See Chapter 12, *Space Commerce: The Inside Story by the People Who Are Making it Happen.* Aerospace Technology Working Group, 2010.

beyond what current fuel and metallurgy technology permit. However, it is clear that the horizontal / horizontal booster / space access systems will be a quantum technology step and the cost of space access and space goods will change dramatically. While that is a bold statement, the *shuttle system proved the cost of up-mass can be compensated for by down-mass when the vehicle system can in fact perform both functions.* Here, we are focused upon the booster system for its up / down-mass capabilities as *a part of a much larger logistical system.* That logistical system is *the space port*, which has connection to terrestrial ports.[11]

ECONOMIC IMPACT OF INTERNATIONAL COMMERCE NOW AND TOMORROW

How severe is this fuel / speed / efficiency / technology "gap" in commerce in our world today? A recent survey shows that the current mega-freighters, also known as Post-Panamax ships, larger ships, are slowing as much as 20% below top speed with the intent of fuel cost savings, averaging over 40%.[12]

While this is a suitable business response to immediate economic conditions, it may also ultimately delay the development of better, more systemic solutions to the same underlying issues. This interplay between profit and research is an ongoing dialog throughout history.

Between 1970 and 2010 the volume of world ocean commerce more than tripled, from 11 million ton-miles to 34 million ton-miles. In 2010, the cargo was comprised of 50% crude oil, oil products and coal, roughly 25% iron ore and grains of all types, and the rest of merchandise consisting almost exclusively of 1 and 2 ton equivalent-unit (TEU) containers of finished goods.[13]

A Gross Domestic Product (GDP) in 1990 dollars of about $41,000 billion across the globe in 2000, and an estimated $80,000 billion by 2010, certainly fueled by the 34 million ton-miles of goods transported for trade purposes.[14] Contrast this with the new wealth of the 1950s America, where the GDP of a post World War II world was a mere $4,000 billion.

---

[11] *Ibid.*
[12] "Modern cargo ships slow to the speed of the Sailing Clippers", by John Vidal, The Observer, July 24, 2010.
[13] *The Wilson Quarterly*, Autumn 2010, "The World Trade Revolution", Martin Walker, Pages 23-27
[14] http://econ161.berkeley.edu/TCEH/1998_Draft/World_GDP/Estimating_World_GDP.html Estimating World GDP, One Million B.C. - Present

This is a 10-fold increase in the interconnection of the world for goods and services in just 60 years, clearly a principal factor in the virtually unbroken growth of prosperity of the period 1950 – 1978.

What will happen, then, when lunar and space assets are added to the mix and the world becomes even more interconnected?

The wealth this represents is potentially staggering, and the international cooperation that is achieved in meeting this challenge is likely to be beyond what the world has ever experienced. While our world is a more dangerous place today than it was 50 years ago, with nearly every nation now participating "inside" the system of trade, no country that wishes to continue to exist can dare seriously disrupting this flow of money.

Our expectation is that sooner rather than later, every nation will also aspire to participate in the space commerce arena for the same reasons, namely national competitiveness. This same dynamic can also be harnessed to promote peace both terrestrially and beyond.

HOW TO GROW THE ECONOMIC RESOURCES NEEDED? TECHNOLOGY MUST BE FUNDED, CREATED AND EMPLOYED

The technical literature is full of examples of the impact of fuel technology and fuel cost on commerce.

The era of "cheap oil," for example, ended with the Oil Shocks beginning in about 1973, and this has had a fundamental impact globally ever since. Airlines that were once profitable entities and highly attractive stocks have seen their profit margins eroded to nothing, and the possibility of earning any profit at all generally depends on fuel prices. Andrew Soare, from the Lux Research Corporation noted, "It's hard to hang your hat on either $150 per barrel or extensive government support because either one can shift within a few months." He noted the average cost of jet fuel went from $3.89 / gallon in January 2008 to $1.26 in February of 2009, then back to $3.14 in July of 2011.[15]

Figure 6 shows that the specific fuel consumption per seat-mile fell exponentially for 15 years,[16] but then went flat precisely when fuel prices

---

[15] Aerospace Engineering", December 14, 2011, "Booking Flights on Bio-fuel", by Patrick Ponticel, Page 20-21.

[16] "Fuel Efficiency of Commercial Aircraft: An Overview of Historical and Future Trends", Peeters, P.M., Middel, J., Hooolhorst, A., Nationaal Lucht-en Ruimtevaartlaboratorium, National Aerospace Laboratory, NLR, NLR-CR-2005-669, November 2005.

rose, profits were squeezed, and research and development budgets were cut.

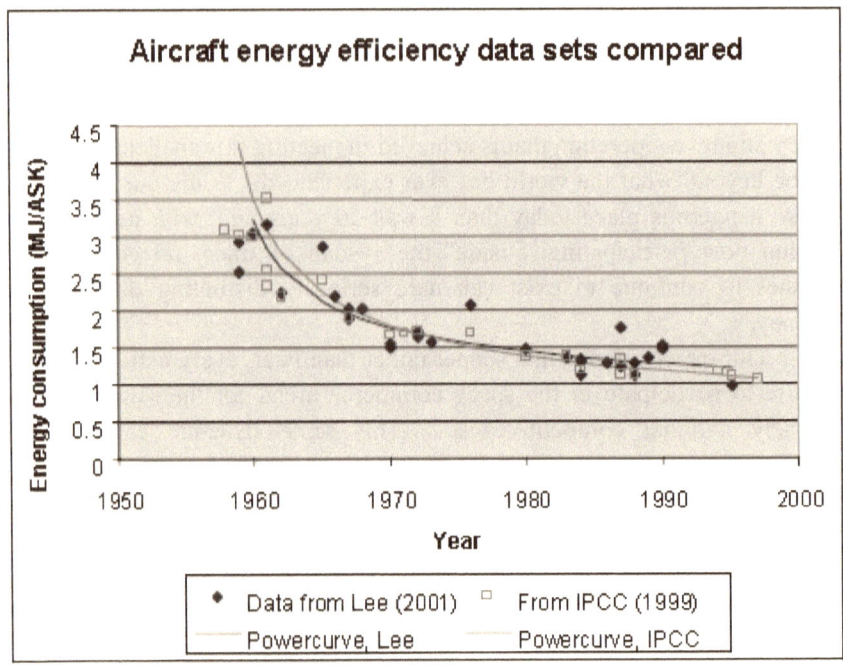

Figure 6: Seat Mile Costs Normalized by Fuel Cost

As we noted in chapter 16 of *Space Commerce*, as ocean cargo ships grow larger on a given technology they tend thereby to have a smaller cost per unit of delivered cargo, clearly driven by business efficiency requirements.

Similarly, Figure 7 shows the downward trend of total fuel consumption as aircraft became larger.[17] But even this curve has flattened out, showing that there is a point where a given approach, such as size alone, cannot yield additional economic benefit.

---

[17] *Ibid.*

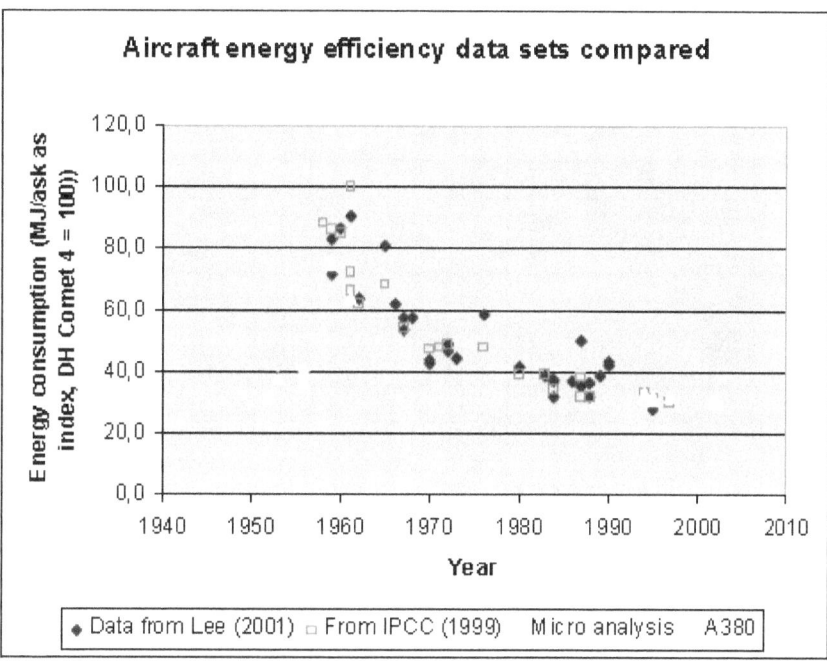

Figure 7: Total Specific Fuel Consumption Including All Technologies

Hence, for breakthroughs to occur, a uniform, continuous, and simultaneous push is required to be exerted on all related technologies to the transportation industry, inclusive of all modes in question.

And each industry will vary, depending on the unique and specific technologies involved. The aircraft industry will have a radically different solution set than the shipping industry.

What must be sought after in the space arena is a phenomenon as depicted in Figure 8.[18] The ultimate solutions might include fuel depots, space-based cryogenic fuel handling / depots, closed cycle food production, as well as solving medical issues, such as the reversing long term effects related to lack of 1 g-force, and reproductive / radiation effects.

---

[18] *Ibid.*

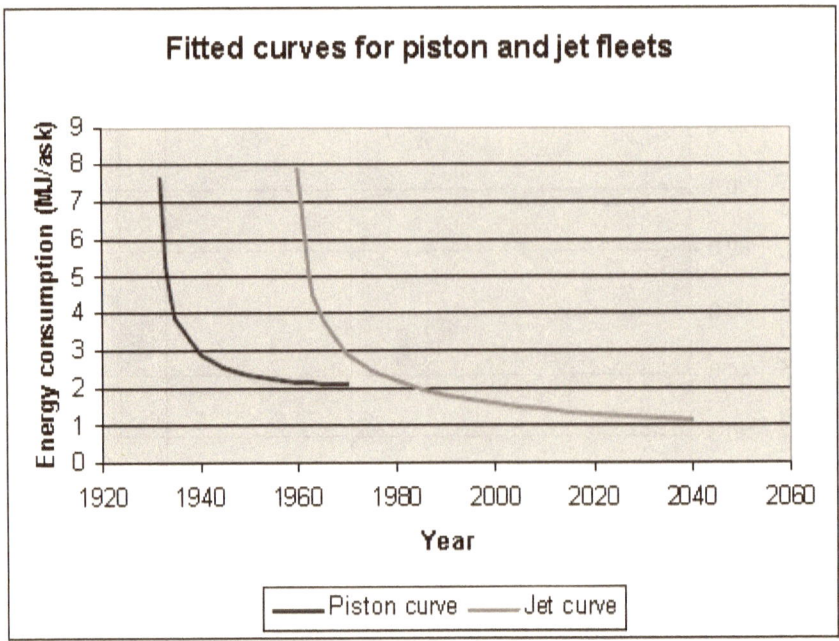

Figure 8: Technology Transition And The Elative Efficiencies Piston Versus Jet

LOGISTICS AND CONTINUOUS FISCAL SUPPORT: LOCAL, REGIONAL, NATIONAL AND INTERNATIONAL POLITICS INTERACT IN UNINTENDED WAYS

Significant change is coming with the opening of the enlarged Panama Canal, and we will use this as an example of how public policy and private commerce interact.[19] In a briefing to a public forum on September 21, 2011, Alex Dreyer, Chief Executive Officer of the Port of Houston Authority (POHA), discussed this rapid and profound change:

> *"The impact of the new, wider Panama Canal is a "game changer" for shipping, especially on the Gulf Coast when it is completed in 2014. Conservative estimates indicate we will see an increase of about 15 percent on the number of TEU's we handle. Those containers will be arriving on ships up to 3 times larger than the present ones transiting*

---

[19] Chapter 16, *Space Commerce – The Inside Story By the People Who Are Making It Happen*, Langdon Morris and Kenneth Cox, editors, Aerospace Technology Working Group, ISBN 978-0-578-06578-6, 2010.

the current canal."[20]

Houston is the second largest port in tonnage in the United States, and yet:

> "The Corps of Engineers estimates that it takes $40-50M per year to maintain the Houston Ship Channel at its authorized depth. For FY2011, the port of Houston needed $40.6M to meet critical dredging requirements, yet only $24M was funded by the federal government. This is despite $768M in customs revenues being collected."[21]

We see from this that when government is the controlling entity or major partner, the result is often mismanagement at the strategic level because the highly useful feedback derived from the profit motive has been removed. In a publically-controlled venture, acknowledging reality does not mean that will be reflected in funding.

> "Roberto Aleman, Chief Executive of the Panama Canal Authority, is on record as saying that Houston stands to benefit from his canal's expansion more than any other port on the Gulf Coast. However, he also recognizes that there is fierce competition among U.S. ports for dredging funds and he candidly admits that not all those ports vying for this funding will be called on by post-Panamax ships."[22]

Government oversight alone might not be sufficient to assure appropriate use funding. Lacking details, it appears to be the arbitrarily imposed limitation on maintenance requirements but in fact has the largest effect upon the enhancement and augmentation areas.

Another issue, also examined in *Space Commerce*, is that the traditional roles of the Navy and the Coast Guard would certainly be replicated in a spaceport because of the role that a spaceport would play not only in commerce but also in national security. Well-maintained ports with adequate infrastructure are stimuli for economic growth and fiscal soundness, and they are also necessary for a strong national defense.

---

[20] Remarks delivered at Baytran Monthly Meeting, 9/21/2011 by Alex Dreyer, Chief Executive Officer, Port of Houston Authority, 1111 East Loop North, Houston, Texas; www.portofHouston.com
[21] *Ibid.*
[22] *Ibid.*

However, strategic needs, certainly international ones, and most national ones, are rarely the same priorities as the expediencies of politics. (And certainly the politics of international space commerce are at best, still emerging.) Hence, we can expect that space commerce policy may be even more disconnected from longer term goals and objectives.

Nevertheless, it is our view that free international trade and trade into space, out of space, and between locations in space will be the key to future wealth. It is strategic, and like the Medieval cathedrals, building these capabilities will be a multi-generational process.

As with any endeavor that must endure over a prolonged period, education will be essential to sustaining the effort not only for the general public, but also for the nation's legislators. (NASA has also done a rather poor job at this.)

In summary, we see the necessity to *clearly segregate* or filter tactical port and commerce issues *into a checklist that requires further evaluation and careful study for applicability to the goal of international space commerce and cooperation in space.*

SOME MAJOR TECHNICAL CHALLENGES OF A SPACE PORT NETWORK: INTERMODAL OPERATIONS

Whether it is ocean vessels or "sky vessels" in the form aircraft or spacefaring vehicles, several axioms become clear from the discussion thus far:

- Fuel cost and speed are directly related for all transportation, terrestrial or space based: speed costs money.
- Speed of transportation is directly tied to cost of goods transported.
- The world is lagging in significant progress in fuel / transportation systems technology to increase speed for all transportation modes, but most notably aircraft and space transportation.
- The integration of aircraft and space vehicles may hold the technology key to economic use of space for commerce.
- Speed of goods transportation and ability for wealth creation are directly related.
- Fuel cost variability may kill technology investment precisely when that investment is needed most due to the scarcity of investment capital and the risk to profits.

- The trade space for economics terrestrially include speed but exclude it in space because of the physics of space transportation.
- Even in "old" modes of transportation such as ocean shipping, technology is evolving and governmental control struggles with strategic planning; the struggle in the "new" modes will be commensurately more difficult.
- Game-changing technologies in any technical area will not immediately nor quickly be embraced in terrestrial or space transportation because they are seen as, *and in fact are*, disruptive.
- Depending upon the extent of the disruption and the attendant conditions, most significantly the funds available to push it into use, these game changers *will force themselves* into the market.
- All policy and allocation is definable at a national level but the national perspective is developed regionally and politically.
- Going to the international level is the integration or confluence of these mixed-message national attempts at strategy.
- All of these factors will be present in space commerce, and each is both a growth opportunity and a potential deal breaker.

There are several identified problems of a technical nature that need to be discussed.

ADDITIONAL TECHNOLOGY CHALLENGES THAT SPACE COMMERCE MUST SOLVE

Figure 8 above shows that for a period of time, jet aircraft were more expensive per seat-mile than piston aircraft in terms of fuel consumption. How then did jet aircraft not only survive, but eventually drive the front line aircraft to be exclusively jets for military and commercial applications? For the same reason people in the 1930's were willing to pay ten times as much to book a transatlantic flight on the zeppelins as on a Cunard or White Star liner. The answer is that there was a strategic proposition at work - *there was no way a piston aircraft could match speed, and all assorted benefits that come with speed.* Piston technology had matured and was at the end of its life-cycle.

It so happens that the performance limit of piston / propeller aircraft wasn't metallurgy or fuel related, but simple physics. At any speed over about 450 mph, the propeller tips are induced to become transonic and then supersonic, with no load damping because they were built for subsonic airspeeds. In fact, the cost disparity led the jet community to aggressively pursue this and *not only to close the gap*, but to reach the same level of cost efficiency at a higher speed. This in turn provided for an increased profit margin because there are more ton-miles achievable by each aircraft.

A review of this case is warranted before the parallels between ocean shipping and space transportation are projected onto the space realm. During the 1950s and 1960s, the airline industry, driven both by domestic and international competition, was obliged to replace their propeller-driven fleets with jet aircraft. The jet engine had less complicated maintenance, cheaper fuel (JP-4 versus 120 octane aviation gas), and was a technology that held promise, seen in the Concorde and the TU-144, for supersonic speeds, limited only by metallurgical technology and avionics.

Yet today there are many propeller-driven aircraft still in use. After the opening of the enlarged Panama Canal the old, lower-capacity cargo ships will be in use and providing useful service for another 20 years. *Technology leaps raise the bar but not all are required by commerce factors to leap over the bar – there are many viable commercial pathways.*

PROPULSION TECHNOLOGY: NEW TECHNOLOGY IS REQUIRED FOR CRUISING THE OCEAN OF DEEP SPACE

An example of a successful technology that failed commerically, even when there was funding and both national and international interest behind it, was the revolutionary ocean shipping technology development of the "Atoms for Peace" initiative of President Dwight D. Eisenhower.[23] Introduced on December 8, 1953, the first nuclear powered merchant ship was the NS Savannah, shown in Figure 9. The program was halted after a completely successful sea trial and demonstration of capability for speed, reliability, and virtually no emissions. In an era more concerned with non-proliferation of nuclear energy into non-peaceful pursuits than carbon footprint or emissions as is the case today, the public did not support the Savannah.

---

[23] "Atoms for Peace" was the title of a speech delivered by U.S. President Dwight D. Eisenhower to the UN General Assembly in New York City on December 8, 1953.

Figure 9: NS Savannah

Fast forward to the post-1973 era, when scenarios for energy availability were radically changed internationally by the "oil shocks." Today when the debate continues to rage over whether global warming is indeed a real issue or a created phenomenon, this concept of clean, non-polluting power is not even given a consideration. How can that be? What happened to nuclear power in the ocean is yet another excellent example of the confluence of conflicting and largely counterproductive protection of the status quo under the banner of risk. An updated design of the reactor, with new metallurgy, placed in a double hulled, post-Panamax size ship would be an interesting economics case study. Clearly, discounting the solvable hijacking issues, this would be a very viable ocean transportation option.

And nuclear power is certainly attractive in space where concerns about accidental radiation release are of little or no consequence. A fission reactor would be a very logical source of power for a moon base, and would also be ideal for a long duration Mars mission, especially when coupled with a mature concept like the Variable Specific Impulse Magnetoplasma Rocket (VASIMR) engine, shown in Figure 10.[24]

Recalling that mass is not an issue for space transportation except when entering or exiting a gravity well, such as a celestial body landing or ascent, it is clear that any reactor capable of being launched and working in microgravity would be an effective power source.

---

[24] *New Scientist Space*: Ion engine could one day power 39-day trips to Mars, July 2009 by Lisa Grossman. http://www.newscientist.com/article/dn17476-ion-engine-could-one-day-power-39day-trips-to-mars.html

VASIMR Laboratory Experiment

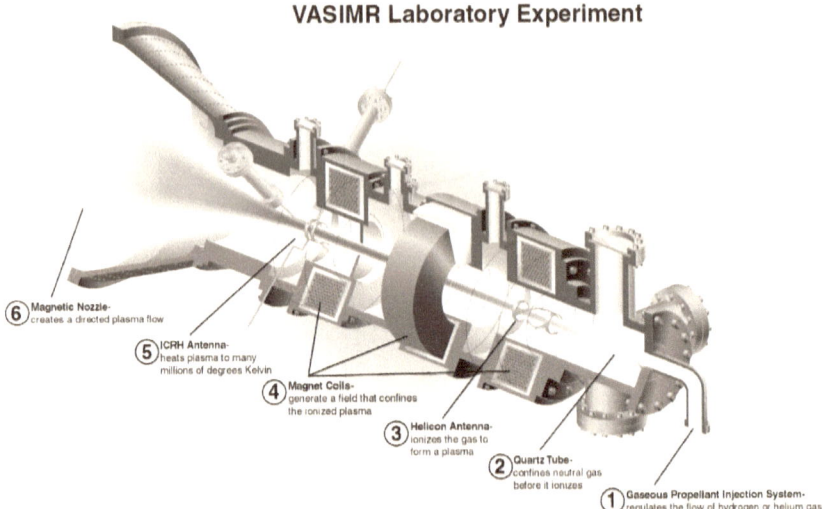

Figure 10: Prototype VASIMR Engine

The vehicle upon which the engine resides looks remarkably like the typical trans-lunar or trans-Mars transfer vehicle.

Figure 11: A VASIMR-Powered Trans-Mars Transfer Stage
Intermodal Shipping Containers and Handling of Commercial Space Cargo: A Key Element in the Profitability of Space Commerce

While the VASIMR is presently a terrestrially bound working

demonstration of the concept, the notional idea clearly is as revolutionary as the steam engine was to ocean travel after 4000 years of sail powered craft. This type of technology breakthrough will be required to economically cruise the oceans of deep space.

THE INTEGRATION OF SPACE TRANSPORTATION SYSTEMS WITH TERRESTRIAL TRANSPORTATION SYSTEMS.

Space based cargo elements, whether up or down mass, look substantially like the training mock-up shown in Figure 12.

Figure 12: A Typical Space Vehicle – Cylindrical with Spherical Appendages

The configuration geometry has mostly to do with the vehicle design, and less dependency upon volumetric or cost efficiency. The "space truck," the Space Shuttle, had the most effective and efficient cargo manifest system of any vehicle to date, yet it too was cylindrical, as shown in the training artifact shown in Figure 13.

Contrast the cargo bay of the shuttle, which itself had a squarish slab sided profile, to the standard 1 and 2 TEU (Ton Equivalent Unit) containerized shipping module for ocean / land shipping as shown in Figure 13. (NOTE: a TEU is the volume of a 20-foot-long (6.1 m) intermodal container, a standard-sized metal box which can be easily transferred between different modes of transportation, such as ships, trains and trucks.)

The first thought is the shuttle cargo envelope is larger than a TEU container. The terrestrial consideration of consequence is mass, while in space it is volume. This is an excellent illustration of the conflicting metrics, and the need for system integration of a workable mission level solution to cargo. The innocent response would be if the shipping container from space is cylindrical, larger than the standard 1 or 2 TEU container, then cargo would have to be repackaged for land and sea shipping back on earth.

Figure 13: Shuttle Cargo Bay model next to 1 and 2 TEU Shipping Containers

That thought highlights the *real challenge* here beyond geometric optimization, incompatibility of materials of construction and properties thereof. Those issues of course must be solved but that is a design consideration and considered a low risk. The *real challenge* is that the cost of ground handling of non-commercial cargo sent to the International Space Station and utilized on the space shuttle for over 135 missions was roughly 40% of the total processing cost for the mission. That would be an insurmountable cost penalty for international space commerce. That number would be of necessity larger than purely terrestrially bound cargo, but certainly would have to be cut by an order of magnitude for space sourced or space bound cargo. A 10% maximum is perhaps a starting point with the goal to be minimized to only fractionally more than terrestrially originated material.

A BRIEF HISTORY OF MODULAR SHIPPING

Containerization is a system of freight transport. Containers are built to standardized dimensions, and can be loaded and unloaded, stacked, transported efficiently over long distances, and transferred from one mode of transport to another, including container ships, rail and semi-trailer trucks, without being opened. The system, developed after World War II, led to greatly reduced transport costs, and supported a vast increase in international trade.

These were prototyped in the mid-1930s, and by 1953, the larger railroads had made the transition to containerized transport. The next step in the process of global intermodalization was purpose-built ships, specifically designed to accommodate containers. The first vessels to carry containers began operation in Denmark in 1951. In the US, ships also began carrying containers in 1951 between Seattle and Alaska. The containers were unloaded to purpose-built railroad cars for transport north to the Yukon, in the first intermodal service using trucks, ships, and railroad cars. This first lower 48 intermodal system initiated operation in late 1955 on the Atlantic seaboard. Full on shipping of containers was initiated April 1956, when a refitted tanker ship sailed from Newark to Houston.

During containerization's first 20 years, many container sizes and corner fittings were used; there were dozens of incompatible container systems in the US alone. The standard sizes and fitting and reinforcement norms that now exist evolved out of a series of compromises among international shipping companies, European railroads, US railroads, and US trucking companies. Four important ISO (International Organization for Standardization) recommendations standardized containerization globally during that time:

- January 1968: R-668 defined the terminology, dimensions and ratings
- July 1968: R-790 defined the identification markings
- January 1970: R-1161 made recommendations about corner fittings
- October 1970: R-1897 set out the minimum internal dimensions of general purpose freight containers

In the United States, containerization and other advances in shipping were impeded by the Interstate Commerce Commission (ICC), which had been created in 1887. The ICC was abolished in 1995 due to these containerized shipping and jurisdictional conflicts, as no longer needed.

Containerization greatly reduced the expense of international trade and increased its speed, especially for consumer goods and commodities. It also dramatically changed the character of port cities worldwide. Crews of 20-22 were replaced by a single crane operator who could still outperform that crew.

Another result was that the location of the port was changed, something that would have been nearly impossible to predict. As a result of inappropriate dockage configuration and an unwillingness to upgrade, the Port of San Francisco virtually ceased to function as a major commercial port, while the neighboring port of Oakland accommodated containers and emerged as the second largest on the West Coast.

In the 1950s Harvard University economist Benjamin Chinitz predicted that containerization would benefit New York by allowing it to ship its industrial goods more cheaply to the Southern United States, but did not anticipate that containerization might also make it cheaper to import such goods from abroad. Most economic studies of containerization merely assumed that shipping companies would begin to replace older forms of transportation with containerization, but did not predict that the process of containerization itself would have a more direct influence on the choice of producers and increase the total volume of trade.[25]

The lesson from this is clear. *While attempting to change the technology, policy control such as the ICC and standards will emerge but the gyrations and false starts are all part of the success.* This is not something that can be centrally planned and then executed.

How quickly can this same transformation be accomplished in space? That depends, of course, upon many factors, including the amount of money involved, the degree of international cooperation, and by which countries and companies. How willing will the terrestrial links to space commerce be to accommodating the newest intermodal player – space?

What is completely clear is that this process will occur but not under control by any government or governments, and the pace will be determined, as was the adoption of terrestrial containerization, by the economic engine of profit.

Two other considerations. A converted container can be used as an office or home. Could space-based units also be used for storage or habitat? How would that affect the logistics from the moon? From Mars? L2 or L5? Could these be like the post Panamax boats, the harbinger of the "super space ship"?

On the ocean, the Emma Mærsk, 396 m long, was launched August 2006. It has been predicted that, at some point, the size of container ships will be constrained only by the depth of the Straits of Malacca, one of the world's busiest shipping lanes, linking the Indian Ocean to the Pacific Ocean. This so-called "Malaccamax" size constrains a ship to dimensions of 470 m (1,540 ft) in length and 60 m (200 ft) wide.

---

[25] Marc Levinson (2006). The Box: How the Shipping Container Made the World Smaller and the World Economy Bigger. Princeton Univ. Press. p. 1. ISBN 0-691-12324-1. http://www.pupress.princeton.edu/titles/8131.html.

What does this mean for space commerce? At best a prediction of radical and rapid change is certain, but how it manifests is conjecture. After all, who would have correctly projected the shoe box sized mobile phone of the 1980's would be a palm sized mobile computing system that is in 2012 an inexpensive commodity?

With the demise of the shuttle program with STS-135 in July 2011, it has been argued that the United States and indeed even the world has lost the bridge to the future of space commerce because we decommissioned the only space truck that ever flew operationally.[26]

While this certainly is true for the International Space Station (ISS), this is also an issue for the space freighter of the future. The bulk of the ISS components are of the "round / cylindrical" variety, as are most of the smaller parts. When there are efforts to set up manufacturing and mining in space environments, there will be a requirement to build in the geometric considerations as part of the program. That certainly was the case for the shuttle program and the round / cylindrical space hardware. The requirement for the cargo bay of the shuttle and its capability of volume – recall the mass versus volume discussion in the Chapter 16 on Space Ports from the previous volume in this series, *Space Commerce* – was instituted *before the Apollo 11 astronauts walked on the surface of the moon!*[27]

Since the shuttle is no longer available to perform the prototype role for this packaging and shipping experience, the timeline is of concern for future design and integration of the shipping system.

In this regard, note that the logistics vehicles that support the ISS were being developed a full 30 years before the station itself was built. The message here is that effective long term planning is essential for successful space commerce development.

And at the same time, initial concepts will evolve significantly as they are being developed. Hence, the original design for the space shuttle and the final design were only in common by 10%. The driving force is the underlying idea, not the resulting product.

With regard to space commerce, the fully formed transport system will not be a single vehicle, but a fleet of them. That clearly compels the

---

[26] "Return to Reality: Why a Space Shuttle Program is Vital to the Survival of the International Space Station," George W.S. Abbey, Baker Botts Senior Fellow in Space Policy, James A. Baker III Institute for Public Policy, Rice University, Houston, Texas.

[27] "Notes on Headquarters Meeting to Define the NASA Space Station Logistics Vehicle – or "Did the Huddle Muddle the Shuttle?", May 9, 1969, Kenneth A. Young, FM6/ Orbital Mission Analysis Branch, JSC, United States Memorandum

space commerce community to unprecedented international collaboration on the design of this fleet of crafts and the "transition" craft that are the final leg from space to the spaceport. It must work for the entire world, as the 1 and 2 TEU containers currently do, or the costs will be prohibitive and commerce will falter. One could think of the current container ship and its evolution, which took less than 20 years from inception to the total demise of non-container shipping, as well as the evolution to the current Panamax ship. That sense of collaboration will be required again, but on *a much more complex problem*, space to ground and ground to space. Manufacturing will occur at near absolute zero temperature, in a perfect, but dirty, vacuum and shipping will take them to a human, supportive environment on Earth, all economically. These considerations cannot be added on; they must be designed into the spaceport system if international space commerce is to be successful.

Work Force Considerations: More Stringent than Terrestrial Considerations

Figures 14 and 15 show that the space shuttle / ISS space suit presents a manufacturing and logistical problem for any spaceport. Every astronaut must have a custom suit made, including legs, torso, gloves, and inner cooling gear. Expensive for sure, and unfortunately not interchangeable with other suits, and therefore subject to difficult logistical/supply issues.

These suits and cooling gaments require a special piece of computing machinery to generate the pattern that the suit maker, by hand, must translate into a durable, space-rated exo-garment. Certainly this machine shop could be put on orbit, but that would also require the artisans to live and work there as well. Perhaps this will be the next quasi-industry in space.

Another group of craftsmen labor over the custom-made glove, shown in Figure 16. While there are adjustments that can be made on all of these components, the adjustments are quite small. The utility of the garment pieces rests primarily upon the component artisan, and the creative interpretation of the estimation computer system.

Figure 14: Space Suits: The Logistical And Cost Challenge

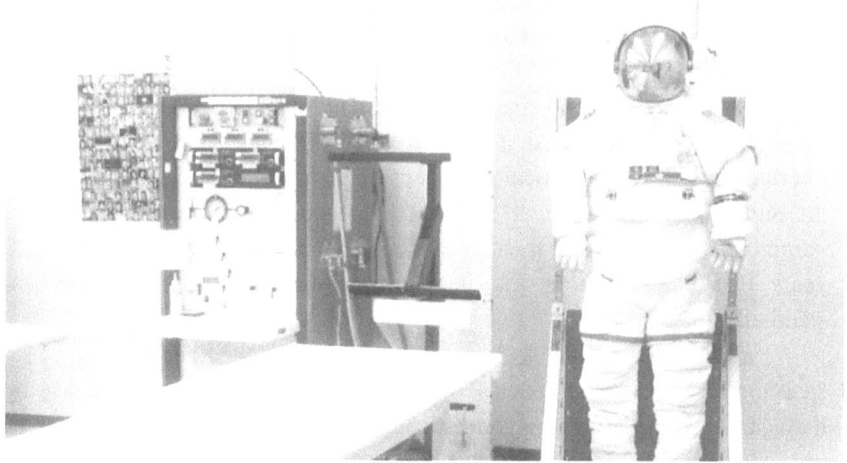

Figure 15: The Space Suit Artisans Craft Hall

With the termination of the Shuttle program, experience gathered during 30 years of shuttle flights has also been dissipated almost completely.

Russian partners wear an Orlon suit, which, while not as robust or capable, is in fact a second source. This situation points to a potential collaboration opportunity to meet space and terrestrial needs for garment tailoring techniques, systems, and technology that might make fitted suits and space suits from an industrial rather than a craft process. The number of people going into orbit will determine the size of the market, and thus the extent of the commerical opportunity.

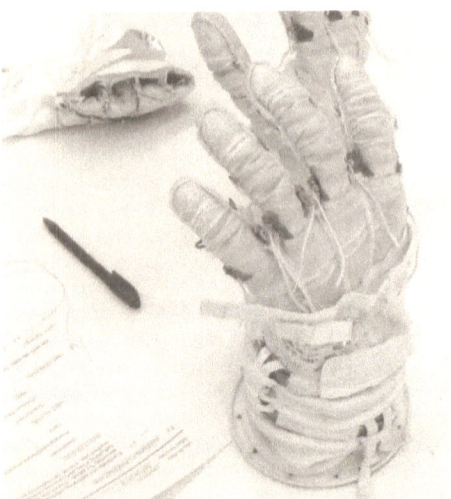

Figure 16: The Art of Space Gloves

If the space suit and space glove system looks like a throwback to the 1960s, it's because indeed it is. Reducing the art to a science on an international level is a critical path item for space commerce, and of course the suit is but one element that will be required *for large scale space commercialization*; recall the discussion of containerized freight and the 20 years that elapsed from the intial implementations in the "rush" to standardization.

Space tools must also become standardized, whether used inside a spacecraft or outside in space. And these tools must be rated for use by humans as well as usable by sophisticated robots.

Today's approach is based on massive customization of everything, which is also massively expensive to achieve, and complicated. Future space commerce systems will be designed from the top level to the smallest level of detail to be produceable at much larger scale, at one or two magnitudes lower cost.

The ability to have a tool box rich on improvising possibilities is essential in any deep space scenario, where dealing with the unknown will be the norm, and the distances involved mean that there is no recourse to Earth in an emergency situation.

The scenarios for training will also be different.

It would be expensive and impractical to remove a human from space service, de-orbit, recondition (to 1 g), train for space tasks, and return to service, which means that training will occur in space. A great deal of the training for ISS crew members is already occurring on board, as much as practically possible, although the Sonny Carter Training Facility at the Johnson Space Center is where ISS EVA training is currently conducted.

Figure 17: Sonny Carter Training Facility at the Johnson Space Center Zero g Simulation

Tools might be of different construction when adapted to space but the design and performance must be consistant with terrestrial tools, or the risk is created that the speciality of the tools, coupled with the lower production numbers and no commonality with terrestrial applications, would force the price, in a market economy, to a completely unsupportable cost.

Today on the ISS, items such as commercial camcorders are being utilized off the shelf (with the exception of the requisite certification for batteries, a safety issue).

It's also worth giving some thought to the technology required to commercialize space food. While current space food is vastly improved from the quality, diversity, and taste of anything available in the past, it is still, produced, processed, and prepared on Earth, repackaged and reconstituted for ingestion. Shelf life is limited, and the ability to stockpile is also limited, making constant resupply a necessity. While mechanical parts, if absolutely necessary, can be logistically inventoried, food is not infinite in shelf life. New technology is required that, once perfected, would most certainly find emergency relief, military, and third world applications where transportation is an issue, such as during emergencies and famines.

But if the perspective of the commercialization of space requires the generation of enormous capital profits, then the lure of spin-off technology has a significant pull in a positive direction. If space food were as tasty, nutritious, and economical as terrestrial grown and prepared food, it would be extensively used in many applications. There is a long way to go to understand and meet the requirements of long term exposure to the rigors of space, and proper and tasty nutrition.

While we're discussing supply and logistics, the issue of standards deserves a final word. The mix of partners, a multitude of participating nations, virtually assures that issues of integration and interoperability will require extraordinary effort and attention.

How competing "standards" for the logistical parts and sub parts is handled will also be a factor in determining eventual profitability. *Standardization and terrestrial collaboration on the associated technologies must be carefully guided by policy that promotes profitability. A future version of the International Standards Organization, ISO, can be expected to emerge, which we might think of as the SSO, Space Standards Organization. Among the many issues that will be discussed and negotiated is how various technical and mecahnical standards will be developed for the varying degrees of gravity that occur on various celestial bodies. Should tools designed for use the Moon also be usable on Jupiter? Or at LEO? Or L5?*

It's likely that we'll depend on entrepreneurs to drive not only the development of these required standards, but also the full scope of the business systems that will make space commerce a reality. We can look to today's and tomorrow's Warren Buffets, as well as Elon Musks, Robert Bigelows, Richard Bransons, and all the other investors, entrepreneurs, visionaries, engineers, and managers to drive these industries forward.

ASSEMBLING THE PIECES: A THOUGHT EXERCISE ON HOW A SPACEPORT COMPLEX MIGHT FUNCTION

Can we construct from the initial conditions of today a speculative but plausible scenario describing how the commercialization of space might emerge?

The ensemble of vehicles and concepts shown in Figure 19 is from a NASA work on that very topic.[28]

---

[28] "A Flexible Path to Mars, With Commercial Opportunities and Public Benefits Along the Way", Lynn Harper, et al, NASA Ames Research Center, Moffett Field, CA, 2011.

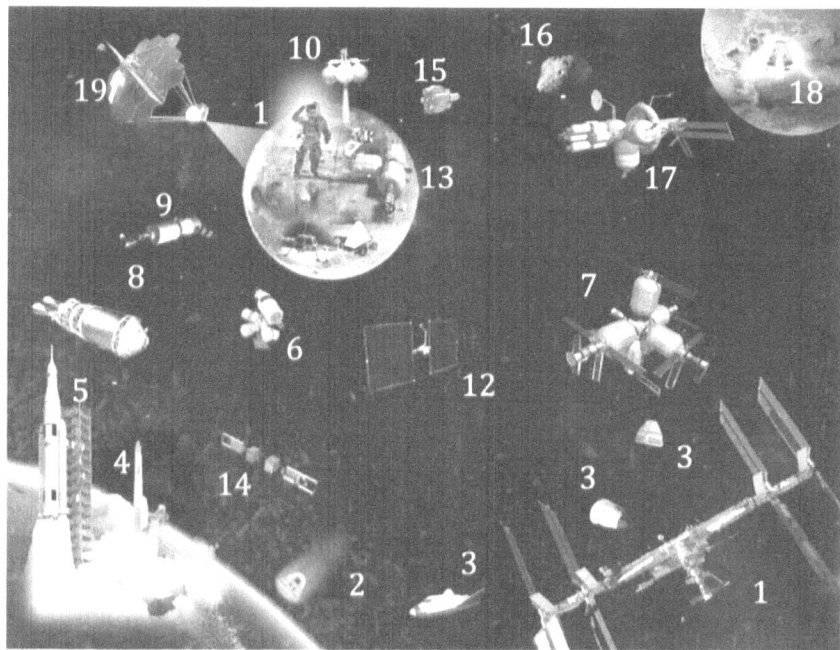

Figure 19: Potential Space Commerce Mixed Fleet Components: 2012 - 2065

1. International Space Station, ISS
2. Orion (and derivatives) – NASA vehicle for space access
3. Commercial crew space access vehicles:
   a. Dragon
   b. HTV
   c. Dream Chaser
4. Ares / Delta / Atlas – near term launch systems
5. HLV – future heavy lift system
6. Cryogenic tank farm in LEO for refueling
7. Commercial space station – with inflatable habitats
8. Trans-lunar stage
9. Robotic lunar / Mars transport
10. Morpheus – Methane-oxygen fueled lander system
11. Crew expedition / lunar outpost
12. Space based power generation station at Lagrange point
13. Lunar inflatable habitat
14. Orbital cargo transfer system
15. Deep space expendable cryogenic fuel tanker
16. Asteroid target / manufacturing resource
17. Mars transport vehicle
18. Mars crew lander system
19. Solar collector / concentrator for power generation and beaming

No credibility is extended to any of these systems, vehicles, or concepts, except to say they are plausible and possible, and have progressed beyond the state of an idea.

This collage of vehicles shown in Figure 19 in no way is optimized or proposed as a viable fleet, or even a desirable fleet. Rather, it takes what is in the inventory, what is currently under development, what is proposed, and what is thought to be a possible next step and welds them together to illustrate the point of this chapter in a story, which is that there will be a mixed fleet of vehicles, expectations, dreams, hopes and beneath all that, a mixed set of regulations, folkways, mores, and inspirations, all of which must, like the discovery of the New World, be melded together. From this forging the next chapter of humankind will be told.

For purposes of a hypothetical scenario, a spaceport hub is shown. Starting on Earth with a vertical – vertical or at best vertical-horizontal launch system, ISS (1) orbits the earth being resupplied by a host of commercial cargo / crew systems (3), competing for the business of science, product, and people ferrying. Launched primarily by medium capability lift vertical-vertical launchers (4), the crew and the return cargo (2) are entered through the atmosphere in the technique of the Apollo era, landing in the water. This allows for gentle handling and facilitates the "use and replace" motif of the transport system. This activity dominates the space transport business through the early 2030s, and only with the advent of the very heavy lift system (5), again vertical-vertical, does the industry grow into the trans-lunar and lunar Lagrange point market for vehicles and crew. This heavy lift system and the medium lift system combine forces to put inflatable commercial space stations (7) and space tourists in orbit.

These habitats require by their nature as tourist destinations routine and frequent launches of both launch systems (4), (5). While shuttling people and completing re-supply, the launch systems also provide the parts and manpower to assemble tank farms for cryogenic propellant storage, as well as the first high orbit communications network intended for space-to-space communications and high-speed celestial internet / web system (12) to service the L2 and L5 Lagrange point assets.

Low cost commercial and perhaps some government sponsored robotic landers (10), orbited by launch systems (4), (5), provide robotic material (9) processing, and the habitat precursor capability on the lunar surface. The literal scrap yard of lander bases is collected for the construction of the permanent human habitat (13). Power is microwaved to the lunar surface to facilitate the construction of lunar habitats from a collection station (19).

Within 5 years of the first landing, construction crews (11) will complete a permanent lunar base in 2035, providing sufficient infrastructure for a long duration crew settlers habitat (13) that operates the mining, refining, packaging shipping and manufacturing of the repair parts for the lunar surface operations. This integrated silicon, germanium, arsenic and iron production provides the materials for the L2 factory and L5 processing (17), (12) that are a major profit center for the international computer chip consortium. These extremely high speed chip sets are utilized to construct directly on the lunar surface the majority of the computing power necessary to continuously operate with precision the cost effective product and people transfer shuttle flights (14) from the lunar city to Earth.

Utilizing the assets of the Lander bases and indigenously produced parts, the necessary fuel production equipment, along with assets from earth launches of the heavy launch system (5), the first pieces of the infrastructure for the Mars Team are launched in 2050 (15). In 2060, the Orbiting Mars Control Center for Exploration (MCCex) (17) is launched into a polar Martian orbit, providing continuous control for the robotic elements of the infrastructure on the Martian surface.

Space tourism is a huge high profit center on this budding inter-celestial highway, as despite the cost, the duration of absence from earth and inherent dangers of radiation, distance, and lack of Earth level medical care in space, demand far outstrips the ability to supply opportunities.

Space workers and their robotic assistants get first selection seating on the available shuttle craft. The shift of assets and interest to permanent Mars habitation in 2065 constrains the entire celestial highway with traffic and resupply needs. The mix of government vehicles and predominantly commercial vehicles, most of which are owned by an international consortium, is opening up a market for "pre-owned" space vehicles, which are slower, less safe, and more prone to delays. This industry will no doubt be a major market force by 2075.

In Earth orbit, where space manufactured goods are placed in a parking orbit, the payload bay of the passenger liner Skylon III is filled with cargo modules after the Skylon III delivers passengers into orbit. This horizontal-horizontal cargo shuttle has such a rapid turn-around in the launch-to-delivery-to-reload-to-reentry that along with its fuel-only launch system, it is likely to capture all but the very large element cargo market in less than a decade. By 2080, a Skylon IV that handles twice the cargo volume will be operational. This will increase the profit margin of celestial manufacturing considerably, and evokes the memories of the 2020 post-Panamax ocean ships, and the dramatic cargo cost reduction achieved in terrestrial commerce as a result of these massive container ships.

The advent of large scale, non-vertical launch, point-to-point, land-based cargo systems has caused the Port of Houston Authority, located only 10 miles from NASA's Johnson Space Center, to expand its operations and become known as the Celestial Orbital Port of Houston Authority (COPOHA). With facilities to ship by unmanned aircraft vehicle (UAV) robotic rail, and robotic highway cargo pods, the complex at COPOHA handles nearly 40% of the dollar volume of goods entering the United States.

It is projected that by 2062, only a short 100 years after the space race began in earnest between the USSR and the USA, spaceports around the world have ceased being organized around dominance in warheads and counts of missiles in inventory. Rather, collaboration in space has shifted competition into a matter of profit centers focused on developing and deploying faster, more reliable, and higher technology strategic cargo systems.

Hundreds of years from now, surely someone will write a book about the parallels of the journey of Columbus, the first transcontinental railroad across the United States, the Trans-Siberian railroad, the Suez and Panama Canals, and other great voyages and transportation firsts, and recognize commerce as the very common human thread uniting all these endeavors. Yes, indeed: Houston we have a profit ....

SUMMARY

In this chapter we have discussed many of the essential commerical and technological issues that must be considered as commerce moves outward from Earth. Some of the principles we have explored include:

- The speed of commerce often dictates the scale of the profits that may be derived.
- Speed is also the connective tissue between terrestrial and celestial vehicles.
- Speed is a function of fuel / propulsion systems terrestrially and in space, with the added factors of gravity wells and Keplerian motion being critical features of space transportation as well.
- Technology and policy, social trends and international market

- forces often pull in different directions, but synergy must be realized in order for commerce to emerge in space at any significant scale.
- Legislative intentions often fail to translate into funding allocations that address intended needs, even in vital areas such as transportation. This suggests that the more control lies with commercial parties, the faster will be the progress. The challenge then becomes balancing speed and safety, and not sacrificing safety in the reckless pursuit of speed.
- Humankind's sorties into space will require national and international focus on the expansion into permanent footholds required for international space commercialization.
- Eventually, none of this will be about flags, footprints, and fame; it will be about market share and profit potential.
- A new ISO will likely emerge, a Space Standards Organization, through which humans – and perhaps robots as well – will negotiate the standards that will enable commerical interoperability and efficiencies for tools and technologies that are used across the solar system.
- All this will unfold with all the strengths and weaknesses of all of humankind, in full evidence and on display.

May the authors and readers alike not only live to see this adventure move beyond the beginning that is unfolding today, but live to see it become a significant reality, and perhaps witness the beginnings of the next chapter in human evolution, commerce-based outposts in the cosmos.

•••

Tom Diegelman

Tom Diegelman has been in the aerospace community for over 35 years, involved in the research, development and operation of training simulators, ground based flight control installations and facility operations. Tom started his career with Cornell Aeronautical Laboratory as a research engineer, working on early versions of shuttle handling quality study simulations and shuttle shock tunnel testing.

Tom moved to Houston in the late 70's to join Singer / Link Flight Simulation and worked in the Shuttle Mission Training Facility (SMTF) as a model developer, and later a manager of simulation projects. In 1988, Tom joined NASA to lead the $170M redesign of the SMTF. Assignments at NASA / JSC include projects in advanced mission control technology, technology development, and facility operations control. He served as Facility Manager for Mission Control for 3 years before accepting an account manager position in the Technology Transfer Office, developing partnerships and Space Act Agreements.

The design of the training facility for the Constellation Program culminated his nearly 30 years of experience at JSC in the Jake Garn astronaut training facility. Tom most recently became the ISS Vehicle Safety Engineer for Communications and Tracking Subsystem.

Tom was elected to Seabrook City Council in 2006, and served two terms, during which he worked closely with the Port of Houston Authority on the Seabrook / Bayport Terminal Facility issues. Tom is a member of Baytran, a non-profit organization promoting inter-modal transportation solutions in the Houston / Bay Area, and continues to be involved in local, state and federal government on behalf of the space technology community.

A dedicated writer, he is coauthor of two chapters in the previous volume in this series, *Space Commerce*, and also coauthor of Chapter 17 in the present volume.

CHAPTER 7

INTERNATIONAL SOCIAL COOPERATION IN SPACE AWARENESS
YURI GAGARIN, YURI'S NIGHT AND *FIRST ORBIT*

DR. CHRISTOPHER RILEY
PROFESSOR, THE LINCOLN SCHOOL OF MEDIA, UNIVERSITY OF LINCOLN

AND

DR. CHRISTOPHER WELCH
DIRECTOR OF MASTERS PROGRAMS, THE INTERNATIONAL SPACE UNIVERSITY

© Christopher Riley and Christopher Welch, 2012. All Rights Reserved.

1. INTRODUCTION

Yuri Alekseyevich Gagarin's flight into space in April 1961 turned a hitherto unknown Soviet Air Force officer into a hero around the world.

Upon his death in March 1968 he became an icon, his image forever frozen. Since then, the date of his flight has been celebrated in the Soviet Union and its successors as Cosmonautics Day. April 12, 2001 marked the 40th anniversary, and April 12, 2011 the 50th.

Between 2001 and 2011, under the banner of *Yuri's Night*, an annual global celebration of the anniversary has been organized. Taking the form of a worldwide grassroots effort mediated via the internet, *Yuri's Night* and associated activities have celebrated humankind's exploration of space, and provided a mechanism for reaching out to the general public, particularly young people, and raising awareness of space exploration. A recent example of this has been the hit 2011 experiential documentary film *First Orbit*, which celebrates Gagarin's flight in a unique and powerful way. Created with the specific intent to be premiered on the 50th anniversary, *First Orbit* became a significant global phenomenon that has brought additional attention to both *Yuri's Night* and to the broader commitment, felt by people around the globe, that humanity's movement into space will continue and will engage more people and more nations in a global cooperative effort.

2. THE GENESIS OF *YURI'S NIGHT*

In 2000, Loretta Hidalgo, Trish Garner and George Whitesides realized that the 40th anniversary of Gagarin's flight was approaching, and determined that his achievements should be celebrated globally. They conceived the idea of a world space party to do this, calling it *Yuri's Night*.[1,2] When they met for the *Space Generation Forum 2000* in Vienna, they used the SGF network to initiate the project. The *Yuri's Night* cofounders gathered email addresses, and in a few days designed a basic web page.

After the summer of 2000, they worked on the website, sought funding, told everyone they knew, wrote letters and sent emails around the world. Enthusiastic replies came back, many more people became involved, local organizers agreed to arrange *Yuri's Night* events, and the project moved into high gear.

A team of volunteers based in California developed additional resources, including posters, press kits, graphics, a party locator, 'how to' web pages, and a *Yuri's Night* chat room.

---

[1] Space Generation Forum, http://spacegeneration.org/index.php/about-sgac/history Retrieved 13-02-2012.

[2] UNISPACE III, http://www.un.org/events/unispace3/. Retrieved 13-02-2012.

With the success of an early grant application, the founders were also able to provide local organizers with $100 seed cash and a 'party pack' containing a *Yuri's Night* T-shirt, publicity material, and a selection of space music.

3. YURI'S NIGHT 2001

On April 12, 2001, sixty-four *Yuri's Night* events took place in twenty-nine countries on seven continents.

About 1200 people attended the *Yuri's Night* flagship event at The Palace Night Club in Los Angeles. In the main lobby was a mini-convention for space, with stands from The Planetary Society, The Mars Society, the International Space University and others. During the five-hour party, lasers spelled "Yuri's Night" and drew pictures of famous spacecraft. Silver-suited go-go dancers adorned two platforms, and above the whole thing was a screen showing classic space moments, such as Gagarin's launch, STS-1's launch and Neil Armstrong's 'small step.' The entire event was webcast and much media attention was generated.

Figure 1: The 2001 Yuri's Night party in Los Angeles

In Vancouver, 150 people attended a space rave that ran from 2200 to 0800 the next day. The event featured a number of DJs together with stunning space visuals.

In Bujumbura, Burundi, 600 young people attended a conference entitled Humans and the Environment that considered topics including, How can we preserve life in the world? How can science and technology contribute to human life without destroying our planet? And Yuri Gagarin as a Pioneer of Technology and environment development. The day concluded with a social evening and dancing.

The Paris *Yuri's Night* event took place at the Cafe de Flore, attended by thirty-five artists, writers and scientists. Representatives of the French space agency, CNES, brought to the party a flight-ready duplicate of their Sputnik model, launched from Mir in 1995, and a 17-second video showing the cosmonaut hurling the little satellite into space during an EVA.

The *Yuri's Night* party in Leiden, the Netherlands was attended by about 110 people, including Dutch space journalists and a future Dutch astronaut. Posters, balloons and spacey frisbees, courtesy of ESA, were used as decoration in an old pub.

The London *Yuri's Night* Party was held at The Rocket Complex from 2000-0100 GMT, attended by 200 people. It started with a space-oriented short film show and some space performance art, and moved into high gear featuring DJ sets combined with stunning space and Yuri visuals.

Figure 2: Flyer for the London 2001 Yuri's Night party

The *Yuri's Night* event in Sydney was run as a special event for members of the Powerhouse Museum Members' Association. It was held in the members' lounge with a backdrop of soaring aircraft and spacecraft suspended from the gallery ceiling. The event attracted over 70 people, including children and senior citizens. A variety of multimedia presentations enhanced the night, including giant video projections of Gagarin's flight beamed onto the Museum's inner walls and ZIA's *Yuri's Night* music.

About 100 students and professionals attended the Adelaide party, hosted by the Stag Hotel. Entry was free. Educational posters decorated the venue, and the DJ asked quiz questions throughout the party based on the information available on the posters, with free drinks for correct answers. Andy Thomas, the Adelaide-born astronaut who had recently returned from the International Space Station where he had performed a spacewalk attended. (Dr Thomas had trained at the Yuri Gagarin Cosmonaut Training Centre, and lived aboard the Russian space station Mir for 141 days.)

The *Yuri's Night* celebration in Lahore, Pakistan featured a number of events attended by about 30 lawyers, scientists, engineers and students. The Pakistan space agency Suparco provided the venue and presentation equipment.

The Cape Town *Yuri's Night* event was a wedding held at a planetarium set to ZIA's music Back 2 the Moon and To Mars! The bride and groom exchanged titanium rings, and all 130 guests adjourned to the planetarium foyer for a buffet with a cake in the shape of the US Space Shuttle.

4. LESSONS LEARNED FROM *YURI'S NIGHT* 2001

Yuri's Night 2001 was the first event of its kind to happen on such a scale within the space (enthusiast) community. It started with an idea and was transformed through widespread effort and enthusiasm into a truly global event. It was, to borrow a phrase, very definitely a matter of "Think globally, act locally."

Electronic communication was vital, and while social media had not yet emerged, simple email and websites were enough to deliver information and resources.

Yuri's Night 2001 was a proof-of-concept that demonstrated that significant global space outreach events were achievable by the space community, using a decentralised approach and utilising electronic communications for co-ordination. Furthermore, the simultaneous nature of all the *Yuri's Night* events lent a sense of excitement and connectedness that generated enthusiasm in organisers and participants, and significant interest from the media.

The informal and celebratory aspects of *Yuri's Night* attracted and educated many people who may not have been be reached by more formal and educational space outreach activities.

Although it was originally conceived as a celebration of the 40th anniversary of Yuri Gagarin's flight, it continues to be celebrated each year, and by the time of the 50th anniversary in 2011, it had become a fixed part of the space calendar. The number of *Yuri's Night* parties and activities increased steadily, boosted by the emergence of social media communications channels such as Facebook and Twitter, to a total 567 events during the 50th anniversary in 2011, involving an estimated 100,000 people in seventy-five countries on all seven continents.

In addition, one particularly successful project, combining both international cooperation in space, international internet-mediated co-operation, and social media is the film *First Orbit*.

5. 2011 - FIRST ORBIT

To mark the 50th anniversary of human spaceflight on the April 12, 2011, a feature-length film, *First Orbit*, was produced and directed by Christopher Riley to bring the story of Yuri Gagarin to a new generation. In collaboration with the European Space Agency, NASA, Roscosmos and the Expedition 25, 26, and 27 crews on board the International Space Station (ISS), a new view of Earth was filmed over several weeks as the ISS passed over the same ground track at the same time of day as Vostok-1 had done on its 1961 flight.

The resulting footage was edited together with the original voice recordings from Gagarin's mission and a new musical score from composer Philip Sheppard, and on April 12, 2011 the film was premiered on YouTube, generating the largest audience for a long-form film release in the website's history.

On the same day the film was screened at over 1600 venues in more than 130 countries around the world, making it one of the most widely released independent films of all time.

5.1 ORIGINAL CONCEPTION

In late 2009, following his previous documentary film projects *In the*

Shadow of the Moon (2007), *Moonwalk One – The Director's Cut* (2009), and the video installation *Apollo Raw and Uncut* (2009), which projected the entire Apollo flight film archive into public gallery spaces in London and Montreal, Riley looked for another archive-based film project to celebrate the 50th anniversary of human spaceflight. It soon became clear that Yuri Gagarin's Vostok-1 flight of April 12, 1961 was the right subject.

Gagarin lifted off from the launch site near Baikonur, not far from the Aral Sea, at 06:07 UTC on April 12, 1961. He flew northeast across the eastern part of the Soviet Union and Siberia, and on across the terminator and into night over the Pacific Ocean. At 07:10 UTC he emerged into sunlight again over the Southern Atlantic and passed over Africa, the Mediterranean Sea, Turkey and the Black Sea, before landing just north of the Caspian Sea, 108 minutes after launch.

The flight was not captured in any significant way on film or video, and only a single TV camera on board Vostok-1 was used briefly to transmit an image of the cosmonaut inside his capsule during his flight over Soviet territory.[3] But as Gagarin headed East over the Pacific Ocean, the transmission signal was lost and no attempt was made to continue recording pictures on board the spacecraft.

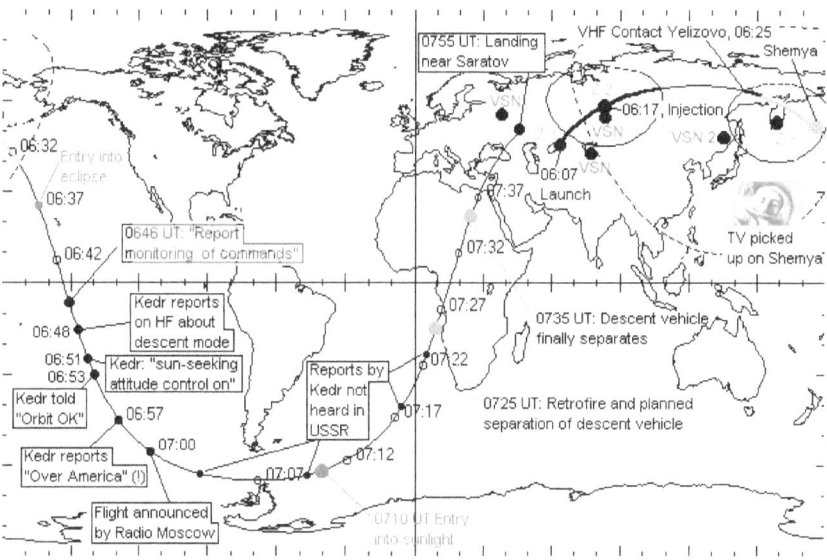

---

[3] Grahn, S. TV from Vostok
http://www.svengrahn.pp.se/trackind/TVostok/TVostok.htm. Retrieved 12-03-2010.

Figure 3: The ground track of Vostok-1 (Courtesy of Sven Grahn)

Audio recordings of the flight loop between Vostok-1 and mission control were made throughout the flight, both on board the spacecraft and on the ground, when communications allowed. Translated transcripts have been widely circulated since then, but the complete audio recordings were apparently never released outside Russia.

In 2009 Riley conceived the idea of creating a new Vostok-1 film, which would include as a central element a view of the Earth that Gagarin could have seen. The initial proposal was to piece together existing archive footage shot in Earth orbit over the same ground track and at the same time of day as he flew. Maps of the Vostok-1 trajectory[4] were used to guide the archive research needed for such a production approach (see Figure 3). A review of the Earth-view footage in the NASA archive that was shot over the past fifty years quickly led to the conclusion that it would be too difficult to utilize existing media of consistently high-enough quality to make the film in this way, and the idea was shelved.

Figure 4: NASA astronaut Tracey Caldwell-Dyson in the ISS Cupola, during Expedition-24 (Courtesy of NASA)

However, in early 2010 when the Italian Space Agency's cupola was

---

[4] Grahn, S. An analysis of the flight of Vostok http://www.svengrahn.pp.se/histind/Vostok1/Vostok1X.htm. Retrieved 12-03-2010.

installed on the International Space Station (ISS), much was made in the media about the unparalleled views of the Earth which it offered (see Figure 4), and the idea to create new, high-definition digital video views of the same ground track at the same time of day that Vostok-1 had flown almost fifty years before was born. The resulting footage would then be edited together into a 108-minute film, and combined with the original voice recordings to create a new video installation for gallery spaces around the world.

5.2 Producing the Film

With the support of Bob Chesson, Head of the Human Spaceflight and Exploration Operations Department at ESA, an initial feasibility study was undertaken.

	Vostok-1	ISS
Orbit	302 x 170 km	350 x 350 km
Inclination	65.0o	51.6 o

Table 1: Comparison of Vostok-1 and ISS Orbits

However, full orbital elements for Vostok-1 proved impossible to track down, and for parameters that were identified, there were small discrepancies in their values between different sources. The figures adopted for *First Orbit* were eventually taken from *Man's First Space Flight – a TASS Report.*[5] Using these values and the map of Vostok-1's orbital ground path (see Figure 3), ESA's Gerald Ziegler performed the initial calculations to see if the Space Station's ground track ever matched that of Vostok-1. Ziegler concluded that a similar ground track would be made by the ISS every 48 hours or so, but when matching the time of day as Vostok-1, the frequency of occurrences dropped to around once every six weeks.

While there was no chance of capturing a complete 'Gagarin view' during a single pass around the Earth, Ziegler recommended breaking up Vostok-1's ground track into a series of five separate segments which would be matched to future ISS ground tracks (see Figure 5). The filming opportunities for each segment could then be identified and the footage captured on different days and then edited together to give the illusion of the single 108-minute flight around the Earth that Gagarin originally took.

---

[5] Yuri Gagarin, et al, 1961. *Man's First Space Flight – A TASS Report*, pg 9. Soviet Man in Space, 2001, University Press of the Pacific. Reprinted from the original edition.

ESA started making plans for Italian astronaut Paolo Nespoli to carry out the filming towards the end of 2010, when he arrived on the ISS as part of Expedition 26.

A draft shot list was compiled, noting the preferred camera positions and directions for each orbit segment. The camera, a Canon G1 HDV, is one of the standard video cameras on board the station. It was set up in a fixed position, the recording started, and the camera left to run for the duration of the ground track segment.

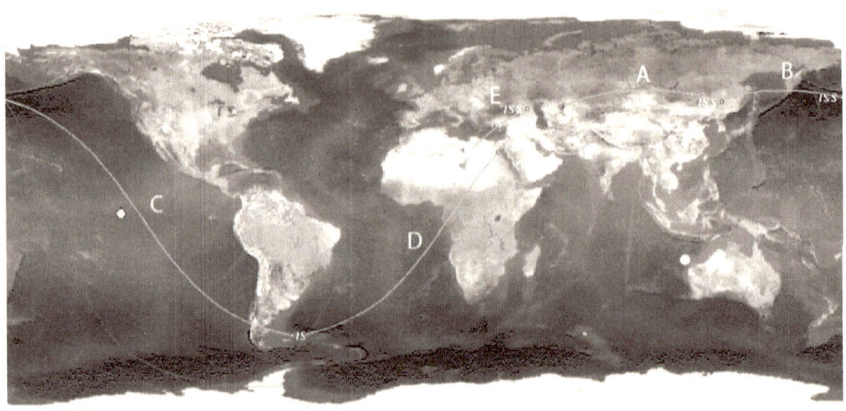

Figure 5: ISS ground track segments A-E chosen for filming purposes to most closely match the ground track and time of day of Vostok-1 (see Figure 3)

5.2.1 FILMING

In October 2010, NASA astronaut and Expedition 25 Commander Douglas Wheelock performed a test shoot inside the cupola. The footage captured was compressed into a single file and transmitted to Houston through NASA's Tracking and Data Relay Satellite System (TDRSS) network, and then on to ESA-ESTEC in Noordwijk, The Netherlands, where it was shared with Riley through an ESA media centre ftp link, co-ordinated by Jean Coisne and Melanie Cowan.

Nespoli was launched to the ISS on Soyuz TMA-20 with NASA astronaut Catherine Coleman and Russian Commander Dimitri Kondratyev on December 15, 2010. Because of ISS crew operational constraints, it soon became apparent that not all the filming opportunities could be accomplished.

To supplement the footage Nespoli was attempting to capture, ESA flight directors recorded additional passes over other orbit segments using remote controlled standard definition cameras mounted on the outside of the ISS and downlinked live to recorders on Earth. This procedure did not require any crew time and was easier to conduct, but the quality of the footage from the external ISS cameras was not as compelling as the crew-captured cupola footage, and in the final film the producers tried to minimise the use of this footage.

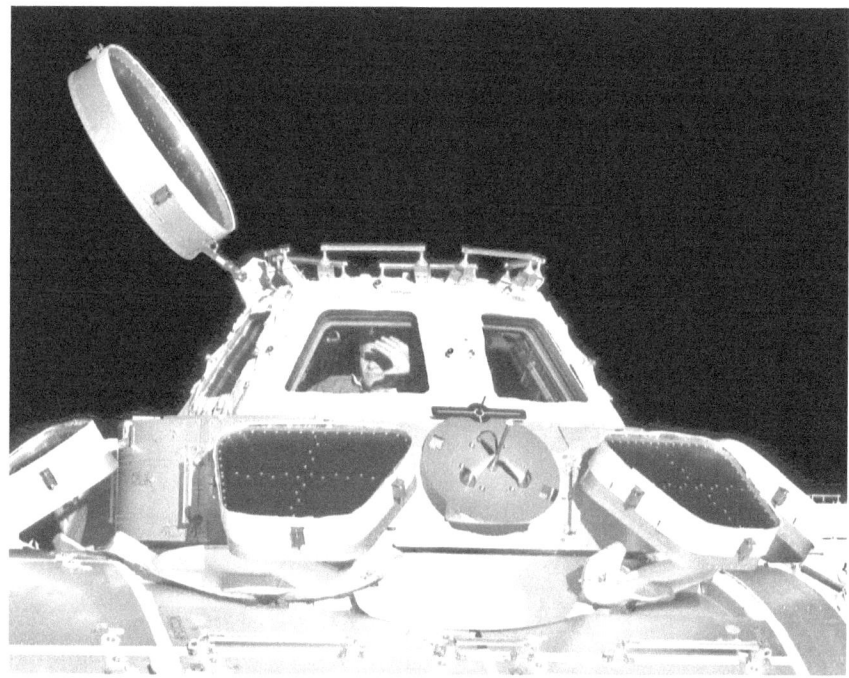

Figure 6: ESA Astronaut Paolo Nespoli inside the Cupola during filming of *First Orbit* in early January 2011 (Courtesy of ESA).

On January 8, 2011 life on the ISS was significantly disrupted by the attempted assassination of US Congresswoman Gabriella Gifford in Tucson, Arizona. Gifford is the sister-in-law of NASA astronaut Scott Kelly, who was serving as Expedition 26 Commander on the ISS at the time. Flight controllers immediately sought to minimise disruption to core ISS activities so as to reduce stress on the crew, and extra activities such as filming for *First Orbit* from the cupola was restricted.

By this time Nespoli had captured four of the five segments, A, B, D, and E, but capturing segment C as a night pass over the Pacific proved impossible, as every opportunity coincided with crew sleep periods which could not be interrupted. To cover segment C, the producers requested NASA archival footage of night passes over the Pacific, which was generously supplied.

The Moon had not been visible to Gagarin during his Vostok-1 flight and he had written in his autobiography, Road to the Stars, that he would '...try to see it next time.'[6] Sadly for Gagarin, there was no next time in space. So as an extra tribute to the pioneering cosmonaut Riley used shots of the Moon from NASA's Pacific night footage to give Gagarin the view of the Moon he never got a chance to see from space.

5.2.2 Editing

First Orbit was edited so that the view of the Earth from space at any point matched the timings from Gagarin's flight. Additional archive footage was donated by Footagevault to construct opening and closing titles. Edited by Tabitha Moore, the opening sequence aimed to set the scene for Gagarin's flight, blending footage of his preparation together with opening credits and a speech Gagarin had made prior to launch. The closing sequence, to simulate Gagarin's re-entry, was constructed from Apollo 10 onboard footage, and his parachute descent was simulated using Project Excelsior III archive and V2 test flight footage from White Sands.

Further shots contributed from Footagevault and the ESA archives helped to construct missing parts of segment B, including the dramatic view of the setting Sun as Vostok-1 passed through the terminator and into the night side of the Earth. A final archive shot from Footagevault's collection showing the famous portrait of Yuri Gagarin holding a dove, which had been fixed to a wall inside the Russian section of the ISS as a final tribute to the world's first space man, still orbiting the Earth in spirit 50 years after his pioneering flight.

5.2.4 Music

Although the original concept for the film involved no music, it became apparent during editing that a musical score was needed. Composer Philip Sheppard, who had worked with Riley on *In the Shadow of the Moon*, was approached. Quite by coincidence it turned out that he'd been working on an album called *Cloud Songs*, inspired by spaceflight, which he generously donated to the project, along with some additional tracks.

In a further coincidence, NASA astronaut Catherine Coleman, a friend of Sheppard's, had carried *Cloud Songs* to the ISS on her Soyuz flight with Nespoli in December 2010. Although it was not known at the time, at one end of the ISS Nespoli had been filming *First Orbit* while at

---

[6] Gagarin, Y. 1961. Pg 153. *Road to the Stars. Notes by Soviet Cosmonaut No. 1.* 2002, University Press of the Pacific, Reprinted from the original 2002.

the other end Coleman had been listening to *Cloud Songs*, the music that would eventually accompany his footage.

5.2.5 Gagarin's Voice Recordings

From the start of the project Riley had enlisted the support of human behaviour performance specialist and native Russian speaker Iya Whiteley to help source the original voice recordings from Vostok-1. Whiteley's search took her from the National Archives to contacts at the Russian Federal Space Agency Roscosmos, NATO, the British Embassy and even the Russian military. Only a few weeks before the film was completed, Whiteley finally tracked down the original recordings at the Russian State Archive of Scientific and Technical Documentation, and the full mission audio from the flight was acquired. Whiteley then painstakingly undertook subtitling the audio into English.

Gagarin is most vocal during the first 20 minutes of the mission, but as he passes out of contact with the Soviet ground stations, he becomes quieter. He speaks very little after passing into the night side of Earth over the Pacific, only commenting briefly on the view of stars. After sunrise over the South Atlantic he makes one more brief comment about the direction of travel of the sea below, and then does not speak again.

With the help from the post-production company Unit in London, the film was completed at the end of February 2011, about five weeks before the anniversary. Including opening and closing titles, its duration is 105 minutes; just three minutes short of the 108 minutes of Gagarin's flight.

5.3 Related Media

In addition to the main film, supporting media were created around the project, including iPhone and Android apps that compressed the entire orbit into a single 100 second video clip (the equivalent of orbiting the Earth at over a million miles an hour).

A Twitter channel (http://twitter.com/FirstOrbit) was also set up to promote the film and to carry live tweets of the mission audio (in English) at exactly the same time as Vostok-1's flight 50 years before, and Facebook film page was also set up to promote the project (http://www.facebook.com/firstorbitfilm).

5.4 Distributing the Film

The strategy for distributing the film enabled anyone to download it from a website and screen it at their own celebratory event. Applications to download the film went live from the March 23, 2011, when the story about the project broke, and the YouTube channel was also launched with three trailers for the film.

On April 12 a short 'making of' film was also added to the project's channel, bringing the total number of *First Orbit* films on the project's YouTube channel to five (three trailers, a making of and the main feature).

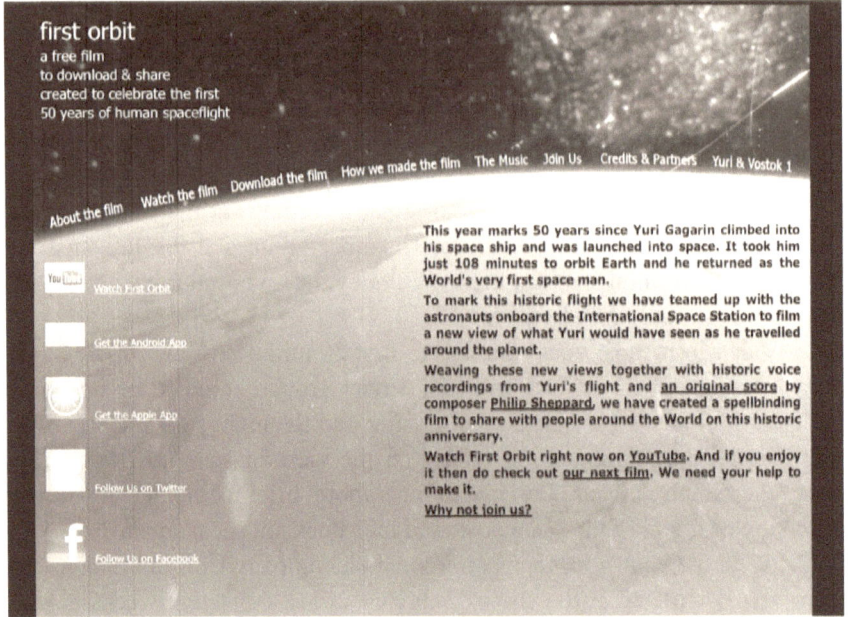

Figure 7: The firstorbit.org home page as it appeared 6 months after the film's release

5.5 How the film was received

The agency Sister (www.sisteris.com) was recruited to promote the project around the world, and their highly effective campaign generated more than 65 major international articles and features on *First Orbit*.

Press coverage of the project was broad and ranged from tabloids to UK broadsheets such as *The Guardian* and other high-traffic sites such as *Wired*, to news and video aggregators. Others linked to the Android platform or the project's Twitter feed (www.twitter.com/firstorbit) or Riley's personal website (www.chris-riley.com). Broadcasters including the BBC, and *Russia Today* recorded reports and interviews for their outlets.

Media coverage was strongest in the United States (66%), followed by the United Kingdom (22%), with significant media coverage in Spain, Russia and New Zealand. The European Space Agency (ESA) website,

Guardian Online, El Pais, the BBC, the MoonAndBack, the Russian site Moscow Nedelia and the British Interplanetary Society all ran more extensive articles of a full page or two in length.

Almost 2000 download requests were received. Internet traffic received on April 12 crashed the project's server, and the volume of registrations also pushed the firstorbit.org GoogleMail account into a spam alert state, which prevented the team from replying to anyone for twenty-four hours. But by this time the film was already available to watch on YouTube.

In the run up to April 12, a total of 1646 requests were approved to download the film in advance for anniversary screening events around the world.

700 of these were for school and university celebrations, 290 requests were for Yuri's Night party premieres, and another 639 were for public premiere events. Notable premieres included the BBC breakfast premieres on their giant screens in 20 city centres around the United Kingdom.

Figure 8: BBC Big Screen breakfast premieres, Swansea

In total, the film premiered 'offline' in more than 130 countries around the world, playing to an estimated 200,000 people.

In Russia, schools were encouraged to hold a special lesson to celebrate the anniversary, and with the help of the British Council and Roscosmos, the film was promoted for use in this anniversary lesson.

5.5.6 THE YOUTUBE *FIRST ORBIT* PREMIERE

In the build up to the film's global YouTube release on April 12, the film's trailers had together attracted over 800,000 views. At 00:00 UTC (01:00 BST) on April 12, the file was switched live, and *First Orbit* became visible to the global Internet population. Within the first six hours YouTube registered only 300 total viewings of the film. Then, at 06:07 UTC, the project's campaign to watch the film at exactly the same time as Yuri's flight 50 years before kicked in, and 24,000 people across the World watched *First Orbit* on YouTube simultaneously.

Correspondence about *First Orbit* on YouTube rose steadily throughout the day, quickly making it one of the most talked about subjects on the website. More than fifteen thousand people subscribed to the channel and left almost 5000 comments, the vast majority of which were positive. As the day passed, the main feature film attracted 24,752 likes, 609 dislikes, and 4445 friends. Google/YouTube helped to drive substantial traffic to the channel by linking to it from their 'doodle' of the day, which depicted Yuri's achievement.

Figure 9: Google/YouTube's Gagarin Doodle which replaced their logo on the April 12th world wide, linking directly to the *First Orbit* YouTube channel.

By 18:00 UTC the viewing count for *First Orbit* reached 600,000 people, and by 23:59 UTC it was over 1.2 million. Within 48 hours of release, *First Orbit* had received just over 2 million viewings, making it the most watched long-form film release in YouTube history. At the time of this writing, over twelve months later, the count online has reached over 3.6 million viewings, with the *First Orbit* YouTube channel as a whole receiving over 4.7 million upload views for the trailers, making of video and the main film combined.

According to the statistics on the YouTube channel, the film was most popular with men aged 45-54, but interest from both sexes in the age range 13-17 was also noted. Globally it was most popular in Russia, with Taiwan, India, North America, the UK, Canada, Brazil and Australia ranking next. According to YouTube it was watched in every country on Earth except five on the African continent and one in Eastern Europe

5.6 SOCIAL AND ORCHESTRATED MEDIA

A number of supporting social and orchestrated media campaigns were carried out as part of the *First Orbit* project, including the establishment of a Twitter channel, a Facebook page and the creation of both Android and Apple Apps designed to interact with Gagarin's orbital path whilst watching the film. An analysis of these extra elements is presented below.

5.6.1 TWITTER AND BLOG INTEREST

Browser window captures for Twitter searches for the hashtag 'firstorbit' were carried out between April 12 and 14, for later analysis. Subsequently, in preparing this paper, a retrospective analysis of the Twitter data relating to *First Orbit* was also carried out using 10 different applications, including topsy.com and peoplebrowsr.com, the latter which claims to access all tweets from the last 1000 days.

Analysis of these results shows three peaks relating to *First Orbit*. The first peak occurred on March 24, the day after the *First Orbit* web site went live and the day Jonathan Amos broke the story on BBC News Online. This single story resulted in 627 tweets and 194 blog posts. The second social media peak occurred on April 12, resulting in almost 15,000 tweets and 1175 blog posts in a single day.

To compliment a campaign encouraging people to watch the film at exactly the same time as Gagarin's flight – starting at 07:07 BST, an orchestrated media live tweet was planned, broadcasting the entire mission audio translation into short 'tweet-length' English phrases. These micro-blog friendly transcripts were prepared by Vix Southgate of YuriGagarin50 and Scott Andrews at the British Council. Tweetdeck was employed to queue their transmission, with plans to start the live tweet with the hashtag 'orbit1' a couple of hours before 07:07 BST. This ran well until just after 'lift-off', when Twitter's automated systems mistook the *First Orbit* feed as a spam attack and locked the account, bringing the stream to a halt.

5.6.2 Facebook

Facebook proved to be the least useful channel of communication for this project. A basic page (www.facebook.com/firstorbitfilm), was created which attracted 420 likes without any promotion. However, Riley failed to interest Facebook in the project despite repeated calls to their UK and Ireland HQs. Problems with the Facebook user interface and a lack of support from the company meant that the team had to abandon further plans for promotion of *First Orbit* through this route.

The Wikipedia entry on the film, which was auto-listed on Facebook, attracted a further 246 likes. http://www.facebook.com/pages/First-Orbit/215819291767403

5.6.3 Android and Apple Apps

Sales of the Apps were low, with only around 680 on the Apple market and fewer on the Android platform. The poor take-up on these, compared to the reception for the main film, is attributed to a shift in the apps market place away from paid apps to free or freemium ones supported by embedded adverts.

5.7 *First Orbit* 2012

Requests from fans of the film to translate Yuri's story into other languages started to come in soon after April 12, 2011 and Riley decided to crowdsource translations.

An English transcript of the film was published as a spreadsheet on the firstorbit.org web site in October 2011, and a call to action was launched at the 2011 International Astronautical Congress in South Africa, during the presentation by Dr. Christopher Welch of a paper on the project by Riley and himself.

Translations into more than 30 languages were received within a couple of months from more than 70 volunteers, and all have been published at www.firstorbit.org/translations. The *First Orbit* translation challenge remains open, with the goal of eventually translating Yuri's story, in his own words, into every language on Earth. Anyone interested in contributing a new language to the project should visit www.firstorbit.org/add-a-language

To make the most of this effort, Riley decided to publish a multi-language version of *First Orbit* on DVD and BluRay in time for the 2012 anniversary, encoding the first 30 languages. Funding for the manufacture of these discs was raised on the crowdfunding platform IndieGoGo, (www.indiegogo.com/firstorbit) by pre-selling hard copies of the film and offering fans the chance to add their names to the end credits.

A new 2012 campaign trailer was added to the YouTube channel and attracted over 7000 viewings during the eight week marketing drive through December and January. About 100 of them donated funds to the project and although only 20% of the requested funds were raised, plans for releasing the film in this way went ahead.

In the Spring of 2012 the film was lightly re-edited and re-mastered in preparation for manufacturing onto disc. The most significant change to the film was the production of new subtitles in 30 languages. To help promote these new multi-language hard copies of the film a new 2012 campaign was run, again by creative agency Sister. The campaign revolved around creating public screenings of the film around the world in all 30 new languages.

In London the British Interplanetary Society hosted a *First Orbit* season, screening the film in all 30 languages, starting with Russian and English on the April 12, 2012. And further campaigns in collaboration with the Yuri's Night community, and other special interest space flight groups saw further foreign language screenings of the film around the world.

To further promote the translations of the film a new Live Tweet of the mission starting at 07:07 UT on the April 12, 2012 was run by Riley. Tracking of re-tweets for #firstorbit and #yurisnight for this event showed a reach of over 400,000 people. On YouTube a further 100,000 people watched the original 2011 version of the film with the English subtitles.

6. Conclusion

Fifty years after his Vostok-1 flight, details of Gagarin's story have been reconstructed in a compelling and original way through the creation of *First Orbit*. ESA's enthusiastic support for the project, and Paolo Nespoli's work on board the International Space Station to bring it to fruition brought significant attention to the ISS during this major anniversary year.

The value of combining the archive mission audio of Gagarin's flight with new high definition color footage of the route he flew has been demonstrated; bringing new life to a previously inaccessible oral record of Vostok-1 and making it more accessible to today's visually driven generation.

The film's release attracted a great deal of interest around the world, and amassed a significant community of fans and followers through a number of channels.

The global reach of YouTube, and Google's willingness to work creatively with content producers in this way makes their platform an unparalleled tool for releasing long-form experimental video content which is unsuitable for broadcast on conventional TV channels. The capacity for human spaceflight to capture interest and to intrigue audiences around the world prevails, and the combination of beautiful views of the Earth from space set to music resonates as deeply with us today as it did when humankind first set eyes upon Earth from above.

The first decade of Yuri's Night celebrations since its inception in 2001 has shown just how strong the story of humanity's first spaceflight still appears to be. Fifty years after Gagarin's pioneering mission, the courage and spirit of adventure which human spaceflight epitomizes still appeal widely to the people who live today on the planet he first orbited.

Yuri's Night 2012 featured 236 events in more than 50 countries involving thousands of people in the celebration of the first human space flight. In the years to come, there will be many opportunities to raise the profile of other memorable spaceflight anniversaries, such as a global Moonlanding Night on the July 20/21 each year to mark the anniversary of Apollo 11. And as more and more people participate in space flight through broader commercialization in the coming years, there will be more occasions to celebrate, and an ever-growing community of those who have been to space, and those who can realistically expect to share this experience at some point in their lives.

...

Pertinent Websites

www.yurisnight.net
www.firstorbit.org
www.youtube.com/firstorbit
www.twitter.com/firstorbit
www.facebook.com/firstorbitfilm

DR. CHRISTOPHER RILEY

Dr. Christopher Riley is a writer, broadcaster and film maker specializing in history and science. He has worked on many of the BBC's iconic science programs from *Tomorrow's World* and *Rough Science* to *Science in Action* and *The Sky at Night*. In 2004 he won the Sir Arthur Clarke award for his work producing the BBC ONE blockbuster series *Space Odyssey: voyage to the planets*. His 2007 feature documentary film *In the Shadow of the Moon*, the story of the Apollo astronauts, was premiered at the Sundance Film Festival, where it won the World Cinema Audience Award.

Chris is visiting Professor at the Lincoln School of Media, the University of Lincoln. He gained his doctorate at Imperial College, where he worked with Metric Camera data from Spacelab-1, flown on Space Shuttle Columbia's STS-9 mission. He is the author of more than thirty articles and books on astronomy and planetary science and regularly broadcasts and lectures on these topics. His book *Apollo 11, an owner's workshop manual*, published by Haynes in June 2009, was an Amazon top ten science and nature book of the year. He is the producer and director of the unique Yuri Gagarin 50th Anniversary film project *First Orbit*, the subject of this chapter.

DR. CHRISTOPHER WELCH

Dr. Welch is Director of Masters Programs at the International Space University (ISU) in Strasbourg, France. He has a PhD in spacecraft engineering from Cranfield University – where he is also adjunct faculty – and an MSc in space physics from the University of Leicester in the United Kingdom. His research interests include space propulsion, space exploration and microgravity physics. In 1989 Dr Welch was one of the final 20 candidates to fly to the Mir space station on the UK-USSR Juno mission, which continues to fuel his passion for human spaceflight and space education. He is former chair of the UK's Space Education Council and is Vice Chair (formerly Chair) of the International Astronautical Federation Space and Education and Outreach Committee. He is on the board of the British Interplanetary Society, the World Space Week Association, the Spacelink Learning Foundation and the Arts Catalyst. In 2009 he won the Sir Arthur Clarke Award for Achievement in Space Education, and was Scientist in Residence at London's National Museum of Childhood in 2007. He is also a frequent commentator on space and astronautics and has made more than 300 television and radio broadcasts and has also advised on a number of space-related television programs and films.

Acknowledgements

The assistance of Dr E. Detsis and A. Ebrahimi of the International Space University in the analysis and production of the *First Orbit* internet statistics is very gratefully acknowledge by the authors. Without the generous support of the European Space Agency (ESA) *First Orbit* would never have been made. Their scientists, engineers and public affairs staff were invaluable partners in this project from start to finish. Most notably Jean Coisne, Roland Luettgens, Gerald Ziegler, Giovanni Gravili, Melanie Cowan and astronaut Paolo Nespoli, who gave up some of his precious time on his mission to the International Space Station to capture the footage used in the film. Additional orbital views of the Earth were sourced and supplied by NASA, where we are particularly grateful to Jody Russell, Gayle Frere, Mike Gentry, Silvia Gederberg, Sheva Moore, James Hartsfield and Kylie Clem.

CHAPTER 8

THE SCIENCE AND TECHNOLOGY OF SPACE EXPLORATION:
LEVERAGING CYBERSPACE FOR GLOBAL COOPERATION IN THE DEVELOPMENT OF SPACE

RITA M. LAURIA, JD, PH.D.
ATTORNEY AND COUNSELOR AT LAW
MEDIATOR

© Dr. Rita M. Lauria, 2012. All Rights Reserved.

I. INTRODUCTION

Cyberspace, the electronic environment that consists of myriad connections between people and information, is also the transition space to outer space, the place where we may learn to productively to live, to work, and to cooperate in preparation for our actual journeys off the Earth.[1] While future space flight missions that go beyond today's International Space Station promise to take us to new frontiers made possible by impressive, new levels of human cooperation, these missions also demand that we consider a new spectrum of preparations for those who will undertake them. These preparations range from mission and operations training to psychological concerns that engage the forward edge of continued evolution of humanity, its cultures, and its societies.

Cyberspace is our contemporary training ground for the cooperative purpose of space development. We train in cyberspace as a transition from our Earth habitat to off-planet habitats in space. Cyberspace can be and should be seen as a training, adaptive space.

We adapt in cyberspace to constructed, alien environments, different from the natural environment of Earth in which we evolved. We learn in cyberspace how to project beyond time and space the embodiment of our personhood,[2] even to the point of teleporting the essence of our humanity, what for aeons has been called our "soul."[3] While this chapter does not address the evolution of humanity *per se* or the need to open the Earth habitat system to support the evolution of our species,[4] it does suggest that leveraging cyberspace to support more effective communication can play an important role in the advancement of human knowledge.

Collaborative international teams can leverage cyberspace for global cooperation in the development of space by working together to build

---

[1] The author extends her recognition and sincere appreciation for the collaboration engaged in over a year with Dr. Jacquelyn Ford Morie, Dr. Gustav Verhulsdonck, and Dr. Kathryn E. Keeton on the benefits of virtual worlds (VWs) for astronaut and ground crew training for long-duration space flight. This work is based on this year-long research effort to investigate the potential of using advanced communication media in training for conflict management, communication, collaboration, team building, and skills proficiencies. See Morie, J.F., Verhulsdonck, G., Lauria, R., & Keeton, K. (August 2010) *Operational assessment recommendations: Current potential and advanced research directions for virtual worlds as long duration space flight countermeasures.* [Final Report] NASA Technical Manuscript NASA/TP-2011-216164, Houston, TX: BHP/Wyle Laboratories.

[2] *See e.g.,* Lauria, R. and Robinson, G. S., (2012). From cyberspace to outer space: Existing legal regimes under pressure from emerging meta-technology. 33 *U. La Verne L. Rev.* 219 (2012).

[3] Robinson, G. and Lauria, R., (September, 2004) Legal rights and accountability of cyberpresence: A void in Space Law/Astrolaw jurisprudence *Annals of Air and Space Law,* Volume XXVIII: 311-326.

[4] *See* Robinson, G. S., "An incomplete species: Unfolding of Space Law to support the survival of humankind and its unique envoys migrating off-Earth," this volume, Chapter 19.

virtual worlds (VWs) in which to train our envoy[5] astronauts who travel, work, and live in space. Jessy Cowan-Sharp, a collaborative web technology developer at the NASA Ames Research Center in Mountain View, California notes virtual worlds are "the seed of something that's going to be one of the major transformational technologies of our species."[6]

II. INTERNATIONAL COLLABORATION

In the past, the US and the USSR competed to become the major power in space. But where these nations once engaged in a Cold War space race, the US and Russia now cooperate in the development of space, with Russia acting as the transporter of US astronauts to the ISS. Other nations, notably China, pursue their own national space programs with significant ambitions. While Chinese leaders insist that China's purposes are peaceful as the nation looks for ways to exert its growing economic strength[7] and to demonstrate its global might, others remain concerned.

International collaboration in the building of virtual worlds for envoy astronaut training suggests a strategy that can help generate cooperative objectives towards becoming global partners in the development of space. An international collaborative model to accelerate the development of space is not a novel idea. Indeed, the purpose of this chapter is to examine how such collaboration can be extended.

Just as NASA uses Space Act Agreements to support commercial space endeavors such as the recent 2012 SpaceX mission to the ISS, a similar initiative could be established to support using virtual worlds to train our envoy astronauts towards enabling them to develop specific competencies necessary for expert teamwork, communication, leadership, interpersonal skills, and other competencies. Such collaboration should engage the international community.[8]

---

[5] The United Nations' space treaties use the phrase "Envoys" to characterize astronauts. See also supra, notes 3.

[6] Pellerin, C. (10 July 2007). NASA harnesses power of virtual worlds for exploration and outreach. *Space Mart.* [Online] Available: http://www.spacemart.com/reports/NASA_Harnesses_Power_Of_Virtual_Worlds_For_Exploration_And_Outreach_999.html [1 June 2012].

[7] Wong, E., and Chang, K., (29 December 2012). Space plan from China broadens challenge to U.S. *New York Times* [Online] Available: http://www.nytimes.com/2011/12/30/world/asia/china-unveils-ambitious-plan-to-explore-space.html?pagewanted=all [1 June 2012].

[8] See *infra*, Section IV. *See also* NASA. International Space Station Human Behavior & Performance Competency Model -Volume I & Volume II, Mission

III. Defining Cyberspace[9]

Cyberspace is a constructed virtual reality that provides alternate environments for humans to use and to inhabit. Although creating virtual realities requires a cluster of interactive technologies and techniques to create an alternative reality through the engagement of one or more sensory channels,[10] virtual reality is defined more by user experience and interactions among participants than by a particular technology.[11] This experience is often characterized by a compelling sense of presence, of being inside an environment created by computer-mediated systems.[12]

In the late 1980s a group of researchers and entrepreneurs popularized the term "virtual reality." Often terms from science fiction literature merge with the idea of VR. For instance, William Gibson coined the term "cyberspace" in his 1982 short story *Burning Chrome*, and continued to use the term in his 1984 novel *Neuromancer*. In *Neuromancer*, cyberspace is described as a connected "consensual hallucination" shared by real people inhabiting the virtual space inside the computer. Science fiction writer Neal Stephenson coined the term Metaverse in his 1992 book *Snow Crash* to describe a similar concept of shared, computer-generated spaces. Stephenson's term suggests such spaces represent a metaphorical universe with spatial dimensions.

Operations Directorate, ITCB HBP Training Working Group – L. Bessone, E. Coffey, N. Inoue, M. Gittens, C. Mukai, S. O'Connor, L. Tomi, V. Ren, L. Schmidt, W. Sipes, S. V. Ark, and A. Vassin. NASA/TM–2008–214775 Vol 1 & Vol 2.

[9] This section is largely extracted from the final report to NASA by the author and her colleagues who engaged in a year-long research effort to explore the possibility and implications of using virtual worlds as countermeasures for long-duration space exploration and habitation. See supra, note 1, Morie, J.F., Verhulsdonck, G. Lauria, R., & Keeton, K. (August 2010) *Operational assessment recommendations: Current potential and advanced research directions for virtual worlds as long duration space flight countermeasures*. [Final Report] NASA Technical Manuscript NASA/TP-2011-216164, Houston, TX: BHP/Wyle Laboratories.

[10] Biocca, F., Lauria, R., & McCarthy, M. (1997). Virtual Reality. In Grant, A. (Ed.), *Communication Technology Update* 6th Ed., Boston: Focal Press.

[11] Steuer, J. (1994). Defining virtual reality: Dimensions determining telepresence. In Biocca, F., & Levy, M. (Eds.). *Communication in the age of virtual reality*. Hillsdale, NJ: Lawrence Erlbaum Associates, 127-158.

[12] Morie, J.F., Verhulsdonck, G., & Lauria, R. (January 2010) *Countermeasure technologies for long duration spaceflight: Using virtual world technologies to enhance physical, psychological, and team performance*. Johnson Space Center JSC 66220, Houston, TX: BHP/Wyle Laboratories at § 1.2.

Figure 1: Avatars Inhabiting a Virtual World
(Courtesy of J.F. Morie, USC Institute for Creative Technologies.)

Most early virtual reality applications were single-purpose environments in which only a few people shared the computer-generated virtual space. These VR applications were often created for virtual travel to famous monuments, or promoted as new game spaces, or perhaps as artistic experiences. These were also referred to as Virtual Environments (VEs) and were explored for their serious potential as venues for therapy to mitigate phobias such as fear of spiders, flying, heights, public speaking, as well as for training purposes. NASA Ames Research Laboratories worked to use VE tools to engage in telepresence to effect results at a distant place.[13] Telepresence refers to the ability to use actions in a virtual environment to manipulate objects in a non-virtual environment, often in remote locations.

Virtual reality and virtual environments afford a sense of immersion and presence in a synthetically generated, computer-mediated space. The objective of VR is to provide an authentic experience of presence in a virtual reality. Early systems featured head-mounted displays that insulated and physically isolated users within a digital space. Various

---

[13] Fisher, S.S., McGreevy, M. Humphries, J., and Robinett, W. (1986). Virtual environment display system, *ACM 1986 Workshop on 3D Interactive Graphics (pp. 77-87)*. New York: ACM Press.

interface tools, such as gloves and tracking devices, enabled users to experience the simulated world as though they were traversing the virtual space, or manipulating virtual objects, and provided a distinct spatial experience. For this reason, virtual environments are often understood as "spatial virtual reality."

A spatial VR typically has limited or no connectivity to other VRs, is created for a single purpose, has a repeatable set of starting parameters (no persistence across usage) and minimal, or no self-representation. If there is self-representation, the representation is usually defined by the program's authors and not by the individual user.

Virtual worlds, by contrast, are highly social virtual realities built to be inhabited, traversed, and manipulated by many concurrent users who interact and communicate with each other through their virtual representations, their avatars.[14]

In virtual worlds, participants model, explore, and interact within an environment that can be similar to our physical surroundings. On the other hand, virtual worlds may present novel, imaginary, or fantastic environments that are not bound by the constraints of common reality. Therefore, virtual worlds allow for a greater range of opportunities and experiences that may help prepare an individual to inhabit even further dimensions than the Earth and cyberspace.

Video games share much with virtual reality and with virtual worlds as they provide common spaces for players to interact with each other in a virtual space. Video games and virtual worlds are popular pastimes whose use is growing steadily. According to GigOm Research, one in eight people in the United States report having used a virtual world.[15] The use of VWs, especially by children from age 5 to 15, seems to be increasing exponentially.[16] Indeed, combining its use of games, virtual worlds, digital communications, and tools like Google, Facebook, and YouTube, the current generation can rightly be called a "digital generation."

---

[14] An avatar representation is literally the "embodiment" of someone in a virtual space. According to Webster's Ninth New College Dictionary (1986), to embody is to give a body to (a spirit), to incarnate or to make concrete and perceptible.

[15] Wagner Au, J. (2009). *Other World Notes: 1 in 8 Americans Virtual World Users.* [Online] Available: http://nwn.blogs.com/nwn/2009/07/other-world-notes-5.html [31 May 2012].

[16] Marsh, J. (2010). Young children's play in virtual worlds. *Journal of early childhood research* 7(3), 1-17. Thousand Oaks, CA: Sage.

The digital generation is comfortable learning, living, and communicating through computer games and computer-based environments.[17] The growth and prospects of the digital generation for NASA means that learning and communicating with computers are familiar and expected ways of being in the world, thus reinforcing the tremendous opportunity that exists to use VWs as digital, learning environments in many contexts.

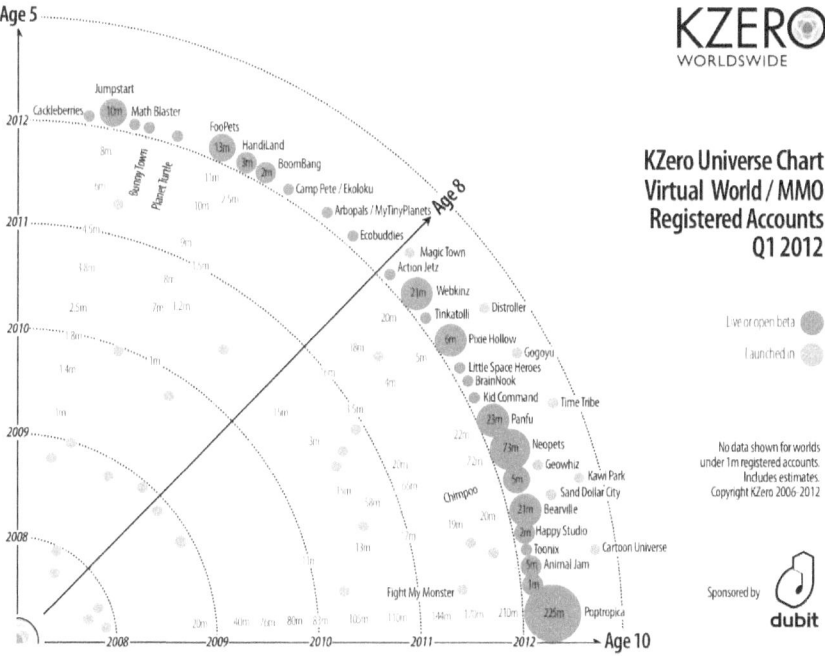

Figure 2: The Digital Generation's Participation in Virtual Worlds Organized by year from the origin, and by age (5, 8, and 10 years old.) Reprinted by permission.

IV. Virtual Worlds for Space Development

Conceiving, planning, building, operating, and using the ISS research platform is a remarkable achievement in international cooperation and collaboration and the latest step in humanity's quest to explore and to live in space. Completion of the space station has already strengthened international cooperation towards the development of space as a scientific

---

[17] Prensky, M. (2005). *Digital-Game Based Learning.* New York: McGraw-Hill.

collaboration among five space agencies representing fifteen countries.[18] These accomplishments illustrate the potential for international cooperation and collaboration to lead to unprecedented technological, political, and human achievement. Indeed, NASA considers the ISS program's greatest accomplishment to be as much a human achievement as it is a technological achievement.[19]

Through 2020 and beyond, the ISS will serve as a test-bed for permanent human presence in space, and as a foothold for long-term exploration. The tools and technologies envisioned and developed today will not only support the ISS through 2020, but also will support programs to the moon, to Mars, and to asteroids.

Training in virtual worlds can augment current mission training for these ventures. As we model the final frontier, engaging in long-duration space flights and space habitation towards developing space, all the participating space agencies engaged in the ISS program agree that specific human behavior and performance competencies encompass the knowledge, skills, and attributes required for successful performance on long-duration missions.[20] Using virtual worlds to train our envoys in these competencies offers great potential.

NASA's Human Research Roadmap indicates that the agency works to identify how virtual reality and virtual world technologies can be used to train crews and controllers to mitigate behavioral health and performance risks.[21] NASA seeks tools, strategies, and technologies to extend our human presence beyond low Earth orbit. To support humanity's journeys into space, new tools and technologies must be identified and developed that can help prevent performance degradation, human errors, and failures resulting from interpersonal conflicts or lack of team cohesion, coordination, and communication, and a host of other factors that can lead

---

[18] NASA: A Global Partnership for ALL Mankind [Online] Available: http://www.nasa.gov/multimedia/videogallery/index.html?collection_id=66481&media_id=97768842 [8, April 2012] The International Space Station is a joint project between five global partners, or participating space agencies: the US NASA, the Russian Federal Space Agency, (Roscomos), the European Space Agency (ESA), the Canadian Space Agency (CSA), and the Japanese Aerospace Exploration Agency (JAXA).

[19] NASA: International Cooperation [Online] Available: http://www.nasa.gov/mission_pages/station/cooperation/index.html [8 April 2012]

[20] L. L. Schmidt (2008), Competency modeling for the final frontier: Supporting psychosocial health and performance in low earth orbit. Performance Improvement, 47: 52–58, p. 54.

[21] TOPIC: X14 Behavioral Health and Performance. [Online]. Available: http: http://sbir.gsfc.nasa.gov/SBIR/sbirsttr2011/solicitation/SBIR/TOPIC_X14.html [3, April 2012]

to performance degradation and human errors. Hence, NASA seeks to develop virtual reality environments as a countermeasure approach to support and/or to enhance team functioning.[22]

One tool NASA has developed and uses is the Human Behavior and Performance (HBP)[23] model, which defines eight psychosocial competencies and associated behavioral markers.[24]

The HBP competences include:

1) Self-care/self management;
2) Communication;
3) Cross-cultural understanding;
4) Teamwork and group living;
5) Leadership;
6) Conflict management;
7) Situation awareness; and
8) Problem solving and decision-making.

NASA identified the Human Behavior and Performance competencies as requirements for those participating in international long duration missions.[25] Virtual worlds, social versions of virtual reality, provide an ideal environment in which to conduct HBP training. Indeed, collaboration in virtual worlds training may not only be practically necessary, it could also support continued cooperation to build a powerful global partnership such as that which constructed the International Space Station.[26]

Future tasks will focus on virtual reality development to utilize efficient and cost-effective approaches to providing support for crewmembers engaged on long duration missions related to in-flight interventions and countermeasures for supporting task performance,

---

[22] NASA: Human research roadmap. [Online]. Available: http://humanresearchroadmap.nasa.gov/gaps/?i=408 [31 May 2012].
[23] Ibid.
[24] For a discussion of each ISS partner's training locations and contributions see e.g., NASA. Behind the scenes: Training locations. [Online] Available: http://spaceflight.nasa.gov/shuttle/support/training/isstraining/locations.html [1 June 2012]. See also infra notes 29 and 30.
[25] NASA, supra note 9 - International Space Station Human Behavior & Performance Competency Model -Volume I & Volume II, Mission Operations Directorate, ITCB HBP Training Working Group – L. Bessone, E. Coffey, N. Inoue, M. Gittens, C. Mukai, S. O'Connor, L. Tomi, V. Ren, L. Schmidt, W. Sipes, S. V. Ark, and A. Vassin. NASA/TM–2008–214775 Vol 1 & Vol 2.
[26] See International Space Station. Available: http://www.nasa.gov/mission_pages/station/main/index.html.

teamwork, and psycho-social performance.

According to NASA's human research roadmap, "The desired state of knowledge includes a suite of tools that span the mission phases (pre, during, and post) that crews can use to support their team functioning for a long duration mission and furthermore, enhance both their performance and well-being. Thus ... it is very important that the tools that are developed are tested in both ground and spaceflight studies for validation purposes."[27]

NASA seeks to use virtual reality tools and techniques to train crews and controllers in areas such as cross-cultural interactions, interpersonal conflicts, team cohesion and coordination, communication, leadership, and other HBP competencies. The agency intends to develop virtual worlds to train individuals for complicated, multi-agent tasks, such as human and robotic collaboration, and wants to identify how virtual reality and world technologies can be used to train for performance readiness, cross-cultural interaction, psychological support, and effective collaboration with other team members or artificial intelligent agents.[28]

V. HUMAN BEHAVIOR AND PERFORMANCE COMPETENCIES

Virtual worlds can be used for training astronaut envoys in various contexts, including mission preparation, mission support, and space flight. The basic training flow consists of instructional units, or blocks, comprised of several subjects that are further broken into individual lessons. Training for ISS missions consists of classroom sessions that bring each astronaut candidate to the required levels of background knowledge in mission critical areas, which includes introducing the astronaut candidate to the training process, fundamentals, ISS Systems, robotics, Extravehicular Activity (EVA), and other matters.[29]

Each global partner who participates in the ISS program also provides advanced training courses in their specialties. The Columbus multifunctional pressurized laboratory is Europe's largest contribution to

---

[27] *Ibid.*

[28] Intelligent agents act autonomously to respond to a person using Natural Language Processing and thus appear as artificially intelligent. Intelligent Tutoring Systems (ITS) use the input of a person's responses to a task, then gauge their performance on the task and ask follow-up questions that provide the person with post-task reflection. An ITS can provide important mentoring opportunities for students. *See generally,* Morie et al, *Operational Assessment Recommendations,* supra note 1.

[29] International Space Station Astronaut Candidate Training Catalog (Baseline, August 2009), Mission Operations Directorate Spaceflight Training Management Office. Johnson Space Center Manuscript JSC-36543 Baseline, Houston, TX.

the International Space Station, where astronauts carry out experiments in materials science, fluid physics, life science, and technology. The European Astronaut Centre near Cologne, Germany is the training center for all European built ISS hardware, including Columbus laboratory systems, subsystems and payloads, as well as for astronaut operations for the Automated Transfer Vehicle (ATV).[30]

Each global partner holds great pride in its contribution to the ISS, including pride in the training courses the partner develops to train the international astronauts to competency in the system or subsystems associated with the partner's contribution to the space station.[31] An opportunity exists to further advance these accomplishments in international cooperation towards the development of space by collaboratively building virtual worlds for mission training in the HBP competencies. Virtual worlds can be built as a collaborative effort to provide virtual environments that can augment traditional training and exposure to learning materials.

Virtual worlds training platforms are especially promising for preparing for and solidifying learning of cognitive skills that have traditionally been difficult to instruct, are often tacit in nature, and require extended exposure to, and experience with learning situations. While these cognitive skills may be difficult to master, they must be acquired to meet the difficult demands of living, working, and traveling in space where "failure is not an option."[32] These cognitive skills are the HBP "soft skills."[33] Developing these skills can readily lend towards accelerating

---

[30] The European Astronaut Centre. See:
http://www.esa.int/esaHS/ESAJIE0VMOC_astronauts_0.html
See also supra, note 9.

[31] Interviews by the author were conducted with a range of individuals within the NASA organization on a range of topics, including building training within a virtual world context. For a selected sample of the content of the interviews with astronauts working in human space flight programs, with group leadership working inside Mission Operations Directorate, with psychologists working in training and management, with senior scientists working in Space Life Sciences, and with personnel working on the operational side of Human Performance and Behavior see Lauria, R., Morie, J.F., Verhulsdonck, G. (June 2010) *Applying countermeasure technologies to augment current training protocols for long duration space flight: Using virtual worlds to enhance team performance, resiliency, and proficiency skills.* [Operational Assessment and Data Analyses] Johnson Space Center Manuscript JSC 66218, Houston, TX: BHP/Wyle Laboratories.

[32] *See* Schmidt, supra note 20, at 53.

[33] Smith-Jentsch, K. A., Zeisig, R. L., Acton, B. and McPherson, J.A. (1998). Team dimensional training: A strategy for guided team self-correction. In J.A. Cannon-Bowers & E. Salas (eds.) *Making decisions under stress: Implications for*

international cooperation.

While training in virtual worlds can teach system proficiency, navigational skills, Extravehicular Activity (EVA), robotics and telepresence control, the soft skills can be taught in virtual worlds towards developing situational awareness, communication, leadership, conflict management, decision-making, and teamwork. Virtual worlds have inherent qualities that may facilitate learning the soft skills by using game or story-based mechanisms. Such game or story-based strategies allow participants a way to play out, for instance, various experiments in learning without the risk of exposure to others.

Intelligent agents embedded in virtual worlds can provide game-based virtual environments that use the pedagogy of guided discovery for cross cultural engagement and sensitivity training.[34]

Such learning experiments can include leadership or team roles[35] through which participants can actively experiment with various roles using embodied avatar representations to interface with intelligent agents rather than through the traditional interface of face-to-face role-playing with another human being. Such role-playing using avatars in virtual worlds allows cost effective, safe, continued rehearsal in crucial communication, intercultural, decision-making, and other tacit skills.

A virtual world not only provides a safe and varied environment to keep these skills up-to-date, it can be fully integrated with embedded testing and benchmarking. Reports can be generated to identify potential problem areas while allowing the participant the freedom to practice independently and in the safety of a personal, simulated environment.[36]

Virtual worlds can be used for training both physical and psychophysical health, and for various therapies. Therapy training could

---

individuals and team training (pp. 271-297). Washington, D.C.: American Psychological Association.

[34] Hill, Jr., R.W., Belanich, J., Lane, H.C., Core, M., Dixon, M., Forbell, E., Kim, J. and Hart, J., (2006) *Pedagogically Structured Game-Based Training: Development of the Elect Bilat Simulation,* Proceedings of the 25th Army Science Conference.

[35] Moire, J.F., Verhulsdonck, G., & Lauria, R., (April 2010) *Applying countermeasure technologies to augment current training protocols for long duration space flight: Using virtual worlds to enhance team performance, leadership skills, conflict management and communication skills.* Johnson Space Center Manuscript JSC 66219, Houston, TX: BHP/Wyle Laboratories.

[36] Morie, J. F., and Lauria, R., (16 February 2012). How virtual worlds can benefit future space travel. NASA Human Research Program Investigator's Workshop. [Online]. Available:
http://www.dsls.usra.edu/meetings/hrp2012/hrp2012.authorindex.html.

include mindful stress reduction, addressing phobias,[37] therapy for post-traumatic stress syndrome (see Figure 3),[38] for physical therapy, and for other medical purposes such as immersive telemedicine for consultation between patients, physicians, and other clinicians.[39]

Virtual worlds can also be used for recreation and re-creation, for gaming and exploration, for worship and reflection, for calming environments, and for re-generation of spiritual health. Finally, research has shown that virtual worlds can be used for resiliency training.[40]

Virtual worlds in these applications and others can significantly advance international cooperation for the development of space by engaging global partners in training the "soft" human behavior and performance skills that our envoy astronauts and support crews will need to

---

[37] Anderson, P., Jacobs, C.H., Linder, G.K., Edwards, S., Zimand, E., Hodges, L., & Rothbaum, B.O. (2006). Cognitive behaviour therapy for fear of flying. Sustainability of treatment gains after September 11. *Behavior Therapy* 37 (1), 91-97.

[38] *See* Post-Traumatic Stress Disorder Assessment and Treatment (PTSD), [Online]. Available http://ict.usc.edu/projects/ptsd/. [2 April, 2012].

[39] *See* SimCoach, [Online]. Available http://ict.usc.edu/projects/simcoach/. [1 April, 2012].

[40] Interviews were conducted with six individuals who operate as contractors within NASA. The purpose of the interviews was to operationally assess the potential of using virtual worlds (VWs) to augment current NASA Training for astronaut candidates, astronauts, and flight controllers. The interviews related to using VWs as countermeasure technologies to augment current training protocols for long-duration space flight. All identifying information related to the sources was removed to protect privacy of the sources and to ensure anonymity and confidentiality. The interviewees were designated as *Interviewee One* through *Interviewee Six* (I-1–I-6). According to *Interviewee One* (I-1) People who work the Rover Mars mission work a Mars' day. A Mars' day is a 25-hour day. Working a Mars' day changes these people dramatically. Imagine, said Interviewee One: "Fall back every day! ... When you turn your clock back in the Fall you do that every day on Mars' time. A lot of people say, 'I just haven't recovered yet.' If you kept that up everyday ... You evolved in a 24 hour day, in 9.8 meters/second^2 - the acceleration for gravity. And virtually all life we know and understand on this planet evolved in that. So we, as best as we can understand, we have no pre-formed design to adapt to a lower gravity. We have some design for hyper G, but not for lower G."
According to *Interviewee One*, conditions imposed on the body during space flight and working a Mars mission necessitate resiliency. Training for these stresses is important. VWs can facilitate training for resiliency during space flight and for working a Mars' day. *See* Lauria, R., Moire, J.F., Verhulsdonck, G. (June 2010) *Applying countermeasure technologies to augment current training protocols for long duration space flight: Using virtual worlds to enhance team performance, resiliency, and proficiency skills.* [Operational Assessment and Data Analyses] Johnson Space Center Manuscript JSC 66218. Houston, TX: BHP/Wyle Laboratories.

survive in the alien environment of space.

Figure 3: Virtual Iraq/Afghanistan PTSD Exposure Therapy Scenarios (Courtesy of Skip Rizzo, USC Institute for Creative Technologies).

VI. BENEFITS AND OPPORTUNITIES OF USING COLLABORATIVE VIRTUAL SPACES

A. CONVENIENCE, TIME, AND COST EFFICIENCY

Space exploration is a relatively new endeavor in the time frame of human history, and although commercial space transportation and the privatization of space have begun to usher in a new era, very few have actually experienced space travel. However, all space agencies believe maximizing returns while reducing risks as much as possible in such an endeavor is the best option.[41] Adding opportunities to practice or to perfect skills without increasing training time is a critical concern for all agencies.

Several advantages exist to using virtual worlds for human and behavior performance training, including simply reducing the constraints associated with the time and distance required to travel to various international training centers. Peggy Whitson, Chief of the NASA Astronaut Office, is responsible for the mission preparation and on-orbit support for all ISS crews and their support personnel, as well as for organizing the crew interface support for future heavy launch and commercially-provided transport vehicles.[42] She noted that, "I don't miss

---

[41] Schmidt, supra note 20, at 53.
[42] Peggy Whitson, NASA JSC Biographical Data (January 2012). [Online]. Available http://www.jsc.nasa.gov/Bios/htmlbios/whitson.html. [31 May 2012].

the training. For the International Space Station we trained in five different countries, and that was a challenge."[43]

Using virtual worlds to augment current training protocols could alleviate much of this type of dissatisfaction by augmenting face-to-face training in virtual spaces. This offers not only time and cost savings, but allows important team bonding to occur by providing engaging ways of interactive and embodied learning that can both supplement and extend "real world" training.[44]

Virtual worlds offer other advantages as well. Virtual world training can accommodate unexpected performance requirements and trainee timelines, rather than constraining training activity to a particular flow with a particular instructor on a particular time schedule. Resources can be shared to maximize cost efficiencies, and anytime, anyplace learning can be conducted without the rigid scheduling of location-based training lessons. And virtual world simulations are readily reconfigurable and can be adapted to changing design requirements.

B. Promotion of Effective Cross-Cultural Understanding and Interactions

The challenges of living and working in space require that all envoy astronauts consistently perform at a high level, which means maintaining a healthy psychological and social environment. The diverse cultures and backgrounds of those who are and will be living and working together in space suggests that effective cross-cultural understanding and interactions are an important concern of participating space agencies. Psychological training is important to promoting successful teamwork and group living. Virtual training worlds can help astronauts develop important interpersonal and teamwork skills.

People who have known each other for a long time generally communicate with greater understanding because of their shared history, potential affinity, and social bonding. Virtual worlds can facilitate such social interactions as increased group cohesion can be developed by applying the capacities of social networking of virtual worlds like avatar

---

[43] Peterson, L., (12 December 2009). *Whitson heads astronaut office.* [Online]. Available: http://nl.newsbank.com/nl-search/we/Archives?p_multi=CANB|&p_product=SHNP4&p_theme=shnp4&p_action=search&p_maxdocs=200&s_dispstring=%28peggy%20whitson%29&p_field_advanced-0=&p_text_advanced-0=%28peggy%20whitson%29&xcal_numdocs=20&p_perpage=10&p_sort=YMD_date:D&xcal_useweights=no/ [31 May 2012].

[44] Morie and Lauria, *How virtual worlds can benefit future space travel.*

co-presence, chat, and messaging to share learning experiences and to get to know other crew members prior to a mission.[45]

As tight training schedules in multiple locations and the international diversity of crews increase, the ability of virtual worlds to effectively provide a high level of interactivity and inclusion in the virtual space allows for the creation of a much needed infrastructure and framework for social interaction and enhancement of group communication and understanding.

C. Resiliency Training

In studying the topic of virtual worlds training for space flight preparation, I had the opportunity to interview several NASA personnel, including a psychologist working in the training management group[46] who trained flight controllers and others. He noted that while flight controllers work eight-hour shifts, the simulations they train on last for only about four hours. He suggested that an eight-hour training simulation would enhance resiliency preparation, which could aid against fatigue in the critical and demanding job the flight controller performs.

He also noted that virtual world simulations for flight controllers' training would increase the number of individuals who could be certified. This number is currently limited due to the limitations of simulator time.[47]

As we move towards the private commercialization of space development, more people will need to be trained in all phases of mission preparedness and operations. Using virtual worlds for various training can help ensure adequate numbers of people are trained for the activities of space exploration and habitation.

VII. Conclusion

A comprehensive model of psychosocial competencies and associated behavioral markers was defined by subject matter experts to consist of necessary competencies for effective space mission performance. The model reflects eight competency categories that must be mastered by astronauts to adequately support psychosocial health and performance during space missions, including 1) self-care/self management; 2) communication; 3) cross-cultural understanding; 4) teamwork and group

---

[45] Morie, Verhulsdonchk, & Lauria, supra note 12, at § 2.2.
[46] See supra notes 31 and 40.
[47] Ibid.

living; 5) leadership; 6) conflict management; 7) situation awareness; and 8) problem solving and decision-making.[48]

Simulated, virtual worlds offer immersive, inclusive spaces in which to train our envoy astronauts who will work and live in outer space. We currently experiment using cyberspace to effectively project human personhood beyond the boundaries of ordinary reality, to the point even of teleporting the essence of humanity through time and space through the use of avatar representations. We adapt in cyberspace to constructed, alien environments that do not necessarily reflect the constraints of the natural environments in which we as a species evolved.

Leveraging cyberspace to help individuals adapt to these constructed, computer-generated virtual realities, humanity moves one step closer to fully transitioning to the alien environment of outer space. Cyberspace thus offers the possibility of enhancing our capacity to adapt to outer space.

One of the essential keys to higher states of cooperation and understanding is communication. Higher states of cooperation can be attained by cooperatively and collaboratively building virtual worlds for the training of our envoy astronauts. The next steps will require agencies, organizations, and entities interested in developing space to identify ways to increase training support.

Enhanced psychosocial training will become even more significant as ISS crews meet private crews who travel to space for tourism, work, and life in space. By innovating together to build virtual worlds for the training of our envoy astronauts, including technical and mission training as well as training in the soft skills like communication, cross-cultural understanding, leadership, conflict management, teamwork, situation awareness, problem solving and decision-making, international cooperation in the development of outer space will be facilitated. The time is ripe to build not only space stations as endeavors in international cooperation, but also to engage in the building of virtual worlds through and in which envoy astronauts can train to master the psychosocial health competencies necessary to effectively perform in the alien environment of space.

•••

---

[48] *See* Schmidt, supra note 20.

DR. RITA M. LAURIA

Dr. Rita M. Lauria founded Metalaw®.US to develop a legal and consulting practice that would build upon her years of experience in media, communications, and the law. Dr. Lauria is licensed to practice law in California and North Carolina. Based in Los Angeles, she specializes in cyber communications and criminal law and serves as Adjunct Professor of Cyber and Media Law at the University of La Verne College of Law. She also teaches at the University of Southern California (USC).

Prior to founding Metalaw®.US, Rita Lauria served the Superior Court of California, County of Los Angeles as judicial clerk extern in Major Crimes. There she assisted the Honorable Kathleen Kennedy on death penalty and other complex capital cases. Dr. Lauria served as Special Advisor for Communications to the Secretary of the Department of Transportation, Communications, and Infrastructure of the Federated States of Micronesia, as Associate Professor of Media Law, Ethics, and Emerging Technologies in the University of North Carolina system and was Director of the Global Virtual University there.

Metalaw®.US continues Rita Lauria's work at the frontier of media and communications science where she investigates the impact of cyber and communications upon legal regimes and other societal systems.

PART II

NATIONS AND REGIONS OF THE WORLD IN SPACE

INTRODUCTION TO PART II

In Part I we explored some interesting themes that show connections, patterns, and opportunities that link the space efforts of many nations and cultures. One of our goals here in Part II is to give the reader a deeper understanding of how various nations view their own space programs, and how they think about the prospects and need for cooperation. Hence, we here present nine chapters that explore in considerable detail the space programs of *Australia*, *China*, *Japan*, *Russia*, the *ESA*, *Africa*, and the *US*. As the combined populations of the nations covered is approximately 3.5 billion people, we have essentially representation of half of the world's people.

Our authors examine the history, successes, challenges, and in many cases the philosophies behind these national (and in some chapters regional) programs. These chapters also give us a glimpse toward the future, with thoughts on goals and ambitions, and suggestions for future policy.

It should be noted that a detailed review of any of these nation's or region's programs would require not one book, but an entire shelf full of them, and indeed our authors have drawn upon their own considerable expertise as well as dozens of these key reference documents, and many are listed in footnotes for readers who wish to explore more deeply. But given the nature of an anthology such as this, it's obvious that we cannot hope to cover each program in exhaustive detail. Our intent instead is to provide a succinct overview that may help us to see points of commonality among nations, and to recognize emerging future patterns which may also lead to insights precisely through the act of synthesis.

A last point here – although we were not able to obtain chapters on the efforts of Brazil, India, or Israel, we close this part of the book with a brief chapter complied from various media sources covering all three.

Budgets

As a further introduction to this section, please consider the list on the facing page showing the 23 largest space programs by budget, organized according to the actual or estimated annual spending. The combined total of the 13 programs listed here is $24,688 billion, or 75% of estimated total world, public sector, civilian investment in space.

On the same page in Wikipedia where from which this was copied you will also find a list of the countries that have national space programs. 58 countries are included, although there are some obvious omissions. If we count the regional and continent-wide space efforts of Africa on the list, which you will read about in depth here in Chapters 12 and 13, then the total population accounted for is about 6.6 billion of the world's 7 billion people, or 94% of us. And since some of the guiding policy frameworks for space are the result of United Nations initiatives, we can say without exaggeration that nearly everyone in every nation is in some way affected by some national or cooperative space efforts.

Country	Agency	Budget (USD)
United States	NASA (National Aeronautics and Space Administration)	$17,700 million (2012)[64]
ESA [show]	ESA (European Space Agency)	$5,430 million (2011)[65]
Russia	ROSCOSMOS (Russian Federal Space Agency)	$3,800 million (2011)[66]
France	CNES (French Space Agency)	$2,822 million (2010)[67]
Japan	JAXA (Japan Aerospace Exploration Agency)	$2,460 million[68]
Germany	DLR (German Aerospace Center)	$2,000 million[69]
India	ISRO (Indian Space Research Organization)	$1,320 million[70]
China	CNSA (China National Space Administration)	$1,300 million[71]
Italy	ASI (Italian Space Agency)	$1,000 million[72]
Iran	ISA (Iranian Space Agency)	$500 million[73]
United Kingdom	UKSA (UK Space Agency)	$414 million[74]
Brazil	AEB (Brazilian Space Agency)	$343 million[75]
Canada	CSA (Canadian Space Agency)	$300 million[76]
South Korea	KARI (Korea Aerospace Research Institute)	$300 million[77]
Ukraine	NSAU (National Space Agency of Ukraine)	$250 million[78]
Belgium	BELSPO (Belgian Federal Office for Science Policy)	$170 million[79]
Argentina	CONAE (Comisión Nacional de Actividades Espaciales)	$148 million[80]
Spain	INTA (Instituto Nacional de Técnica Aeroespacial)	$135 million[81]
Sweden	SNSB (Swedish National Space Board)	$100 million[82]
Pakistan	SUPARCO (Space and Upper Atmosphere Research Commission)	$82 million[83]
Netherlands	SRON (Netherlands Institute for Space Research)	$26 million[84]
Switzerland	SSO (Swiss Space Office)	$10 million[85]
Mexico	AEM (Mexican Space Agency)	$8.34 million[86]
World	All space agencies (Total of listed budgets)	$32,894.34 million

Figure 1: National and ESA Civilian Space Budgets[1]

THE MILITARY AND THE PRIVATE SECTOR

A couple of additional factors to be aware of as we attempt to get a sense of the scale of humanity's engagement in space activities are the following. First, space-oriented military budgets are rarely published, and in many cases are understood to be greater (and in some cases much greater) than the civilian ones. These budget figures shown here also do not include investments by the private sector, which is also considerable.

---

[1] http://en.wikipedia.org/wiki/List_of_space_agencies – Accessed August 3, 2012.

The private satellite industry is quite substantial in terms of aggregate revenues, as according to Futron Corporation research conducted for the Satellite Industry Association, aggregate industry revenues for the industry will reach $177 billion in 2012, which they account as 61% of total space industry revenues of $290 billion. That's a large number and a significant percentage, an indication of the role that communication plays in our modern world. We will explore this more deeply in Chapter 21.

The report further notes that there are presently 994 satellites in orbit around Earth, of which 494, or 49%, serve government and commercial communications functions.[2]

...

With all that in mind, let us turn our attention now to a review of the history, prospects, and thinking behind some of the world's most influential and productive national space programs.

---

[2] Futron Corporation, "State of the Satellite Industry Report, May 2012." Sponsored by the Satellite Industry Association. Available at http://www.sia.org/satellites-101/

CHAPTER 9

Australia's Place in Space:
Historical Constraints and Future Opportunities

Brett Biddington AM
Chairman, Space Industry Association of Australia

© Brett Biddington AM, 2012. All Rights Reserved.

Australia occupies an unusual place in the story of human activity in space. Australia's approach to space has been and remains intensely pragmatic and collaborative, leading to outcomes that have puzzled many observers and commentators over the years. Principles of necessity and sufficiency have been dominant. Grand plans for space exploration and space industry development have failed to attract sustained interest or investment from any quarter, public or private.

This article outlines the drivers for the situation that exists today and suggests some possibilities for the future. There are four principal drivers for the approach of successive Australian governments to space:

Strategic Geography
Alliance Relationships
Broader International Obligations under the Outer Space Treaty
Cost and Risk.

The first three have been and remain enduring drivers. The fourth addresses a perception, strongly held by many Australian politicians and officials, that space investments entail high technical and financial risks for uncertain returns or for returns for which broader community benefit is difficult to quantify in terms of jobs and votes. Successive governments have reasoned that Australia's essential national security interests with respect to space have been met through the extended deterrence offered by allies in return for their use of Australian soil to pursue their national interests. Ergo, substantial and sustained investment in a local space industry has simply not been a policy or investment priority.

Above all, space has been dealt with by Australian governments as a strategic and national security question within the context of alliance relationships. Investments, activities and interests have been largely driven by this perspective, and much of what has occurred has been and remains highly classified, and thus well beyond the view of many politicians, public servants and the broader population.

THE IMPACT OF GEOGRAPHY

A former representative of the US National Aeronautics and Space Administration (NASA) to Australia, Dr Neal Newman, used to describe Australia as being "there, bare and fair."[1] This epithet neatly captured the essence of NASA's interest in Australia and, for that matter, the essence of the interest of others as well. From the perspective of support to space activities, each of the points warrants brief expansion.

THERE

Australia, in terms of longitude is more or less equidistant from Europe and the United States. This means that satellites, no matter where they are relative to the Earth's surface, are likely to be in view of a ground station in North America or Europe or Australia, making continuous monitoring and downloading of data possible. Satellites in geo-synchronous orbits which are located above the Equator from the western

---

[1] Dr Neal Newman was NASA's Representative in Australia from 2002 to 2006.

Pacific to the mid Indian Ocean can be seen from ground stations in Australia. Australia, in short, fills the coverage gap between North America and Europe.

BARE

Continental Australia is roughly the same size as the continental United States but with a much smaller population (22 million as opposed to 360 million) which is largely confined to the coasts and the south-eastern corner of the continent. The continent is therefore radio quiet and ideal for hosting satellite ground stations and radio telescopes.

FAIR

Australia is a liberal democracy and an ally of the United States, and although less important today, the United Kingdom as well.

USA Australia size comparison. Source: popular tourist postcard, copyright unknown - downloaded from Google Images

When these factors are combined they provide attributes not available anywhere else in the world.

The Joint Defence Facility at Pine Gap near Alice Springs, in the middle of the Australian continent, was located there as a direct consequence of geography. The site was radio quiet and likely to stay that way for a long time. It was far from the coasts and international waters in which the electronic intelligence collectors of the Soviet Union operated during the Cold War. This was vitally important because the first generation of satellites supported by Pine Gap are reported to have had an unencrypted communications link, and the United States was very keen to protect the strategic advantage obtained by having these satellites in orbit.[2]

THE IMPACT OF ALLIANCES

Since the mid 1940s, Australian space activity has been bifurcated. On the one hand Australia has been a staunch supporter of the space ambitions of "our great and powerful friends," to use the phrase coined by Robert Menzies, Australia's longest serving Prime Minister.[3] In the 1940s and 1950s this "friend" was the United Kingdom. Since the 1960s it has been the United States. Only a handful of senior Ministers, the members of the National Security Committee of Cabinet, are comprehensively briefed about the functions and capabilities of the so-called "joint facilities" which have been established by the United States on Australian soil and which are jointly staffed and managed by Australian and American personnel. This means that junior Ministers, backbench Members of Parliament, and the wider Australian public have been forced to depend on sometimes scant, incomplete, and even incorrect information in their efforts to appreciate the importance of these facilities to national, regional and global security. Key decisions about alliance relationships have been made by senior Ministers, including the Prime Minister, the Attorney General, the Treasurer and the Ministers for Defence and Foreign Affairs.

On the other hand, responsibility for investment in space science and a domestic Australian space industry, especially in launchers and satellites, has been the preserve of junior Ministers who, although well-meaning, have failed to win the support of their more senior colleagues for enduring and systematic investment in space activities. The efforts of junior Ministers have lacked direction and focus, have been constrained by lack of funds and characterized by opportunism not connected to Australia's implicit approach to space in which alliance relationships and international civil collaboration have been and remain central pillars.

A consequence of bifurcation is that Australia has not developed a coherent national space narrative that makes sense to politicians and to the

---

[2] Richelson, J.T., *The Wizards of Langley: Inside the CIA's Directorate of Science and Technology*, Westview, Cambridge, US, p 112.
[3] Menzies, R.G., *The Measure of the Years*, Cassell, Melbourne, 1970, p 44.

wider community. Indeed, serious discussion about the role Australia might and should play in space is hindered by ignorance on the one hand, and an unfortunate "giggle factor" on the other. There is also a well-entrenched economic rationalist view that governments have no business in investing in a domestic space program, any more for example, than they should invest in the creation of a biotechnology industry. Politicians and senior officials have come to approach space matters with considerable caution and skepticism, and there is a widely held view "space is expensive and run by geeks."[4] The sector has a major image problem when questions deserving of serious and informed debate arise, although there is evidence that this may be changing, which is discussed in the latter sections of this chapter.

Many commentators both in and beyond Australia express puzzlement and even dismay at Australia's apparent lack of interest and involvement in space matters. Most fail to understand the bifurcation that has occurred since the late 1940s, and its influence and implications.

Australia's core strategic interests in space have been met through alliance relationships. Since the 1960s, the United States has underwritten Australia's security in return for Australia hosting satellite ground stations which were essential elements in the global Cold War intelligence and surveillance system developed by that superpower. These facilities remain as important today, if not more so, than they did when they were established.

Core operational level interests, especially access to Earth observation data for numerous applications including meteorology, mineral exploration, environmental monitoring, vegetation mapping, land use planning and ocean monitoring, have been achieved largely through access to public good data made available through international collaborations. With regard to the sciences of space, astronomy has been favored over space science (the latter taken to include planetary geology, solar physics and space weather), except where space science has been relevant to national security, such as ionospheric physics, which makes an essential contribution to the operations of Australia's Jindalee Over the horizon Radar Network (JORN). The JORN system is arguably the most advanced high frequency sky-wave radar in the world, providing wide area coverage of Australia's northern approaches, monitoring aircraft and shipping

---

[4] This phrase was used by a now retired Minister from the Government of South Australia in a private conversation with the author in March 2011.

movements.[5]

Turning the clock back to the dawn of the space age, before World War II had ended, the United Kingdom was considering what it needed to do to acquire atomic weapons and the missiles to deliver them. By this time, there was no dispute or doubt that Australia was an independent sovereign state within the British Empire.[6] The ties of Empire, however, were strong, certainly strong enough for Australia to enter into a "Joint Project" with the United Kingdom to use remote and sparsely populated areas of the Australian outback to develop and test these weapons. This led to the creation of the Woomera Test Range, 650 km northwest of Adelaide, and later sites as well, notably the Monte Bello Islands off the coast of Western Australia and Maralinga near the border between Western and South Australia, for the testing of nuclear devices.[7]

WOOMERA

Woomera was a very active place in the 1950s and into the early 60s as the Joint Project progressed. However, when the United Kingdom recognized it could no longer afford the costs of Empire, and in the face of increasingly persistent anti-colonial movements, it abandoned its ambitions to be a self-reliant nuclear power which was the death knell of the Joint Project. The European Launcher Development Organisation (ELDO) was active at Woomera for a time, but the French, partly in response to the British change in policy, and determined to develop an independent launch and nuclear capability of their own, left Woomera as well. ELDO was eventually absorbed into the European Space Agency (ESA) and the launch program was moved by the French to Karou in French Guinea where it provided the basis for the successful Ariane family of launch vehicles.

---

[5] For a brief description of the JORN project see:
http://www.defence.gov.au/dmo/esd/jp2025/jp2025.cfm

[6] The relationship between Australia and the United Kingdom is frequently misunderstood by foreign observers due to the fact that the Queen of England is also, formally, the Head of State of the Commonwealth of Australia. This is an artefact of history. Any doubt as to Australia's sovereign independence was removed with the passage of the *Statute of Westminster Adoption Act* by the Australian Parliament in 1942.

[7] The definitive account of the Joint Project is Morton, P., *Fire Across the Desert: Woomera and the Anglo-Australian Joint Project, 1946-1980*, Department of Defence, Canberra, 1989, reprinted 1997.

Woomera's main launch site in the 1960s showing a fully-assembled Europa launch vehicle. (Photographed from a helicopter at an altitude of about 90 metres by Max Ryan, then Officer in Charge of the Still Photography Section, DSTO Salisbury). See http://homepage.powerup.com.au/~woomera/ for further information. Copyright, Commonwealth of Australia.

THE SPACE TREATIES

In the 1960s Australia was an active participant in the international dialogue concerning the regulation of outer space. Australia has signed and ratified all of the five main space treaties including the Moon Treaty, unlike most other countries in the world, and places high value on being regarded as a good international citizen where the regulation of space is concerned. During the Cold War, Australians chaired the S&T Committee of the United Nations Conference on the Peaceful Uses of Outer Space (COPUOS) continuously for 33 years from 1962 – 1995. However, Australian official commitment and interest in the international governance of space waned following the retirement of Professor John Carver from this and other positions in 1995.[8] As will be discussed below, there are signs this may be changing and that Australia, once more, is beginning to take an active interest in space governance.

---

[8] Crompton R W, Dracoulis G D, Lewis B R, McCracken K G and Williams J S. "John Henry Carver: Biographical Memoir", *Historical Records of Australian Science*, Vol 22, No 1, June 2011, CSIRO, Australia, pp 71-72.

THE US ALLIANCE AND THE JOINT FACILITIES

From the earliest days of the US civil, military and classified space programs, the Australian Government has hosted a number of ground stations to support these activities. Some, notably the Defence Surveillance Program (DSP) ground station at Nurrungar near Woomera (operated from 1970-1999), and the Joint Defence Facility at Pine Gap (JDFPG) near Alice Springs in Central Australia (1970–present), have been at times contentious and at the forefront of political arguments and debate in the context of Australia's alliance with the United States, and the extent to which Australian sovereignty was impugned or diminished by the bases being located on Australian soil.

This was especially the case at the start and finish of the Whitlam Labor Government in the 1970s. When the Australian Labor Party (ALP), led by Gough Whitlam was elected to office in December 1972, ending 23 years of conservative rule, deep concerns were held in the US that the incoming government might compromise the security of the joint facilities, or possibly move to shut them down.[9] In fact this did not happen, and the Whitlam Government, although harbouring reservations about the levels of secrecy applied by the United States, supported their continued operations. At the very end of the Whitlam Government's tenure, which closed in controversial circumstances when the Governor General removed the Prime Minister's commission in order to break a parliamentary deadlock, the joint facilities were again at the fore.

A conspiracy theory gained credence to the effect that the Secretary of the Department of Defence, Sir Arthur Tange, possibly assisted by the Chief Defence Scientist of the day, John Farrands, had, in effect, tipped off the Governor-General just days before he dismissed the Whitlam Government, that the United States held grave concerns for the security of the joint facilities if a Labor Government were to remain in office. This theory persisted for more than a decade although no compelling evidence has ever been produced.[10] What these events do point to is the exceptional care taken by the United States and those Australians who are briefed about the joint facilities not to reveal, or have revealed by others, the technical capabilities (and limitations) of the space-based surveillance and intelligence gathering systems of which the Australian hosted facilities are

---

[9] Richelson J T. *America's Space Sentinels: DSP Satellites and National Security*, Kansas University Press, 1999, p 139.

[10] Edwards P. *Arthur Tange: Last of the Mandarins*, Allen and Unwin, Sydney, 2006, pp 271-290.

an important part.

There was a strongly held belief in some sections of the community, and also by some parliamentary members of the ALP as well as Left leaning politicians and commentators, that these facilities would be nuclear targets in the event of war between the Union of Soviet Socialist Republics (USSR) and the United States. This possibility, although remote, was explicitly acknowledged in the 1980s, including in the 1987 Defence White Paper.[11] Although critics of the joint facilities remain, the concerns and reservations that are expressed today are generally more restrained than in the past, a consequence of the passage of time and the end of the Cold War.

In 1985 Professor Paul Dibb reviewed Australia's defense capabilities and his report, published in March 1986, provided the basis for the Defence White Paper published a year later, March 1987 (DOA 87). Both documents refer to the need for the Australian Defence Force to embrace satellite communications and both emphasize the importance of the joint facilities. Dibb wrote, in this instance by inference to the joint facilities, in the following terms:

"We have access to United States intelligence resources, which can provide technical military intelligence coverage beyond the comprehension of previous generations of military planners. These resources contribute not only to our capacity for strategic analysis, but also to the potential effectiveness of our forces in circumstances of combat. They could not be duplicated from national resources."[12]

Both documents were silent, in any explicit way, about outer space. Dibb summarized Australia's principal national security interests under six headings of which space was not one.[13] DOA 87 summarized Australia's principal national defense interests as being:

- The defense of Australian territory and society from threat of military attack;
- The protection of Australian interests in the surrounding maritime areas, our island territories, and our proximate ocean areas and focal points;
- The avoidance of global conflict;
- The maintenance of a strong defense relationship with the United States;

---

[11] Government of Australia, Department of Defence, *The Defence of Australia, 1987*, Canberra, 1987, para 2.11, p 12.
[12] Government of Australia, Department of Defence, *Review of Australia's defence capabilities*, Canberra, March 1986 (the Dibb Report), p 46.
[13] Dibb Report, p 37.

- The maintenance of a strong defense relationship with New Zealand;
- The furtherance of a favorable strategic situation in South-East Asia and the South-West Pacific;
- The promotion of a sense of strategic community between Australia and its neighbors in our area of primary strategic interest;
- The maintenance of the provisions of the Antarctic Treaty, which ensures that continent remains demilitarized.[14]

Both Dibb and DOA 87 also placed considerable emphasis on the concept of 'self-reliance.' Dibb called self-reliance a "slogan" until it was "related to credible and practicable situations."[15] Kim Beazley, the Minister for Defence at the time, in his introduction to DOA 87, explained the implications of self-reliance, writing that self-reliance aimed to achieve the four fundamental objectives of Australia's national and international defence policy:

(1) capacity for the independent defence of Australia and its interests;
(2) promotion of regional stability and security;
(3) strengthened ability to meet mutual obligations especially with main allies (the United States and New Zealand); and
(4) enhanced ability to contribute to strategic stability at the global level.[16]

He also cautioned that self-reliance does not mean self-sufficiency.[17]

Nowhere is there any direct reference to outer space as being relevant to Australia's strategic circumstances, or as an area in which Australia should seek some capacity at all, let alone self-reliance or, even more ambitiously, self sufficiency. It simply did not rate.

A turning point with regard to broad acceptance of the joint facilities came in 1988 when Bob Hawke, the Labor Prime Minister of the day, announced in Parliament that the agreements covering the joint facilities with the United States had been extended for a further 10 years. His statement contained the most fulsome official explanation of the purposes of both Nurrungar and Pine Gap provided before or since. Hawke said:

---

[14] DOA 87, para 2.69, p 22.
[15] Dibb Report, p 107.
[16] DOA 87 p ix.
[17] DOA 87 p x.

"Nurrungar is a ground station used for controlling satellite (sic) in the United States defence support program (DSP). The DSP satellites provide ballistic missile early warning and other information related to missile launches, surveillance and the detonation of nuclear weapons. Few if any elements of the strategic systems of either superpower make such a decisive and unambiguous contribution to keeping the peace as the defence support program.

"Pine Gap is a satellite ground station whose function is to collect intelligence data which supports the national security of both Australia and the United States. Intelligence collected at Pine Gap contributes importantly to the verification of arms control and disarmament agreements."[18]

The Pine Gap complex near Alice Springs.
Image from the cover of Rosenberg, D., Inside Pine Gap, Hardie Grant Books, Melbourne, 2011.

Other ground stations, including the NASA Deep Space Tracking Station at Tidbinbilla and the European Space Agency (ESA) stations in Western Australia, manifestly support space exploration and space science and are an integral part of the Australian space landscape. They are

---

[18] Parliament of Australia, Hansard, H of R 22 Nov 1988, p 2940.

unobtrusive, do good work, provide a small number of jobs into local communities, are the source of occasional good news stories, demonstrate Australia's bona fides as an international partner and cost the Australian Government and taxpayer almost nothing. "Why invest when this is the sort of deal the nation has been able to negotiate in the past?" is a question that few critics and commentators have sought to ask.

AUSTRALIAN SATELLITES

Three satellites have been designed and built (mostly) in Australia. The first was WRESAT in 1967, making Australia the fourth country to launch a satellite made in its own territory from that territory. The second was Oscar Australis 5 in 1970, and the third was FEDSAT in 2002. All were one-off projects and none led to the follow-on activity which was essential if industry was to have the confidence to invest.

CIVIL DEVELOPMENTS

In 1985 the Hawke Labor Government established an Australian Space Office (ASO) following a seminal review of Australia's performance in space, which was conducted at the request of the government by the Australian Academy of Technological Science and Engineering. The review was headed by a distinguished mining company executive, Sir Russel Madigan. Madigan recommended that modest funds ($100 million over five years) be provided to promote space science and to assist space industry developments, including private initiatives to establish launch facilities on the Cape York peninsula in northern Australia and also at Woomera.[19] Successive governments never came close to the investment levels recommended and none of the industry initiatives materialized.

Madigan recommended that the work of the ASO be overseen by a Space Council comprising worthy and experienced people who would guide the ASO and keep the Minister informed as to progress. The Council was initially an advisory group which reported to the responsible Minister. Eventually it was established under its own Act of Parliament in 1994.[20] The functions of the Council were outlined in S6 of the Act:

---

[19] Australian Academy of Technological Sciences, *A Space Policy for Australia*, Melbourne, June 1985 (The Madigan Report).
[20] Australian Space Council Act, No 27, 1994.

6. (1) The Council's functions are:

- To inquire into, and report to the Minister on, such matters affecting the application of space-related science and technology by the Australian public and private sectors as are referred to the Council by the Minister; and
- To recommend to the Minister a national space policy (the "National Space Program") that encourages the application of space-related science and technology by the Australian public and private sectors; and
- To co-ordinate the involvement of representatives of the Australian public and private sectors in developing, and reviewing, the National Space Program; and
- To keep under review, and report to the Minister on, the outcomes of the National Space Program; and
- To consult and co-operate with persons, organisations and bodies in relation to matters affecting the National Space Program; and
- Such other functions as are conferred on the Council by this Act or any other Act.

(2) In performing its functions, the Council must comply with any directions given by the Minister [under section 8.]

(3) The Council may perform its functions in or out of Australia.

The Council produced reports in 1994 and 1995. Emphasis was on launch and focussed on the establishment of a space launch facility on Cape York which was described in the second report as a "Project of National Significance."[21, 22]

In 1996, when the Liberal Coalition Government led by John Howard came to power, the Space Office was scrapped together with the Space Council and their lofty ambitions. Investment in space development activities and space science funding ceased to have any special status in Cabinet, in research funding, or in any other claims to government attention.

Some years later a further attempt to establish a space launch

---

[21] Australian Space Council, National Space Program: Five Year Plan, 1994, Australian Government Publishing Service, Canberra, 1994.
[22] Australian Space Council, National Space Program: Five Year Plan, 1995, Australian Government Publishing Service, Canberra, 1995.

capability on Australian territory was made by the Asia Pacific Space Centre (APSC) which planned to launch satellites from Christmas Island beginning in 2004. Christmas Island is a small Australian island territory south of the Indonesian island of Java and located at 10.5 degrees South. It is close to the Equator and well placed to support heavy lift payloads into geo-stationary (GEO) orbits. APSC planned to use Russian launch vehicles and had an aggressive and optimistic launch schedule. The Australian government offered considerable assistance to APSC mainly in the form of infrastructure upgrades at Christmas Island. However, the launch market dried up, the investors needed to bring the spaceport to life walked away and the project failed.

An enduring legacy of these failed launch initiatives is the Australian Space Activities Act, No 123 of 1998. This Act and its associated regulations is one of the more comprehensive pieces of domestic space legislation in existence anywhere in the world. Perhaps most importantly it defines a lower, if arbitrary, limit of outer space for the purposes of the Act as being 100km.[23]

In 1979 the Australian Government established a new communications company called AUSSAT the purpose of which was to acquire and operate a small number of capable satellites to considerably improve communications services to regional and remote areas of Australia. The satellites were acquired from the Hughes Corporation and launched in 1985 and 1986. Shortly afterwards, AUSSAT was sold to a commercial company, Cable and Wireless Optus, which was later purchased by the Singapore based Singtel corporation. Optus is currently the ninth largest communications satellite company in the world.[24] Although a wholly owned subsidiary of Singtel, Optus is an Australian company and the Australian government has imposed strict licensing conditions to ensure Australia's sovereign interests are adequately protected.

In 2000 the Howard Government released the first Defence White Paper since the one published by the Hawke Government in 1987.[25] Most importantly, the Cold War had ended and alliances and other relationships around the world had been re-shaped. The 2000 document, not unlike its predecessor, was virtually silent on space matters. The word 'space' in the context of outer space does not appear, and the word 'satellite' is

---

[23] Parliament of Australia, *Space Activities Act*, No 123 of 1998.
[24] Space Foundation, *Space Report, 2010*, Exhibit 3k, USA, p 79.
[25] Government of Australia, Department of Defence, *Defence 2000: Our Future Defence Force*, Canberra, 2000.

mentioned only twice – once in the context of high-resolution imagery and once in the context of communications satellites. There is no direct reference to the joint facilities, only higher-level phrases about intelligence cooperation and privileged Australian access to a range of US technologies.

In 2001 the Howard Government established an International Space Advisory Group (ISAG), chaired by the first person born in Australia to fly in space, Dr Paul Scully-Power. The ISAG, according to the media release, was an ambitious undertaking. Among other things, it was to:

- Lay the foundations for the development of a strategy for Australia's engagement in key international space programs,
- Identify opportunities for Australian involvement in the International Space Station (ISS) and other international space programs, and
- Assess the potential scientific and commercial benefits in pursuing such opportunities.

The responsible Parliamentary Secretary of the day, Warren Entsch, was quoted in the same media release as saying, "This could include such things as Australia taking a direct role in testing and providing landing sites for NASA's new X-38 International Space Station crew rescue vessel, right through to possible collaboration in the development of habitation modules for the International Space Station and space based research in the life sciences."[26]

No such activities eventuated. The legacy of the ISAG, which reported in 2002, is that it created an inventory of civil and commercial space activities with which Australia and Australians were involved.[27] Its principal recommendation was that Australia should develop a 'demand' driven strategy for space engagements. Contrary to the hyperbole of Mr Entsch, the report also concluded that:

"Australia should not sign onto international space programs, including the International Space Station, before determining its own national priorities for space engagement through a demand driven strategy."[28]

The ISAG report set the scene for the Howard Government's Australian Government Space Engagement: Policy Framework and

---

[26] Media release, *Australian International Space Advisory Group Meets in Canberra*, Office of Warren Entsch, MP., Parliamentary Secretary to the Minister for Industry, Science and Resources, Canberra, dated 5 June 2001.
[27] Report of the International Space Advisory Group, Canberra, 2002 (ISAG Report).
[28] ISAG Report, p i.

Overview.[29] It would seem to have reinforced the view of Government that space was not an area which called for any special consideration where investment or intervention was concerned.

The apparent lack of interest and investment in space activities in Australia for 25 years from 1980 – 2005 has been a source of conjecture at home and abroad. One paper, describing and lamenting this state of affairs, uses the phrase "punching below its weight" to describe Australia's seemingly passive and confused approach to involvement and investment in space activities.[30]

Some writers point to Canada and wonder why Australia did not choose a similar path and develop a niche, local space industry. Both nations, after all, have vast geographies for which satellites present obvious practical solutions. There are two answers to this question. The first is that the Canadian space industry, for all intents and purposes, is an integral component of the US space industry. Canadian companies supply niche components to the major US satellite manufacturers in a single market. Perhaps more importantly, the drivers of the alliance relationships with regard to space between Canada and the United States and Australia and the United States are quite different. During the Cold War Canada and the United States had common cause in the air and missile defense of North America. They faced a common threat. Canadian servicemen and women served alongside their US counterparts in the organizations that were established to conduct nuclear war with the Soviet Union, such as the North American Aerospace Defense Command (NORAD). This was for Canada a survival imperative, but not so in Australia. Although the joint facilities may have been targets in the unlikely event of global nuclear war, they were in remote locations well away from major population centers. A different risk calculus leading to different investment priorities applied.

CHANGING TIMES

From the mid 2000s a series of events and activities have occurred in and beyond government that have started to develop the narrative about space that was previously absent. At the same time, the political and physical

---

[29] Department of Industry, Tourism and Resources, *Australian Government Space Engagement: Policy Framework and Overview,* Canberra, November 2006 (Space Engagement Framework).

[30] Kingwell J. "Punching below its weight: Still the future of space in Australia", *Space Policy,* Vol 21, Issue 2, May 2005.

circumstances of space have changed substantially, and finally, and perhaps most importantly, there is increasing awareness, certainly by governments, of the increasing reliance that the economy, and social life more generally, has on assured and secure access to space-based services, including position, timing and navigation, communications, and Earth observation.

The closest the Howard Government came to enunciating a national space policy was a document called Australian Government Space Engagement: Policy Framework and Overview. This document, first released in 2003, following the ISAG review, was considered by some critics to be a policy not to have a policy. Its central tenet, that Government saw no special need to provide favored treatment to the space sector over any other sector of the economy in organizational terms, or in support for industry development and research, was roundly condemned by small but vocal space science and space advocacy groups. Space Engagement stated:

"There is no strategic, economic or social reason for the Australian Government to pursue self-sufficiency in space."[31]

Few would have disagreed with this statement. However critics of this policy, such as Jeff Kingwell, made the point that there were opportunities in space science and space industry development, short of self-sufficiency, which, if pursued, offered benefits.[32] Unfortunately, papers such Kingwell's failed to comprehend the impact of the US alliance relationship on space-related discussions in government. Comparisons with countries including Mexico, Venezuela, Egypt and others simply cut no ice with Australian governments because none of those nations had a relationship with any superpower which bore any resemblance to that which existed (and exists) between Australia and the United States and at the core of which are space capabilities.

In 2005 Senator Grant Chapman, a Liberal Senator from South Australia, established a Space Policy Advisory Group (SPAG) which drafted a report, Space: A Priority for Australia, that the Senator forwarded to the Prime Minister in November.[33] This report noted the growing dependence of many sectors of the Australian economy and social fabric on space-based services and argued that the Government did not have a clear view either of these dependencies or their associated vulnerabilities. The report argued that space-based services were, in effect, part of the nation's

---

[31] Space Engagement Framework.
[32] Kingwell, p 161.
[33] Chapman, G., *Space: A Priority for Australia,* Senate Printing, Parliament House, Canberra, 2005.

critical infrastructure and recommended that the Government take a series of steps to allow it to better understand and mitigate the vulnerabilities which were emerging. The Chapman Report, as it became known, led to a formal review of the Space Engagement document and a revised version was released in early 2007.

THE 2008 SENATE INQUIRY

In November 2007 the ALP led by Kevin Rudd won a clear victory at the general election. John Howard, the incumbent Prime Minister, even lost his seat. The new Prime Minister, who had been a diplomat and had an abiding interest in foreign affairs, set a frenetic pace, (which ultimately led to his undoing in June 2009). Shortly after the election, three Senators, all from South Australia (Grant Chapman, Liberal; Natasha Stott-Despoja, Democrat; Annette Hurley, ALP) worked in concert to persuade the Senate, through its Standing Committee on Economics, to conduct an inquiry into Australia's space sector. This was the first time either House of the Australian Parliament had deliberately inquired into Australia's performance in space.

The Committee received 88 written submissions and held six public hearings. It produced an interim report in June 2008 and a final report in November 2008.[34] Its six recommendations were cautious but directed towards much better coordination of space activity in Australia, notably in the civil and commercial sector. The report viewed the creation of an Australian space agency as a goal to pursue and recommended that an Advisory Council be established which would:

- Conduct an audit of Australia's current space activities within six months of the establishment of the Council;
- Analyze the strengths, weaknesses opportunities and threats to Australia's emerging space industry;
- Focus on the key "workhorse" space applications of Earth observation, satellite communications and navigation as the most practical and beneficial initial priorities;
- Systematically evaluate the medium/long-term priorities for a space agency, including the national benefit of defense related

---

[34] Senate Standing Committee on Economics, *Lost in Space? Setting a new direction for Australia's space science and industry sector*, Canberra, November 2008 (Senate Report).

activities, Earth observation, environmental, land management, exploration, national disaster prevention and management, treaty monitoring, e-commerce, and telemedicine;
- Examine the benefits to Australia of improved international collaboration, including membership in international space groups;
- Develop a draft strategic plan for the establishment of a space agency and the most appropriate form of that agency, including public/private funding, budget and staffing priorities; and
- Identify critical performance areas such as research, technological development, development of the skill base, effective partnerships, delivery of new services, and financial management.[35]

The Government formally responded to the Senate recommendations in November 2009, well after the May 2009 budget, which allowed the Government to record considerable progress against the Senate report's recommendations. The 2009 events are discussed later in this chapter.

OTHER FACTORS

There were also other factors at work.

First, the Global Financial Crisis of 2008 prompted the Australian Government to devise an economic stimulus package in order to protect jobs and to help the Australian economy ride the storm that engulfed stock markets and the economies of North America, Europe and elsewhere. The stimulus package provided an unexpected source of income, some of which was used to fund some of the space development activities recommended in the Senate Report.

Second, Prime Minister Rudd made the first ever national security statement to the Parliament on 4 December 2008.[36] This wide-ranging document was completely silent about space, except by inference where discussing the importance to Australia of the US alliance. Perhaps the authors of the document did not know quite what to say. Certainly, in late 2008 and extending into 2009, the Prime Minister's own department conducted a review of Australia's space activities which was intended as

---

[35] Senate Report, Recommendation 5.
[36] Rudd, K., The First National Security Statement to the Australian Parliament, 4 December 2008, Hansard, 2008.

the basis of a briefing to the Prime Minister in April 2009.[37] The outcomes of this review may well have helped to inform decisions announced in the May 2009 budget about civil and science investment in space. It may also have influenced the focus on space in the 2009 Defence White Paper which was also released ahead of the May 2009 budget.

Third, the global strategic environment, of which space forms an integral element, was rapidly changing with the rise of China and the relative diminution of status of the United States. What Australia, as a middle power, might, could and should do to influence the developing relationship between China and the United States may well become the defining question for Australian foreign policy over the next 50 years and as the regulation, use of and behaviour in space may become one of the most contentious areas with which China and the United States will have to deal in this century, Australia finds itself precisely in the middle of the equation.

THE AUSTRALIAN SPACE INDUSTRY

Since it was formed in mid 2009, the Space Policy Unit has commissioned two surveys of the Australian space industry in an effort to gather data about its size and wider contribution to the Australian economy. The second report, which was an update of the first, concluded:

It [the report] reveals that Australian space activities are substantial in their own right in terms of the number of organizations involved (631), the number of people employed (8,418), and the revenue generated ($1 billion - $2.2 billion).[38]

Leaving aside questions about methodology, the word "substantial" hardly seems warranted to describe a collection of companies and research groups in universities and government agencies which contribute directly somewhere between 0.1 and 0.2% of Australia's GDP (which, in aggregate, is approaching one trillion dollars annually). No matter how these numbers are cut, they point to a sector populated by small, even tiny companies and research groups which, using simple arithmetic, employ 13-14 people each. This is not to denigrate the efforts of those involved, far from it. Rather it

---

[37] The author of this paper was an invited participant at one of the workshops held as part of the review.

[38] Asia Pacific Aerospace Consultants, *A Review of Current Australian Space Activities: Executive Summary*, p1. June 2011. At http://www.space.gov.au/SpacePolicyUnit/Pages/DevelopmentandConsultation.aspx, accessed 26 Mar 2012.

is intended to put some realistic measures around the industry, at least as defined by the surveys.

Only one Australian company, Silanna, is known to produce components (radiation hardened computer chips) that fly in space.[39] The industry overall is dispersed and most participants self-identify with other sectors of the economy such as communications, remote sensing or surveying, rather than with space per se. Their space business often comprises a very small element of their business overall.

SATELLITE COMMUNICATIONS

In Australia, satellite communications is a mature, commercially viable market. The Australian government, as noted earlier in this chapter, invested in two communications satellites in the 1980s through AUSSAT, which was the forerunner to OPTUS. The Australian Government, through the National Broadband Network (NBN) initiative, has recently announced its intention to purchase two more highly advanced broadband satellites to deliver high capacity communications services primarily to Australia's remote and regional areas.

The Australian Department of Defence has purchased satellite communications in ways that are innovative and could well serve as a model for others to follow. Defence has used five approaches:

1. A shared satellite approach on the OPTUS C1 satellite where the one satellite bus is shared by a commercial payload operated by Optus and a military payload operated by Defence and the satellite is jointly owned;
2. A constellation approach, evident in Australia's investment in the US Wideband Global System (WGS), whereby Australia has paid for one satellite in a constellation of six satellites which is operated by the USAF and, in return, has gained access to the capabilities of the entire constellation;
3. A hosted payload approach, adopted for an agreement reached with INTELSAT, whereby Australia has paid INTELSAT to place an additional military communications payload onto an otherwise commercial satellite (IS-22, launched on 25 Mar 2012);
4. Negotiated access to allied communications satellites to support specified activities; and

---

[39] Silanna: see http://www.silanna.com/

5. Ordinary commercial leases.

The key point to note is that Australian Governments and military planners have judged that the risks of accepting considerable and growing dependence on space-based services, and over which the nation has little or no direct control, are worth taking. Assured and secure access to the space environment on a sustained and continuous basis has been assumed or taken as a given. A corollary has been that the mere possession of satellites that bear the Australian flag is unlikely to alter the extent of Australia's influence in international conversations around the security and stability of the space environment. The attitude seems to have been that if the United States is prepared to take the risk, then why would Australia not follow suit? The Australian Government has made substantial investments, in both space and ground segments, via the approaches outlined above, a point not always acknowledged by those who argue for more active and explicit space engagement by Australia.

Australia's direct investments in space communications have been achieved in such a way that up-front capital exposures have been shared or minimized and access has been gained, in the case of WGS especially, to a global constellation for the cost of investing in only one part of it. Operationally, providing that the nation remains confident in the strength and endurance of the alliance, this seems like a very good deal. Strategically, such dependencies may become limiting because Australia's options for independent action could be reduced at some point in the future. There is every reason to be confident that the Australian Government is aware of these risks which partly explains why commercial leases with various providers are also in place. Defence would seem to be making considerable effort to ensure path diversity thereby avoiding single points of failure in satellite communications systems which could have adverse strategic, operational and tactical consequences, especially for deployed elements of the Australian Defence Force.

TIMING AND NAVIGATION

Australians, like most other peoples on Earth, are reliant on the US Global Positioning System (GPS) for precision timing and navigation that supports myriad applications. GPS has become a global utility funded by the US taxpayer, and life without access to GPS is difficult to imagine. Many economic and social systems that are highly integrated and tightly coupled would stop. Others might slow to a crawl until alternative methods for determining time and location were established. Several nations, including Russia, China, Europe (as a whole), Japan and India have developed or are developing their own precision timing and navigation systems which are independent of GPS. All of these systems converge in Asia including across the northern parts of Australia, forming what some have called the Global Navigation Satellite System (GNSS) Hotspot.

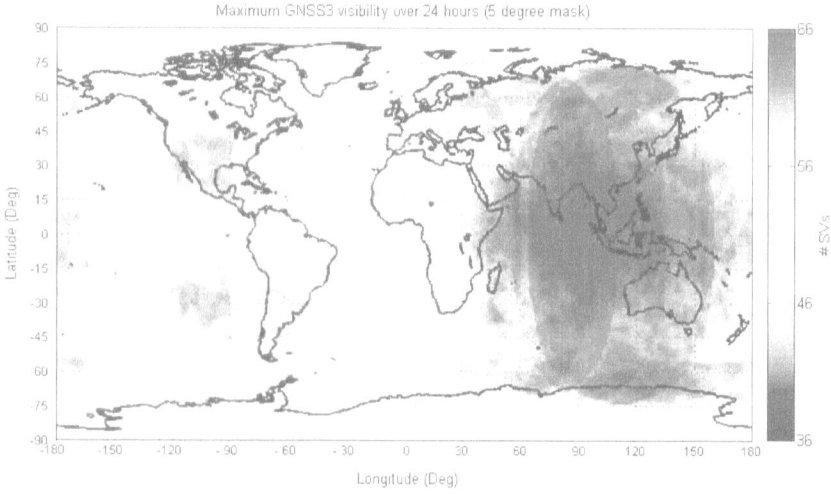

The multi-GNSS "hot spot" over South East Asia (University of NSW, 2011), published in Australian Spatial Consortium, Australian Strategic Plan for GNSS, Canberra, 2012. p 12.

The alternative systems to GPS have been developed in part to mitigate concerns that the US may seek national strategic, economic, or military advantage by selectively denying others access to GPS in circumstances of international tension, disagreement or war. But all of these systems nevertheless assume that access to space will remain assured and secure for all.

The key vulnerability may not be aberrant behavior by a nation on Earth, but rather solar activity of such magnitude that much of the near space environment, if not all of it, could be rendered unusable. Satellites, especially those operating beyond the Van Allen belts, including the GPS satellites, may well be damaged or knocked out. Solar storms of sufficient magnitude to damage and possibly disable satellites have been recorded in the past. Advance warning of such solar activity is short, and although operators can take actions to mitigate the adverse effects of solar storms and cosmic radiation, a situation can be envisaged where the sheer intensity of a storm could be expected to overwhelm the defenses of many satellites, leading to reduced performance and, in extremis, destruction.

Certainly the possibility of such an event is remote, but the possibility is greater than zero, and the global community would do well to consider the impact of a world without GPS or, at best, with severely degraded GPS for an extended period of time. Australia is well placed to frame this discussion for the space disadvantaged nations and to develop realistic mitigation strategies.

The Australian GNSS community, following the lead of space science and Earth observation communities, has recently published a strategic plan for precision location for Australia looking towards 2030 and beyond.[40] This report places dollar values on the economic benefit to be gained by systematic investment in Continuously Operating Reference Systems (CORS) and related capabilities that provide precision timing and location data to all parts of the nation and all sectors of the economy.

EARTH OBSERVATION

Australia is a sophisticated user of Earth observation data especially for climate modeling, weather prediction, and mineral exploration. This status has been achieved without the nation having had to invest, to this point, in satellites of its own. Australia, through international agreements and investments in ground reception stations, data processing infrastructure, appropriately trained and qualified people and modest but capable research infrastructure, has developed a critical mass of capability which is unique among non-spacefaring nations.

A substantial component of Australia's foreign aid program is allocated to the provision of meteorological and related service, from data derived from satellites, to nations in SE Asia and the SW Pacific. Success and sophistication notwithstanding, there are shortcomings and there is considerable room for improvement.

There are several reasons why Australia has not embraced satellite remote sensing in a more fulsome way than might otherwise have been the case.

GOOD WEATHER

Australia's skies, except north of the Tropic of Capricorn during the monsoon season, are mostly clear of clouds, which means that airborne

---

[40] Australian Spatial Consortium, *Australian Strategic Plan for GNSS*, Canberra, 2012.

remote sensing is a well-established industry.[41] Users, both government and private, are familiar with the products and understand their strengths and limitations. Users control and, in some instances, own and operate these aircraft and their sensors. "Why fix a system that is not perceived to be broken?" is a pervasive attitude. Those who seek to persuade Australia to invest in Earth observation satellites rarely acknowledge the efficacy of existing arrangements and fall into a trap of assuming that, almost by definition, a satellite-based solution is in some way superior to anything that might have gone before.

TECHNOLOGY PUSH VERSUS USER PULL

Many companies and government funded agencies from spacefaring nations have come to Australia to demonstrate advanced and seductive satellite-based remote sensing technologies which are assumed to fill pressing information gaps or deficiencies. Invariably these organizations have sought to raise capital or to sell data on an ongoing basis. But few have bothered to learn about the scope, scale and structure of the Australian market. For example, Australian fire and emergency management organizations are favored targets even though the amounts of money to which they have access is small and is held by the States and territories. Until recently there has been no single buying authority for satellite data in Australia. None of the relevant State and Territory authorities have access to the (usually) multiple millions of dollars sought by proponents of satellite-sourced data. More importantly for responder organizations such as fire fighting agencies, their organizational culture seeks to keep to an absolute minimum any investment that is not directed towards response assets. Somewhat simplistically, fire trucks, not information systems, put out fires, and firemen do not want to be satellite operators. This point is often not well-appreciated or sufficiently acknowledged by satellite data vendors and others seeking to collaborate with Australian users.

An additional dimension is that partnership and vendor proposals are often couched in terms of what a particular sensor can 'see' – based on experimental data, much of which has been gathered outside Australia. Invariably, the marketing focus has been placed on the sensor and not the overall system – to include revisit rates, satellite replacement and upgrade plans, calibration and verification, validation and assurance (VV&A) schemes. How the inclusion of satellite data into risk assessments and decision making might allow end users to achieve substantially better

---

[41] Available data indicates that the mean is in the order of 3.2/8 or 3.2 octas. See http://www.bom.gov.au/climate/change/cloud.shtml

operational outcomes is rarely discussed. This involves changing long-established business processes and building trust and confidence in new and unfamiliar sources of data.

There is a tendency for proponents of satellite-derived data to assume that potential customers in Australia will grasp the benefits, essentially by definition, due to the brilliance of the technology. Such attitudes are arrogant, disrespectful and indicate that little or no market analysis has been done ahead of the technology being exposed. Invariably organizations that assume or assert that Australian agencies will derive benefit from access to systems and data on offer leave disappointed and disillusioned. This cycle of disappointment is unlikely to change unless and until vendors seriously invest in understanding and developing the Australian market. This is a long-term effort that is not capable of being met by fly-in/fly-out sales and marketing efforts which is the usual approach.

In summary, vendors reinforce quite strongly held attitudes by politicians and officials that many involved in space endeavours are self-serving, obsessed with technology for its own sake and not focused on providing end user benefit. Until they and potential collaborators come to Australia to listen, rather than to tell, they are not likely to succeed in selling their own products and services or building the relationships they seek to build.

Poor industry behavior and Politicians

Australian politicians have long been wary of proposals to invest in space capabilities. The failed commercial launch proposals of the 1980s and 1990s followed by a succession of ill-conceived, from a business perspective, remote sensing projects have left many politicians skeptical of all things space. Once again, the "giggle factor" comes to the fore and needs to be overcome through evidence and a narrative that emphasizes individual and community end-user benefit.

This evidence, especially with respect to the economic value of satellite based remote sensing, is starting to accumulate and a defensible narrative, told in terms of end-user benefit, is beginning to emerge. In 2008, the Cooperative Research Centre for Spatial Information (CRCSI), commissioned a report on the value of spatial information to the Australian economy. The report concluded that:

". . . industry revenue in 2006-07 could have been of the order of $1.37 billion annually and industry gross value added around $682 million."

The economic footprint of the spatial information industry is considered to be larger than this. Spatial information is increasingly being used in most sectors of the economy where it is having a direct impact on productivity.

The study found that in 2006-07 the accumulated impact contributed to a cumulative gain of between $6.43 billion and $12.57 billion in Gross Domestic Product (GDP) – equivalent to 0.6% and 1.2% of GDP respectively.[42]

Geoscience Australia commissioned a second report by ACIL Tasman in 2010. This found that Earth observation from space contributed at least $3.3 billion to Australian GDP in 2008-2009 with prospects for significant growth. The report also concluded that if Australia were even partially denied access to data from Earth observation satellites there would be a significant negative impact on the Australian economy.[43]

2009: A TIPPING POINT

2009 may well come to be regarded as a "tipping point" in Australia's involvement in space. For the first time, domestic developments and investments were spoken about by Government as being important factors in Australia's commitment to space. An inference is that relatively passive and acquiescent reliance on the Unites States is no longer considered to be necessary or sufficient policy position to meet Australia's sovereign needs and interests in space.

THE 2009 DEFENCE WHITE PAPER

Unlike its predecessors, the 2009 Defence White Paper placed considerable emphasis on space, and on cyberspace.[44] With respect to space, it proposed the acquisition of a Synthetic Aperture Radar (SAR) Earth observation satellite, nominated Space Situational Awareness a "priority" for Australia, indicated that Australia may need to consider "hedging strategies" in Ballistic Missile Defence (BMD) to counter emerging threats and also proposed to establish a dedicated cadre of space

---

[42] ACIL Tasman, *The Value of Spatial Information: The impact of modern spatial information technologies on the Australian economy*, Melbourne, 2008.

[43] ACIL Tasman, *The economic value of earth observation from space: A review of the value to Australia of Earth observation from space*, Melbourne, 2010, p vi.

[44] Government of Australia, Department of Defence, *Force 2030: Defending Australia in the Asia Pacific Century*, Second Edition, Canberra, 2009, (Defence White Paper, 2009).

specialists within Defence. The dependence of the Australian Defence Force on satellite-based services was underlined.

A simple metric illustrates the increased awareness of the importance of space to Defence and to national security more generally. The word "space" appears 32 times in the 2009 White Paper. In its predecessor released in 2000, the word does not appear at all. Similarly, the word "satellite" appears 14 times in the 2009 document and only twice in its predecessor.

The paragraph from the 2009 White Paper, which discusses the SAR satellite, is worth quoting in full.

"As a significant new measure, the Government places a high priority on assured access to high-quality space-based imagery to meet Defence's needs for mapping, charting, navigation and targeting data. It has decided to improve Australia's intelligence collection capabilities by acquiring a satellite with a remote sensing capability, most likely to be based on a high-resolution, cloud-penetrating, synthetic aperture radar. This important capability will add to Australia's standing as a contributing partner within our alliance framework with the United States, which will be given access to the imagery collected by this system."[45]

The paragraph underscores the importance of the US alliance and the concept of self-reliance. Unusual for a white paper, this paragraph is quite specific and prescriptive. A not unreasonable assumption is that the SAR satellite may well represent the beginning of an enduring national Earth observation program for Australia which includes a space element as part of the overall capability. No mention is made of the 'dual use' possibilities of the system but the choice of sensor (SAR) and the explicit link to the United States may imply that this satellite, if ever built and brought into operation, will be optimized to cover Australia's northern approaches and may not be well-suited, or readily available, to meet civilian needs across the Australian continent and surrounding oceans.

DEVELOPMENTS IN THE CIVIL DOMAIN

In the non-defense domain, the 2009 Budget provided funds to establish a Space Policy Unit within the Department of Innovation, Industry Science and Research (DIISR) and to establish a modest Australian Space Research

---

[45] Defence White Paper, 2009, para 9.80, p 82.

Program (ASRP).[46] A Space Industry Innovation Council was also established to provide high level advice to the Minister. Perhaps most importantly, the Prime Minister of the day, Kevin Rudd, invited the Minister for Innovation, Industry, Science and Research (DIISR), Senator the Hon. Kim Carr, to develop and bring forward to Cabinet a proposed national space policy to include defense and national security equities as well as civil, commercial and research requirements as well.

A two-step process was adopted by the Space Policy Unit. The first and most difficult step was to draft and gain acceptance across Government of a series of principles that would guide the development of the actual policy. The real purpose of the principles document, which was released by Minister Carr in September 2011, was to provide a basis for discussion, within and between departments and Ministers, about the increasing dependence on space based utilities of virtually all areas of government.[47] Seven principles are enunciated in the document:

1. Focus on space applications of national significance
2. Assure access to space capability
3. Strengthen and increase international cooperation
4. Contribute to a stable space environment
5. Improve domestic coordination
6. Support innovation, science and skills development
7. National security and economic well being.

The principles document fulfilled, in its drafting, an educative role. At the 11th hour, and this point reinforces that made earlier about the "giggle factor," the word "industry" was inserted into the title of the document. Concerns were raised in senior levels of Government that in a period of fiscal constraint any unqualified reference to a national space policy may well cause media and wider public speculation that the Australian Government was about to embark on some speculative space adventure which could be represented as profligate and out of step with broader economic realities.

The national space policy, hopefully with the "industry" qualification removed, is due for release in the latter part of 2012.

---

[46] Department of Innovation, Industry, Science and Research, *Budget 2009-10, Super Science – Space and Astronomy*, May 2009.

[47] Department of Innovation, Industry, Science and Research, *Principles for a National Space Industry Policy*, Canberra, September 2011. Available at www.space.gov.au

THE AUSTRALIAN SPACE RESEARCH PROGRAM (ASRP)

The ASRP, announced in the May 2009 budget, has achieved considerable success. Fourteen projects, four with an emphasis on education and 10 focusing on industry development outcomes, have been funded. The initial funding for all projects is due to end on 30 June 2013 and follow-on funding remains uncertain. However, several of the projects have already identified sources of continuing support independent of the Commonwealth and others are working to achieve a similar outcome. The ASRP sought to achieve two principal outcomes: enhanced national CAPABILITY and CAPACITY in space science and engineering which would strengthen Australia's international CREDIBILITY in space matters. The principal means of achieving this was through a program that emphasised COLLABORATION between Australian and international participants in the ASRP.[48]

Such practical support has been essential to the building of confidence and credibility around the national space policy development process.

ISSUES AND OPPORTUNITIES

SPACE SCIENCE AND ASTRONOMY

Australia has a rich heritage in astronomy, and indeed European settlement on the east coast of the continent followed the discovery and charting of the coast by Captain James Cook following his voyage to the South Seas (Tahiti) to observe the transit of Venus across the face of the sun in 1769.

Australia is especially well suited to host radio telescopes, being large, flat and, above all, radio quiet. Irrespective of whether the Square Kilometre Array telescope is built in Australia or in southern Africa, Australia, almost certainly, will continue to invest in radio astronomy as a national research strength.

In the context of this chapter, the important point to make is that astronomy is considered by politicians and funding agencies to be an integral part of Australia's space narrative. Funding for space science other than astronomy generally competes directly for funding that astronomers

---

[48] A summary of the ASRP program together with fact sheets on the funded projects may be found at www.space.gov.au

seek to support their research.[49] The Australian astronomy community is well-organized, reasonably coherent and, over many years, has persuaded governments to invest in telescopes both in Australia and overseas. In contrast, the space science community, which covers many diverse disciplines from human performance in minimal gravity environments to planetary geology to ionospheric physics, is dispersed and disorganized. Individual researchers, within their own disciplines, have tended to form their own bilateral links with international collaborators and agencies including NASA, ESA and JAXA.

Australian Square Kilometre Array Pathfinder telescope at Boolardy in Western Australia
Copyright: ANZSKA accessed via Google Images as "Square Kilometre Array."

In September 2010, following a long gestation, the National Committee for Space Science of the Australian Academy of Science released an inaugural Decadal Plan for Australian Space Science.[50] Aspects of the plan were and remain contentious, notably its recommendations to establish a coordinating body seemingly distinct from the fledgling Space Policy Unit, its proposals to collaborate with China on

---

[49] An exception to this general principle was made for the Australian Space Research Program (ASRP), a program that was funded in the May 2009 Budget from Economic Stimulus funds. Astronomy projects were expressly excluded from competing for ASRP funds.

[50] National Committee for Space Science, Australian Academy of Science, *Decadal Plan for Australian Space Science, 2010-2019: Building a National Presence in Space*, Australian Academy of Science, Canberra, 2010.

aspects of ionospheric research and its plans for relatively large investments to support solar research when more immediate priorities to do with water management, especially across the Australian continent, remain to be properly addressed. However, the plan is a stake in the ground that did not exist previously and it has provided a focal point for disparate research groups with diverse space research interests across Australia.

DEVELOPMENTS IN EARTH OBSERVATION

In parallel with development of the 2009 Defence White Paper, the Defence Imagery and Geospatial Organisation (DIGO) and Geoscience Australia (GA) which is Australia's principal civil geo-spatial agency, have worked increasingly closely to avoid unnecessary duplication and to formalize their relationship in a way that has been unprecedented. Evidence for this cooperation includes the establishment of a government panel to rationalize purchases of data and services from commercial providers. Called the Optical, Geospatial, Radar and Elevation (OGRE) data and services procurement panel, the aim is to streamline and improve the way the Australian Government acquires Earth Observation imagery and services by using open frameworks and a cooperative approach.[51]

Also, and alongside the independent reports prepared by ACIL Tasman referred to earlier in this chapter, the Earth observation community produced a strategic plan in 2009[52] and a more narrowly focused policy document in 2010.[53] These reports strengthened the evidence base from which the Space Policy Unit and others have been able to explain to Government that Australia's remote sensing capability, across all elements of capability (people, organizations, sustainment, training, equipment and documented processes) is fragile and, without new investment, will be unlikely to satisfy national requirements with any degree of certainty beyond the short term (the next five or so years).

In 2011, Geoscience Australia released a report called Continuity of Earth Observation Data for Australia: Operational Requirements to 2015 for Lands, Coasts and Oceans (CEODA).[54] The report, based on published

---

[51] See: http://www.ga.gov.au/earth-observation/ogre.html for information about the OGRE panel.
[52] Australian Academy of Science (AAS) and the Australian Academy of Technological Sciences and Engineering (ATSE), *An Australian Strategic Plan for Earth Observations from Space*, Canberra, 2009.
[53] Geoscience Australia, *A National Space Policy: Views From The Earth Observation Community*, Canberra, 2010.
[54] Geoscience Australia, *Continuity of Earth Observation Data for Australia: Operational Requirements to 2015 for Lands, Coasts and Oceans*, Canberra, 2011. (CEODA)

worst-case data about when Earth observation satellites on which Australia currently depends are scheduled to cease operation, painted a grim picture. The report made eleven key points which are reproduced in full below.

1. Earth Observations from space (EOS) data have become pivotal to most environmental monitoring activities being undertaken by federal and state governments in Australia.
2. Australia is totally reliant on foreign satellites for EOS data.
3. Of the 22 EOS sensors currently being used for operational programs in Australia, 19 (89%) are expected to cease functioning by 2015.
4. Australia has not secured access to any future space-based sensors that are relevant to observing the Australian land mass and its coastal regions.
5. Alternate, non satellite-based sources of data do not exist for most types of space-based EOS data, especially those used for environmental monitoring programs.
6. In contrast with the projected rapidly decreasing access to EOS data, Australia's EOS requirements are expected to increase significantly over the next decade. To support a sample set of 91 operational government programs, the total annual EOS data storage requirements in 2015 were conservatively estimated at 1.2PB per year. This represents a twentyfold increase on current annual EOS data storage. These estimates do not include meteorological applications, research and development activities, or new sensor technologies.
7. Two data types, medium resolution optical and Synthetic Aperture Radar (SAR), are most at risk of data gaps before 2015 for land and marine applications.
8. Data continuity for low and high resolution optical data, and for passive microwave data, is also of concern, but improved access to these data types has a lower priority due to the availability of alternative data sources and/or current levels of data usage in land and coastal applications.
9. Urgent action is needed to ensure that the imminent and potentially damaging EOS data gaps are not realised.
10. Australia's participation in the Landsat Data Continuity Mission (LDCM) and ESA's Sentinel missions would significantly reduce the risk of the projected EOS data gaps in the high priority data types and should be the focus of immediate action.

It should also be a priority to encourage an on-going Landsat program.
11. As a matter of priority Australia needs to formalise agreements with several upcoming EOS missions, and formulate a decadal infrastructure plan to safeguard the supply of EOS data.[55]

This report was a further spur to action by the Australian Government and in late 2011 work began on two internal reviews, one focused on precision location and the other on Earth observation. The intention is that both documents are considered by Cabinet in the latter part of 2012, possibly in conjunction with the national space policy.

The precision location review is being led by Geoscience Australia and the Earth observation review by the Bureau of Meteorology. This underscores the critical dependencies that these agencies have on data from satellites to produce timely and accurate products and services on which the broader community places great reliance and trust to support all manner of economic and other activity. The reviews are intended to provide a basis for investment proposals for inclusion in future budgets with the aim of strengthening Australia's precision location and Earth observation capabilities. Almost certainly emphasis will be on education and training, and ground processing and dissemination infrastructure in the first instance.

WHY NO SPACE AGENCY?

A question, often asked, is why Australia does not have a space agency. This is asked both inside Australia and by non-Australian observers as well. The questioners often have in mind a NASA-model, but many of them seem to have no appreciation of the relative sizes of the US or the Australian economies, and still have in mind a launch/manned space flight paradigm as a necessary condition for a space agency. The assumption is that if Australia has a space agency, almost by definition, it would also have launch capabilities and would be involved in programs such as the International Space Station (ISS). This argument is to put the cart before the horse. Successive governments have made abundantly clear by not investing in such activities that neither is considered to be a public policy or spending priority for Australia.

In fact, Australia was invited to join ESA following the withdrawal of the ELDO consortium from Woomera but in 1983 the government

---

[55] CEODA, p ix.

formally notified ESA that it would not be pursuing this option.[56] (ESA, however, maintains ground stations in Australia under treaty level agreements with the Australian Government.)

Departmental and agency structures and funding also play a part in the decision not to have a dedicated space agency. The Australian Government, like many around the world, routinely seeks to harvest "efficiency dividends" typically by cutting the funds of ALL departments and agencies by a fixed percentage of their operating costs. Large organizations are much better able to withstand such austerity measures than are smaller, specialist agencies. The latter can rapidly lose critical mass and become unsustainable. The coordinating and other functions presently performed by the Space Policy Unit within the Department of Industry, Innovation, Science, and Tertiary Education (DIISTE), formerly DIISR, are far less exposed than would be the case if these same functions were being carried out by a small, specialized space agency.

In the past decade, with 2009 representing a "tipping point," Australia's approach to space matters began to change in response to five factors:

1. The space environment and the radio spectrum needed to support space activities, formerly uncluttered, became congested and contested;
2. The international regulatory regime of space which was established in the 1960s, essentially to meet the needs of the United States and the Soviet Union, showed signs of not being able to cope with the changing environment;
3. The pre-eminent position of the United States in space has come under challenge, especially from China;
4. The Australian Government has started to take a more active role in international space affairs:
    a. At the strategic level to work for a new space regime which protects and advances broadly western interests whilst also meeting 'common heritage of mankind' ideals; and
    b. At the operational level to become more self-reliant within the broader settings of the US alliance; and
5. The Australian Government and the wider community has begun to understand its dependence on space-based services, especially the GPS timing signal, and the associated vulnerabilities.

---

[56] Dougherty, K and James, M L., *Space Australia: The Story of Australia's Involvement in Space*, Powerhouse Museum, Sydney, 1993, p 106.

Space: Contested and Congested

The Chinese ASAT antimissile test of January 2007 in which a Chinese satellite was destroyed in orbit, creating a large in-space debris field, was as much a wake-up call in Canberra is it was in many other capitals. The Chinese Ambassador was quickly called in and made formally aware of Australia's concerns and displeasure. Two inferences may be drawn from this response:

That Canberra and Washington consulted closely on the event itself, its implications for the space environment, and how to respond; and

That within the Australian Government there is a cadre of officials who are well-informed about space matters and who maintain a close watch, in particular from the perspective of Australia's strategic interests.

The US destruction of a crippled spy satellite in early 2008 followed by the Cosmos/Iridium satellite collision over Siberia in early 2009 reinforced to policy makers the fragility of the space environment and the vulnerability of satellites, especially those operating in low Earth orbits.

With respect to radio spectrum, Australia has been an active member of the International Telecommunication Union (ITU) for more than a century. The nation's geography compels it to rely, more than many nations, on disciplined use of spectrum to optimize the competing demands of terrestrial and satellite communications systems to ensure assured coverage across the continent and the vast surrounding ocean areas. There is also a need to keep some frequency bands as free as possible for science, notably radio astronomy.

The Regulatory Regime

All spacefaring nations are increasingly aware and concerned about the limitations of the declaratory principles of the Outer Space Treaty and the follow-on treaties and other instruments which were drafted and came into force in the 1960s and 1970s. As more nations develop their own space capabilities, an increasing need is seen for more explicit norms to be developed and embraced by spacefaring nations in order to preserve the space environment itself. The fundamental difficulty with space regulation of any sort and by any name is that the initiative rests with the offense and the distinction between accidental and deliberate damage can be difficult, if not impossible, to establish. This leads to problems with the concept of proportionality when considering how to respond to an aggressive, damaging or destructive act.

Not yet evident in discussions about future space regulation is how the voice(s) of nations that are dependent on secure and assured access to space-based services, but which are not spacefaring, will be heard. This writer argues that Australia is well placed by geography and experience to demonstrate leadership as a space dependent but not spacefaring nation in the negotiations to come that will determine the future regulation of human activity in space. In an increasingly connected world in which space-based services play an important and not easily replaced role, mechanisms will be needed by which these people and their governments can be heard as the code of conduct discussions proceed. The "common heritage" principles outlined in the Outer Space Treaty seems worthy of protection, and indeed, of reinforcement.

The United States and China

The 21st Century has been called by some the Pacific Century, as power shifts from Europe and North America to North and South Asia. China and India are emergent great powers and how they will accommodate each other and the United States remains to be seen. Clearly, there will be elements of cooperation and competition and clearly Australia, by dint of its geography, alliances, and trading relationships, will be both involved in and affected by the ways in which these nations negotiate their future relationships. Space is one of the elephants in the room, with the second being cyberspace.

China is well aware that the US military is dependent on space in a way not yet matched by the People's Liberation Army resulting in an asymmetry that at present is favorable to China. As the PLA itself becomes more dependent on space, the temptation for China to act pre-emptively may be expected to reduce, leading to a situation not dissimilar to the mutually assured destruction achieved in the Cold War when both the United States and the Soviet Union achieved a second strike capability.

There is an active debate among a group of Australian strategic thinkers about the type of role and the extent of any influence that Australia might expect to play in defining the new Asian order. Broadly speaking there are two camps, those who argue for Australia to become even more closely tied to the United States, in anticipation of a serious disagreement with China at some point in the future, and others who argue for a more independent position which acknowledges the importance to Australia of

China in every respect and not simply as a market for Australian raw materials.[57] From a military point of view, the resolution of this debate is fundamentally important because of its implications for force structure. For example, should the Royal Australian Navy be equipped with fast, long endurance submarines, or with slower boats with less range and time at sea before needing to be replenished?[58]

The 2009 Defence White Paper would tend to suggest that its authors and the Government of the day, which approved and released the document, is inclined to the former view. Whether such a position is sustainable and tenable in the longer view remains to be seen.

A More Active Role in International Space Affairs

There is strengthening evidence that Australia is seeking to take a more direct and active role in space affairs in future than has been the case in the past. The cadre of space aware and competent Australian officials is growing in size, capability, and confidence. Australian officials, for example, are becoming more prominent participants in the workings of the United Nations Committee on the Peaceful Uses of Outer Space (COPUOS) and other multi-lateral forums devoted to space matters. Space security is also being spoken about with increasing frequency by senior Ministers and officials, both within and beyond the context of the US alliance. Each year the Australian and US Governments hold the Australia/US Ministerial talks, known as AUSMIN. These are the peak bilateral talks which discuss the state of the relationship and the alliance.

The communiqué for AUSMIN 2010 placed considerable emphasis on space security, notably through improved Space Situational Awareness

---

[57] Professor Ross Babbage and Mr Mike Pezzullo, are among those whom this writer would place in the first camp. Mr Pezzullo was the principal author of the 2009 Defence White Paper and Professor Babbage was closely involved in the shaping of that document as well. Professor Hugh White is representative of the second camp. He wrote a seminal essay in 2010 which confronts these questions squarely and caused something of a stir at the time. See, White, Hugh, *Power Shift: Australia's Future Between Washington and Beijing*, Quarterly Essay, Issue 39, Griffin Press, Melbourne, August 2010.

[58] The Kokoda Foundation has published several papers in the past two years which address Australia's requirement for new submarines. The most recent is Pacey B., *Sub Judice: Australia's Future Submarine*, Kokoda Paper No 17, Canberra, 2012. This paper is available for download on the Kokoda Foundation's website: www.kokodafoundation.org [Declaration of interest: the present author is a director of the Kokoda Foundation.]

(SSA), and under the heading "21st Century Challenges," stated in part:

"Building upon a long history of defence space cooperation, Australia and the United States signed a Space Situational Awareness Partnership Statement of Principles, which should enable further close cooperation on space surveillance to the benefit of both countries.

"Australia and the United States shared a deep concern about the increasingly interdependent, congested and contested nature of outer space and acknowledged that preventing behaviors that could result in mishaps, misperceptions or mistrust was a high priority. Australia welcomed the US decision, reflected in the June 2010 US National Space Policy, to consider space arms control measures that are equitable, verifiable and in its and its allies' national interests. Australia intends to work with the United States to progress their shared goal of enhanced space security, with a particular focus on transparency and confidence building measures. The two Governments endorsed a Joint Statement on Space Security highlighting their shared views and resolve to cooperate with like-minded countries to ensure free and safe access to space."[59]

The SSA Statement of Principles included a commitment by both nations to:

". . . investigate the potential for jointly establishing and operating space situational awareness facilities in Australia to support the United States space surveillance network and to support the development of Australia's space situational awareness and mission assurance capability."

This was an oblique reference to locating, in the first instance, a ground-based SSA radar possibly at North West Cape in Western Australia. Formal announcements about this facility have yet to be made. There has also been comment in the public domain that one of the next generation Space Fence ground-based SSA radars being developed by the US could be located in Australia. Whether this occurs could depend as much on the state of the US defense budget as on operational factors.

In January 2012, the Australian Foreign Minister, the Hon Kevin Rudd, announced that Australia would endorse European efforts to develop a space code of conduct and Australian officials were active participants in the 2012 meetings of COPUOS in Vienna.[60]

---

[59] AUSMIN 2010 Communiqué, Nov 2010.
[60] Hon Kevin Rudd, MP, Minister for Foreign Affairs and Trade, Media Release, *Australia joins the fight against space junk* 18 January 2012, See http://foreignminister.gov.au/releases/2012/kr_mr_120118.html.

Advocacy for Space-Disadvantaged Nations

This writer considers there are 15 nations to which the label spacefaring may be applied. 'Spacefaring' is understood to mean demonstrated current capacity within the nation to design, build, test and operate satellites. The reality of global supply chains means that some components will be sourced from outside these countries, including from non spacefaring nations. However, the bulk of design, integration and testing is done within the spacefaring nation itself.

Some spacefaring nations have launch capabilities while others purchase launch services from a small number of government and commercial providers. The 15 nations are, in alphabetical order: Brazil, Canada, China, France, Germany, India, Israel, Italy, Japan, Russia, South Korea, Taiwan, Ukraine, the United Kingdom and the United States.[61] Together these nations, which comprise 8% by number of the nations on Earth, represent more than 75% of global GDP and 54% of global population.

The remaining 185 or so nations, their peoples and their economies, also have dependencies on services delivered from space including precision timing and navigation, remote sensing data and communications.

A question for spacefaring nations is thus how to ensure that space disadvantaged nations have necessary and sufficient access to space-based services to allow them to participate, as effectively as possible, in economic and other activities which are increasingly globalized.

A question for the international community is how to ensure that the legitimate interests of these nations and their peoples, with respect to the safety and security of space-based services, can be articulated and heard. The principle involved is not dissimilar to that which applied during the discussions that occurred in the 1970s in the conferences which eventually produced the Law of the Sea convention. Landlocked states and others with coasts that did not generate significant maritime zones and obligations argued that, in accordance with the principle of the common heritage of mankind, they had some entitlement to a share of the mineral wealth said to be present on and in the deep seabed. These states formed a loose association that became known as the "blocking third."[62] They did not

---

[61] Nations that might soon be added to this list include Iran, Malaysia, Mexico, North Korea, and South Africa.

[62] Rothwell D R and Stephens T. *The International Law of the Sea*, Hart, Oxford, 2010, p 190.

have the numbers to ensure that particular articles were adopted but they did have the numbers to prevent the inclusion of particular clauses in the convention as it was developed through extraordinary diplomatic effort.

As a middle power, Australia played an active role in the law of the sea negotiations in the 1970s for reasons that are obvious by glancing at a map. Australia has the third longest coastline of all of the world's nations, after Canada and Russia. It generates a vast Exclusive Economic Zone and has search and rescue responsibilities for approximately 13% of the Earth's surface under the Safety of Life at Sea (SOLAS) convention. More than 98% of its exports are carried by ships to markets, mainly in Asia and, in reverse, most imports, which comprise manufactured goods, come from factories, many in China and other countries in Asia, also by sea.

Similarly, this writer sees the possibility of Australia taking a leading role in the international discussions which must occur about the future governance and regulation of space, to ensure that competition between nation states is bounded, that congestion, especially in the Low Earth Orbits (LEO) is managed wisely and fairly and that the finite resource of spectrum is also allocated optimally, taking account of global, regional, national and commercial interests and also the potentials for more effective and efficient use and re-use of spectrum which advances in technology may permit. Norms must be developed, defined and agreed upon and a regulatory regime developed which is sufficiently transparent for satellite owners and operators to comprehend and accept, or mitigate as necessary, investment, commercial and operational risks.

Concluding Remarks

Australia's space journey is quite different from that of any other nation on Earth and this has never been well explained or contextualized. The impact of the bifurcation that has occurred between the nation's strategic and operational goals and interests in space only now is being recognized and its impacts assessed. In 2009 the Australian Government recognized that Australia, like all other nations, had critical dependencies on space-based utilities. It also came to understand that the space environment itself is fragile and easily disturbed, to the detriment of all who depend on space-based services for their work, their safety, their security and even their leisure. The Government has now taken steps to create a coherent space narrative and to develop a critical mass of space qualified individuals within and beyond the public sector who can provide a necessary and sufficient mass of talent and expertise capable of lifting the profile of space in Australia and of Australia's profile in space internationally. This process will take a decade to mature but evidence of progress since the investments made by Government in 2009 is already evident.

The 21st Century has been described as the Asian Century, and space may be expected to play an important role in how broader relationships between China, the United States and other Asian nations develop. Australia, due to its location, its alliance with the United States, its trading relationships with China, Japan and Korea, its vital interests along with India in the security of the Indian Ocean and of the sea lines of communication more generally, has vital national interests at stake. Noting that the space environment is a global commons on which all nations and peoples increasingly depend, Australia standing, as it does, at the cross roads of Asia and the West and of the wealthy North and disadvantaged South, has an opportunity, as a middle power, to provide leadership and be a force for good. In securing its own national interests, Australia is well placed to ensure that the space disadvantaged nations of the world can also enjoy the benefits of secure and assured access to services from space for national, regional and global benefit.

...

Brett Biddington AM

Brett Biddington is the principle of Biddington Research, a Canberra-based consulting company which specializes in space and cyber security matters. Between 2002 and 2009 he was a member of Cisco Systems' global space team which developed the Internet Router In Space (IRIS) program and promoted merged space-ground communications systems.

Brett chairs the Space Industry Association of Australia and sits on several boards and advisory committees that are concerned with the governance of Australian space activities. He is an Adjunct Professor in the School of Computer and Security Science at Edith Cowan University in Perth, Western Australia.

From 2004-2009 Brett was closely associated with the governance of radio astronomy in Australia. He has also been involved in Australia's commitment to astronomy in Antarctica. In 2002 Brett left the Royal Australian Air Force (RAAF) on completion of almost 23 years of service. He was an intelligence and security specialist before moving into capability development where he sponsored command and control, intelligence, surveillance, reconnaissance and electronic warfare projects including the Jindalee Over the Horizon Radar project and classified and unclassified space initiatives.

In June 2012 he was admitted as a Member of the Order of Australia (AM) for services to the Australian space sector.

References

ACIL Tasman, *The Value of Spatial Information: The impact of modern spatial information technologies on the Australian economy*, Melbourne, 2008.

ACIL Tasman, *The economic value of earth observation from space: A review of the value to Australia of Earth observation from space*, Melbourne, 2010.

Asia Pacific Aerospace Consultants, *A Review of Current Australian Space Activities, Executive Summary*, p1. June 2011. At http://www.space.gov.au/SpacePolicyUnit/Pages/DevelopmentandConsultation.aspx, accessed 26 Mar 2012.

Australian Academy of Science (AAS) and the Australian Academy of Technological Sciences and Engineering (ATSE), An Australian Strategic Plan for Earth Observations from Space, AAS/ATSE, Canberra, July 2009.

Australian Academy of Technological Sciences, *A Space Policy for Australia*, Melbourne, June 1985 (The Madigan Report).

Australian Spatial Consortium, Australian Strategic Plan for GNSS, Canberra, 2012.

Crompton R W, Dracoulis G D, Lewis B R, McCracken K G and Williams J S. "John Henry Carver: Biographical Memoir", Historical Records of Australian Science, Vol 22, No 1, June 2011, CSIRO, Australia.

Dougherty, K and James, M L, *Space Australia: The Story of Australia's Involvement in Space*, Powerhouse Museum, Sydney, 1993.

Edwards P. *Arthur Tange: Last of the Mandarins*, Allen and Unwin, Sydney, 2006.

Geoscience Australia, *A National Space Policy: Views From The Earth Observation Community*, Canberra, 2010.

Geoscience Australia, *Continuity of Earth Observation Data for Australia: Operational Requirements to 2015 for Lands, Coasts and Oceans*, Canberra, 2011 (CEODA).

Government of Australia, Department of Defence, Review of Australia's defence capabilities, Canberra, March 1986 (the Dibb Report).

Government of Australia, Department of Defence, The Defence of Australia, 1987, Canberra 1987.

Government of Australia, Department of Defence, Force 2030: Defending Australia in the Asia Pacific Century, Second Edition, Canberra, 2009.

Government of Australia, Department of Innovation, Industry, Science and Research, Budget 2009-10, Super Science – Space and Astronomy, May 2009.

Government of Australia, Department of Innovation, Industry, Science and Research, Principles for a National Space Industry Policy, Canberra, September 2011.

Kingwell J. "Punching below its weight: Still the future of space in Australia", Space Policy, Vol 21, Issue 2, May 2005.

Menzies R G. *The Measure of the Years*, Cassell, Melbourne, 1970.

Morton, P. *Fire Across the Desert: Woomera and the Anglo-Australian Joint Project, 1946-1980*, Department of Defence, Canberra, 1989, reprinted 1997.

National Committee for Space Science, Australian Academy of Science, Decadal Plan for Australian Space Science, 2010-2019: Building a National Presence in Space, Australian Academy of Science, Canberra, 2010.

Parliament of Australia, Hansard.

Parliament of Australia, Space Activities Act, No 123 of 1998.

Report of the International Space Advisory Group, Canberra, 2002 (ISAG Report).

Richelson J T. *America's Space Sentinels: DSP Satellites and National Security*, Kansas University Press, 1999.

Richelson, J.T. *The Wizards of Langley: Inside the CIA's Directorate of Science and Technology*, Westview, Cambridge, US, 2002.

Rothwell D R and Stephens T. *The International Law of the Sea*, Hart, Oxford, 2010.

CHAPTER 10

CURRENT STATUS AND FUTURE DEVELOPMENT IN CHINA'S SPACE PROGRAM:
INTERNATIONAL SPACE COOPERATION

LE WANG
SHAANXI NORMAL UNIVERSITY,
CHINESE ACADEMY OF SCIENCES (CAS)

© Le Wang, 2012. All Rights Reserved.

INTRODUCTION

There is no doubt that China has become a major space power, with its launch technologies, many satellites in orbit, manned space flight program, and its lunar exploration program. China considers the aerospace industry as "a significant symbol of the nation's strength," as remarked by President HU Jintao on the 50th Anniversary of Aerospace Industry (XNA, 2006). Over the last 50 years, China's space program has improved rapidly and advanced through innovation initiated in China and through international cooperation.

This chapter traces the current status and future development in China's space program, with particular focus on studying China's international space cooperation, and reforms on organization structure and policy-making process to promote stability of future development for the space program.

The chapter consists of four parts:

- A synopsis of China's space program including recent reforms by the Chinese government concerning its policy, organization, and budget, to enable the development of its space program to the next stage;
- A review of major achievements;
- Frameworks for international co-operation from the Chinese perspective, including positive steps on international space cooperation with different models under the self-reliance and innovation policy;
- Expectations for future development, in which we describe how China plans to move forward in the development of space technology, space applications, and space sciences, and to strengthen international space cooperation.

SYNOPSIS OF CHINA'S SPACE PROGRAM

ORGANIZATION AND STRUCTURE

According to the National Guideline for Medium- and Long- term Plans for Science and Technology Development (2006-2020) (CNSA, 2006a) (Zhu, XNA, 2006) the goals for the current period in China's space program call for:

> Reform the current scientific and technological management system to combine and coordinate the military and civilian research organizations in order to promote its scientific development;

> Establish a new mechanism to coordinate military and civilian basic research, and integrate research and development for high technology.

In recent years, the government has eliminated redundant organizations and augmented the functions and responsibilities of the central organizations to enhance the space program's effectiveness. At present, there are ten primary organizations responsible for carrying out the functions of space policy making, implementation and supervision, R&D, manufacturing, space exploration, and applications development. China's space program is managed by various departments under the State Council of People's Republic of China's Macro-administration, as shown in Figure 1.

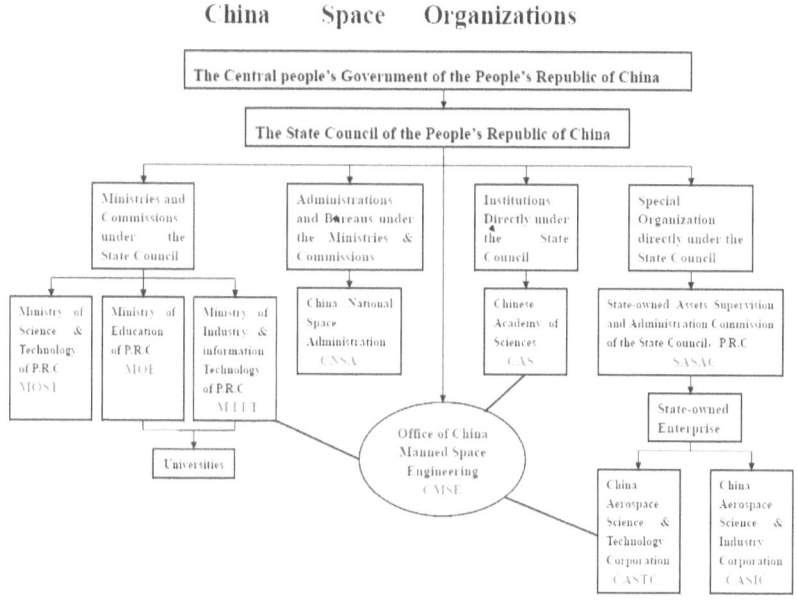

Figure 1: China's Space Organizations

The top level organizations are:

- The Ministry of Science & Technology (MOST) funds and manages some of the R&D efforts of space technology and space science
- Ministry of Industry and Information Technology (MIIT) is responsible for proposing budgets for space programs and space-related large projects
- China's National Space Administration (CNSA) is responsible for space-related policy making and coordinating international cooperation
- Chinese Academy of Sciences (CAS)
- Ministry of Education (MOE)

- CAS and MOE, with their research institutions and educational programs, are largely responsible for proposing space strategy, advising on the techniques, carrying out R&D efforts of space technology and space science, making scientific payloads and instruments, executing space exploration and applications, and doing all related fundamental research
- State-owned Asset Supervision and administration Commission of the State Council (SASAC), with its two large state-owned enterprise groups CASTC and CASIC, builds the infrastructure of China's space program, makes launch vehicles and most of the spacecraft, and carries out all required R&D, and
- China's Manned Space Engineering (CMSE) coordinates China's manned spaceflight program.

Each undertakes their own tasks as assigned by the State Council.

CNSA and CMSE were set up to execute the functions of space activities in 1992 (CMSE Homepage) and 1993 (CNSA Homepage) by the central government. The Commission of Science, Technology and Industry for National Defense (COSTIND) was eliminated, and part of its functions were transferred to MIIT in 2008 (XNA, 2008).

This structure provides macro-guidance by the government, while cooperation and harmonization are important for each organization to jointly implement space missions effectively. The main roles of each organization can be briefly summarized as follows (although there are some overlaps).

SPACE POLICY IN THE NEW DEVELOPMENT STAGE

China's space policy had been interrelated with other policy areas until November 2000, with the publication of the White Papers of China's Space Activities in 2000 (CNSA 2003), issued by the Information Office of the State Council of the People's Republic of China. It is the first formal space policy in China, and it was the first time the government addressed China's space activities on a national level.

Overall, the Chinese government has issued these policy statements as related to the space program:

- The White Paper of China's Space Activities in 2000 (CNSA, 2003);
- The White Paper of China's Space Activities in 2006 (IOSCS,

2006);
- The White Paper of China's National Defense in 2010 (IOSCS, 2011);
- The National Guideline for Medium- and Long-term Plans for Science and Technology Development (2006-2020) (CSC, 2006a);
- Space Science & Technology in China: A Roadmap to 2050 (Guo & Wu, 2010);
- Report on the Work of the Government in 2011 (Wen 2011);
- The Outline of the 12th Five-Year Program for National Economic and Social Development (CSC, 2006b);

To address the curiosity of the international community concerning China's space program, two of the goals of the white papers were: (1) to unveil the development of China's space program to the world; and (2), to eliminate misunderstanding and enhance international space cooperation.

The objectives of China's space policies are expressed as "making innovations independently, making leapfrogging development in key areas, shoring up the economy and leading future trends, to promote independent innovations in space science and technology, make space activities create more economic and social benefits, ensure the orderly, normal and healthy development of space activities, and achieve the set goals in the policies." (IOSCS, 2006)

As described in the 2006 white paper (IOSCS, 2006), China's space policies address these key ideas:

The Chinese government intends "to strengthen its administration and macro-guidance concerning space program;"

China "maintains that international space exchanges and cooperation should be strengthened on the basis of equality and mutual benefit, peaceful utilization and common development;"

As an important part of China's overall development strategy, the Chinese government will give more support and create better environment for the space development;

"The Chinese government attaches great importance to the significant role of space activities in implementing the strategy of revitalizing the country with science and education and that of sustainable development and social progress."

These documents also present future plans and strategies in the short, medium, and long-term concerning the development of the space program step by step.

STRATEGY AND VISION

China's fundamental interests are in developing its economy and realizing its modernization, and the priority aims and principles of China's space program are to serve China's national interests, and to implement China's development strategy.

China emphasizes self-reliance and self-innovation as its national strategy. Premier Wen Jiabao emphasized this at the conference celebrating the 50th anniversary of the country's aerospace industry in 2006. He noted, "China has made great achievements including developing the nuclear bomb, missiles, artificial satellites and manned space flights, and has established an independent and complete system for the aerospace industry and scientific research, and China relies on its own strength to tackle key problems and make breakthroughs in space technology." (XNA 2006) In addition, the aerospace industry is regarded as a significant symbol of the nation's strength (XNA 2006), as stated by President Jintao Hu.

Within these goals, the Chinese government adheres consistently to the well-accepted principle that the exploration and utilization of the outer space should be for peaceful purposes, only for non-military and non-aggressive purposes and benefit to the whole of humanity.

FUNDING AND BUDGET

In China, five-year plans are carried out in all the fields. Space funding is mainly provided by the government, while organizations including MIIT, MOST, MOE and CNSA share responsibility for managing China's space program. However, data on the Chinese national space budget are difficult to obtain, and estimates vary widely. Meaningful figures for historical or projected spending on the Chinese space program are virtually impossible to obtain, and there are no officially published figures about the Chinese government's space budget.

According to the estimated data from *Space Foundation* Reports in 2008, 2009, 2010, and 2011 (Space Foundation 2008, 2009, 2010, 2011), the Chinese government's estimated space program budgets from 2007 to 2010 are shown in Figure 2.

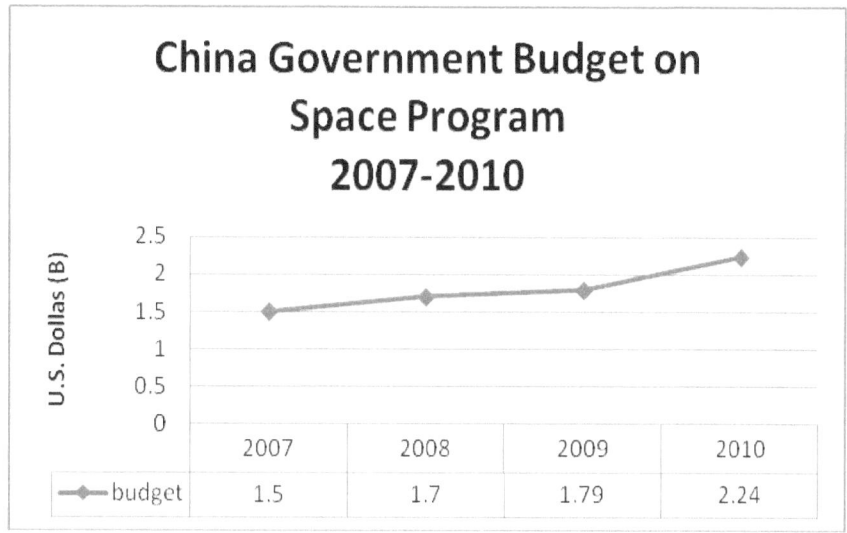

Figure 2: Chinese government space program budgets (2007-2010)
(All data are from the Space Foundation Reports 2008, 2009, 2010 and 2011.)

Estimated government expenditures on the Chinese space program increased from US$1.5 billion to US$2.24 billion since 2007, a growth of 49.3%. The budget is mainly devoted to Earth observation, satellite navigation, launch vehicles, planetary missions, human space flight, and satellite development.

In the 12th-five year plan, covering the period 2011 to 2015, China continues to deploy the BeiDou satellite-based navigation system,[1] with notable developments in its exploration and human spaceflight program Chang'e 2 and Chang'e 3, the spacelab Tiangong program, and also the development of a new space launch facility in the city of Wenchang located in the Hainan province (Wen 2011).

This indicates that the government has begun to devote more resources to the key projects in its space program, as expressed in the National Guideline for Medium- and Long-term Plans for Science and Technology Development (2006-2020) (CSC 2006a).

With the commercialization of space activities in China, some technology companies have also invested in space activities, including telecommunication satellites and small satellites. These investments are expected to become one of the more important funding sources for China's

---

[1] Beidou refers to the seven stars arranged in the spoon-shaped in the Great Bear constellation, which is used to identify the direction. As the result, the Chinese navigation satellite system is named Beidou.

space program in the future.

According to the national guidelines on medium- and long-term program for science and technology development (2006-2020) (CSC 2006a), China will increase its investment in science and technology in the coming years, with its total research and development (R&D) accounting for 2 percent of gross domestic product (GDP) in 2010, and 2.5 percent of GDP in 2020, in its effort to increase its capabilities in high technology and to support international competitiveness of its industries (Yan 2006).

Given the relative importance of China's space program within its overall R&D activity, it seems likely that China will also accelerate investment in its space program, particularly in light of the country's projected overall long term economic growth rate.

Since the long term economic growth of the developed countries of North America and Europe is currently restrained by the recessionary environment, it is possible that the space budgets of these countries may remain flat, which implies that China's role in the international space activities may become more important internationally in the coming decades.

Major Achievements of China's Space Program

China's space program was initiated in 1956 and has evolved through four stages:

- Stage 1, Pioneering (1956-1968): The space program and space physics research were proposed and began implementation;
- Stage 2, Development (1968-1978): The space program was initially developed in all pertinent fields;
- Stage 3, Revitalization (1978-1992): The Chinese government reformed the policy and organization to accelerate the development of space program;
- Stage 4, Cooperation (1992 until today): The space program has been expanded in all respects, including international cooperation in all relevant fields.

China's first man-made satellite was orbited in 1970, Dong Fang Hong (DFH). On October 15, 2003, China launched Shenzhou 5 successfully to Earth orbit with one Chinese astronaut – Yang Liwei, China's first manned launch. China's first lunar probe, Chang'e-1 was launched successfully on October 24, 2007, and its first space laboratory, Tiangong-1 was launched on September 29, 2011.

China has now set up a professional and comprehensive system for space development, including research, design, production and testing. China's space infrastructure has the capability to launch many kinds of satellites and manned spacecraft. The system includes a Telemetry Tracking and Command network consisting of ground stations across the country, and tracking and telemetry ships are in place as well.

Launch technology

The Long March rocket group's capability, shown in Figure 3, permits launching satellites to low-Earth, geostationary (GEO) and sun-synchronous orbits. The launch capability of the Long March rockets is 9,200 kg for low-Earth orbit, 5,100 kg for geo-stationary transfer orbit, and 500 kg for sun-synchronous orbits.

China continues to develop the Long March 5, 6, and 7 family of launch vehicles, with different engines and less toxic fuel than their predecessors. A new generation of rockets is intended not only to meet China's domestic launch needs, but also to compete in the commercial market with American and European launch systems. The Long March 5 has core stages fed by liquid hydrogen and liquid oxygen, a technology route to high specific impulse. Kerosene-fueled modules with YF100s serve as its boosters. (China-Defense-Mashup.com, 2011).

The goal of the largest launch capability Long March rocket is 25,000 kg for low-Earth orbit, 1,500 kg to 14,000 kg for geostationary transfer orbit, and 500 kg for sun-synchronous orbits (sina.com 2011).

China's launch capability has increased significantly since the reform period initiated in 1978. Up to July 29, 2011, China had conducted 152 launches (shown in Figure 5) including 136 successes and 16 failures. This includes 152 satellites and seven spacecraft for manned flights. A total of seven astronauts have been sent to space and returned to the Earth safely.

Figure 4 shows China's launch sites in Jiuquan, Xichang and Taiyuan, sparsely populated areas with flat terrain and broad field of vision and good weather for launching. (Space Launch Report, year unknown). A fourth launch site, Wenchang in Hainan, will be put into service between 2014 and 2015.

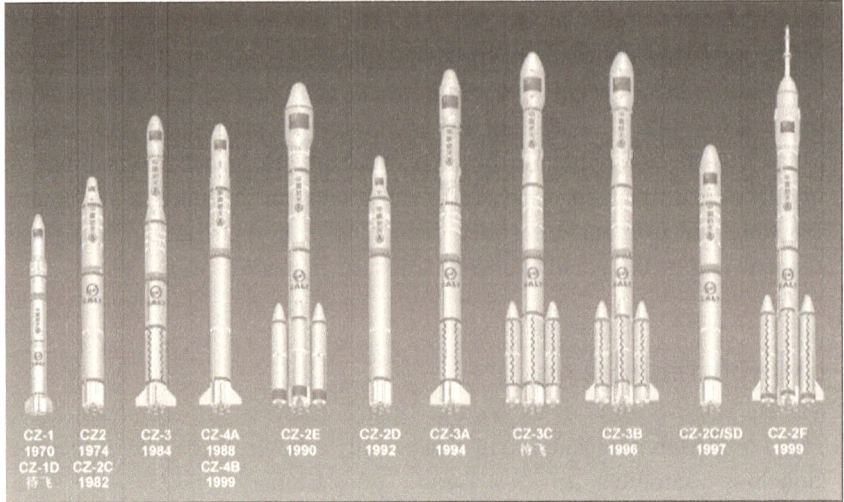

Figure 3: Long March Family
(Source: www.peoplenet.com.cn, *People's Daily* Online.)

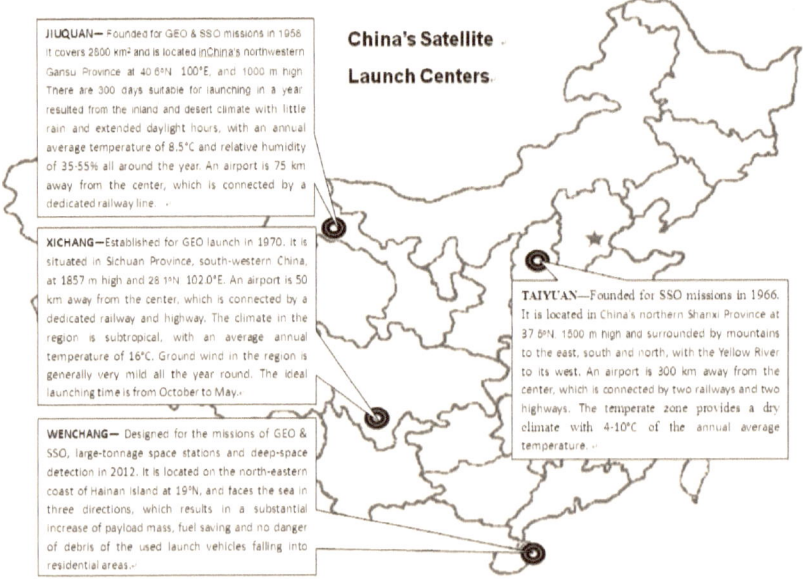

Figure 4: China's Satellite Launch Centers

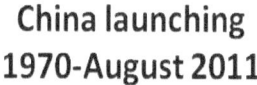
China launching
1970-August 2011

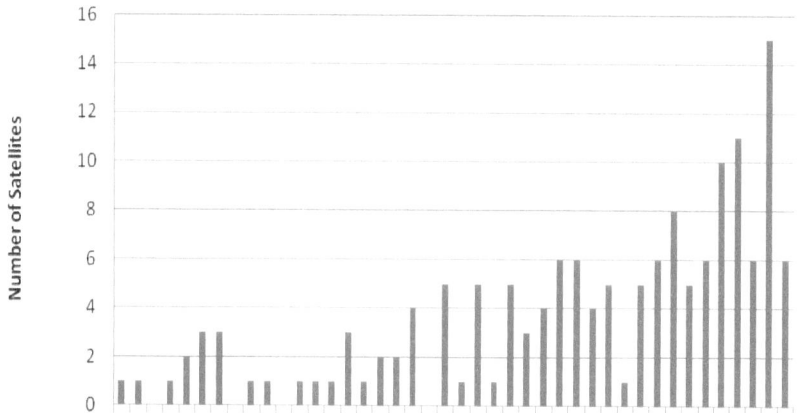

Figure 5: Satellites launched each year in China (1970-July29, 2011)

SATELLITE SYSTEM AND APPLICATIONS

China has successfully developed its own satellite payloads for telecommunications, navigation and positioning, meteorology, Earth observation and space science. By 2011, China's flight success rate was 89%, and a number of satellite application systems have been established, including the six satellite series-DFH (DongFangHong) telecommunication and broadcast satellites, FY (FengYun) meteorology satellites with weather monitoring and forecasting capability, SJ (ShiJian) scientific experiment satellites with recoverable payload capability, ZY (ZiYuan) Earth resource satellites, BeiDou navigation satellite system, and small satellite series for environment and disaster monitoring and warning capability. China is one of the few countries in the world having both GEO and polar orbit meteorological satellites.

These satellite series have yielded useful social and economic benefits, as well as scientific results. (IOSC, 2000).

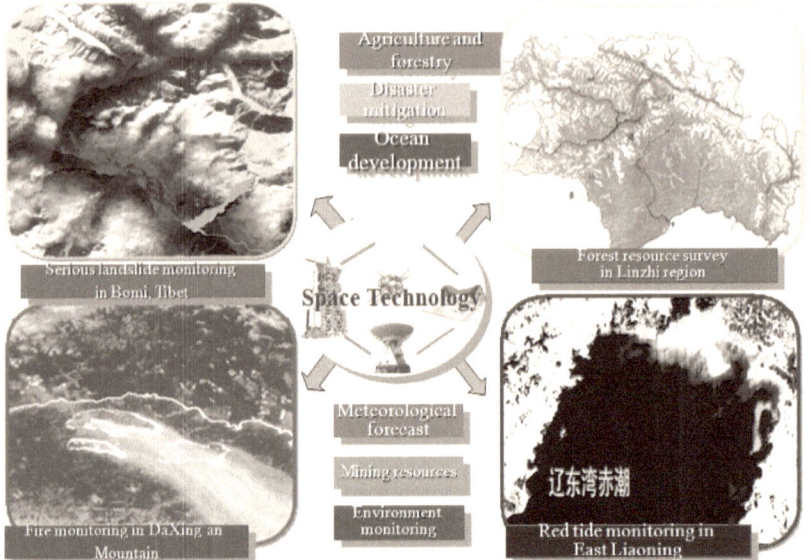

Figure 6: Additional examples of applications of space technology in China.

For example, remote sensing satellites played an important role in construction of major projects, such as the West-to-East Gas Transmission Line, Three George Dam Project, South-to-North Water Diversion Project, Land resource survey and Ecological protection.

China has conducted research on crystal growth in space micro-gravity, and has conducted growth experiments of semiconductor crystals and nonlinear optical crystals such as GaAs and HgCdTe. China has also obtained high quality protein crystals under the space micro-gravity environment, and has worked to master space biology and pharmacy methods with broad prospects.

RECENT ACHIEVEMENTS

LUNAR EXPLORATION PROGRAM

In February 2004, China's Lunar Exploration Program was initiated to pursue objectives and "three-phase" medium- and long-range planning for lunar exploration (Figure 7): Orbiting the moon, landing, and returning samples.

Lunar flyby in 2007   Lunar soft landing in 2012   Sample return in 2017
Figure 7: China's Lunar Exploration Program (Courtesy of CLEP).

On October 24, 2007 China's first lunar orbiter, Chang'e-1, was launched successfully, and on November 26, 2007 China published its first lunar surface image. On November 12, 2008, the full-moon image mapped by CH-1 was published as shown in Figure 8 (CLEP 2008).

Figure 8: Full-moon Image Mapped by CH-1 (CLEP 2008).

On October 1 2010, Chang'e–2 was launched, and completed its designed objectives successfully, as by the end of 2010, the first phase was finished, mapping the lunar surface in three dimensions, analysis of the content and distribution of useful elements on the lunar surface, measuring the density of lunar soil, and monitoring the near-moon space environment (in Figures 9 and 10).

The second stage of China's Lunar Exploration Program aims at achieving a soft lunar landing to make on-site exploration of the lunar surface, which is expected in 2012–2013. The third stage will be another soft landing around 2017, with a goal of bringing back China's first lunar samples.

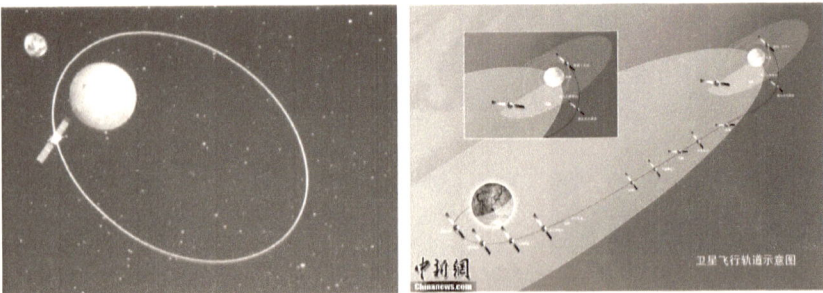

Figure 9: CH-2 in an elliptical orbit   Figure 10: CH-2 Mission Profile
(Courtesy of China News Network)

BEIDOU PROGRAM

The BeiDou Program is China's independent satellite navigation and positioning network, or Compass system.

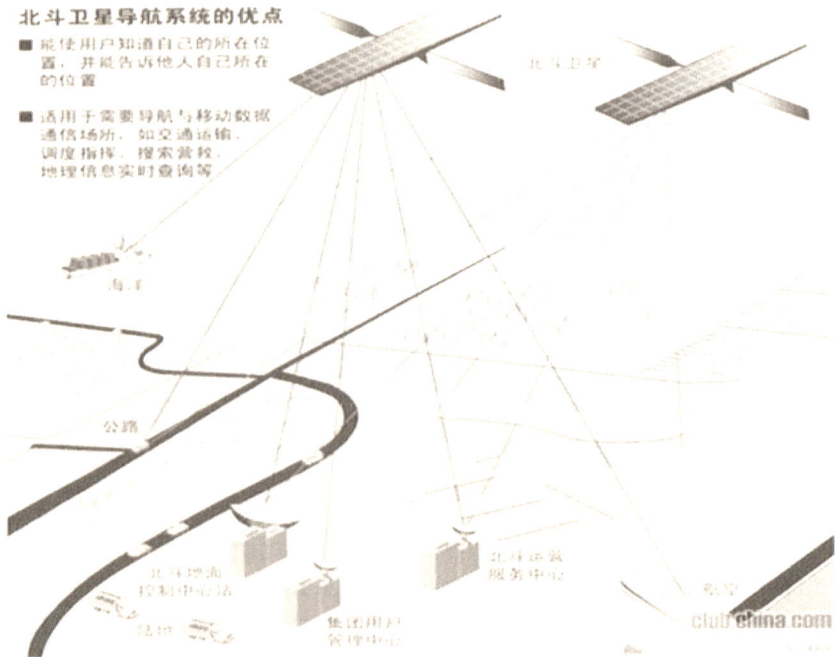

Figure 11: The BeiDou Program
(Source: http://www.china.com/, 2011)

It is a second generation navigation satellite system, designed to provide real-time uninterrupted 3D GPS service, eventually consisting of 35 satellites to provide global coverage, with an accuracy to 10 meters for civilian users.

The program was kicked-off when two orbiters were launched as a double-satellite experimental positioning system in 2000. Besides civilian use, the satellites also provide global positioning data for China's military services, and it is planned that it will be updated to provide global service by 2020 (IBT 2010).

DOUBLE STAR PROGRAM

The Geospace Double Star Exploration Program (DSP) was China's first dedicated scientific satellite with solid scientific objectives, approved by CNSA in 2001. DSP is also the first major Sino-European space science collaboration, consisting of two satellites that explore the Earth's magnetosphere, the magnetic bubble that surrounds and protects Earth, as shown in Figure 12. The diagram shows all 11 spacecraft of the Cluster, Double Star and THEMIS missions orbiting the Earth. All three missions are studying the environment of Earth's magnetosphere. (ESA, 2011).

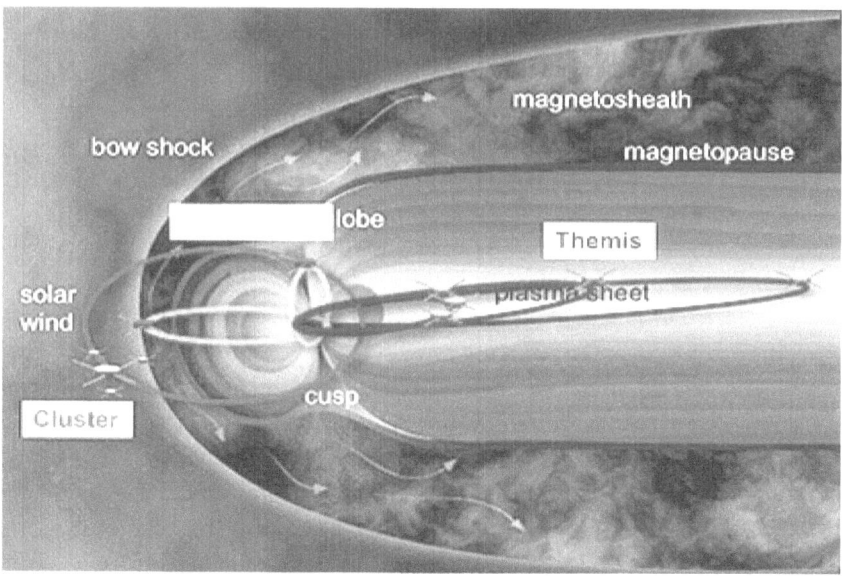

Figure 12: Double Star, Cluster and Themis (Courtesy of ESA)

FRAMEWORKS FOR INTERNATIONAL CO-OPERATION

SPACE POLICY ON INTERNATIONAL COOPERATION

According to the 1996 Cooperation Declaration adopted by the

General Assembly of the United Nations, international cooperation for space activities should comply with the following principles (ESA 2007):

- Accordance with the provisions of international law;
- The interest of all states;
- Fair and reasonable;
- Modes that are considered most effective and appropriate;
- Taking into particular account the needs of developing countries;
- Considering the appropriate use of space applications and the potential of international cooperation;
- Strengthening the role of the Committee on the Peaceful Uses of Outer Space;
- Wide participation.

The Chinese government considers international cooperation as its national strategy, as expressed in Part IV of the "White Paper: China's Space Activities in 2006," and states that "China persistently supports programs involving the peaceful use of outer space, and maintains that international space cooperation should be promoted and strengthened on the basis of equality and mutual benefit, mutual complementarity and common development." (IOCSC 2006)

At the meeting celebrating the 50th Anniversary of the China Aerospace Industry, Premier Wen Jiabao stressed that China adheres to peaceful development and actively extends communication and cooperation in the aerospace field (Wen 2011).

Since China first participated in international space cooperation in the middle 1970s, China has continued to carry out cooperative efforts. Attention has been paid to inter-governmental bilateral and multilateral cooperation, as well as to cooperation with the UN.

China supports international exchanges and cooperation in space technology, space applications and space science, with the following priority cooperation areas, as stated in Part V in "White Paper: China's Space Activities in 2006" (IOCSC 2006):

- Scientific research into space astronomy, space physics, microgravity science, space life science, lunar exploration and planet exploration;
- Data sharing and services of Earth observation satellites, and application and research in the areas of resources investigation, environment monitoring, prevention and mitigation of disasters,

and global climate change monitoring and forecasting;
- Sharing of space TT&C network resources, and mutual provision of space TT&C assistance;
- Design and manufacture of communications satellites and Earth observation satellites;
- Manufacture of ground facilities and key components of satellite communications, remote sensing, and navigation and positioning;
- Application of satellite communications and broadcasting in tele-education and tele-medicine, and expansion of application scope of satellite broadcasting and TV, and related services for satellite navigation and positioning:
- Commercial satellite launching services, export of satellites and their components and parts, and construction and services of satellite ground TT&C and application facilities;
- Exchanges and training of personnel in various fields of space activities.

ACHIEVEMENTS IN INTERNATIONAL SPACE CO-OPERATION

According to the white paper "China's National Defense in 2010" (IOCSC 2011), China has "conducted bilateral cooperation and exchanges with Russia, France, Brazil, Ukraine, the United States and the European Space Agency (ESA) in the fields of space technology, space exploration and space science," in accordance with the principle of peaceful use of outer space.

China also "supports the work of the United Nations Committee on the Peaceful Uses of Outer Space (COPUOS) and Asia-Pacific Space Cooperation Organization (APSCO), and plays an active role in making use of outer space technologies to conduct multilateral cooperation in Earth science research, disaster prevention and reduction, deep space exploration, and space debris mitigation and protection." (IOCSC 2006)

MULTILATERAL SPACE CO-OPERATION

China attaches great importance to multilateral cooperation under the framework of the UN as a member of the UN and standing member of the Security Council. China supports and has participated in the UN space applications program.

In June 1995, for example, CNSA acceded to the Inter-Agency Space Debris Coordination Committee, indicating that China is willing, together with other countries, to continuously make efforts to explore the ways and means to mitigate and reduce space debris, and promote international cooperation on this issue.

In COPUOS China actively promoted the establishment of space debris working group and the draft of Space Debris Mitigation Guidelines. In June 2007, China supported the adoption of the Guidelines submitted by the Scientific and Technical Subcommittee, and in 2010, China National Space Administration worked out the country's first Regulations on Space Debris in an effort to standardize and regulate space debris in China's space activities by administrative measures.

At present, about 20 technical standards have been worked out to guide and regularize the design and on-orbit operation of launch vehicles and spacecraft. Typical of this effort is CZ-5, China's new generation launch vehicle, which included in the early stage of design for research and development the intent to control space debris. During the next five years from 2011 to 2015 it is planned that the Chinese government will continue to increase funds on space debris research.

China continues to promote the Asia-Pacific Region Multilateral Cooperation in Small Multi-Mission Satellites Project, and has started joint research, manufacture and application of small multi-mission satellites together with Bangladesh, Iran, the Republic of Korea, Mongolia, Pakistan and Thailand for international collaborations on the planned Environmental and Disaster Monitoring Satellite Constellation (IOCSC 2006).

In 2010, the IAF General Assembly in Prague selected Beijing as the host city for the 64th International Astronautical Congress (IAC 2013), to be hosted and organized locally by the Chinese Society of Astronautics, another sign that China is opening its space industry and space program to the international community, and expanding its internal space cooperation.

BILATERAL SPACE COOPERATION

According to the IOCSC 2010 report, over the past several years China has signed cooperation agreements on the peaceful use of outer space and space project cooperation agreements with Argentina, Brazil, Canada, France, Malaysia, Pakistan, Russia, Ukraine, the ESA and the European Commission, and has established space cooperation subcommittee or joint commission mechanisms with Brazil, France, Russia and Ukraine.

It has also signed space cooperation memoranda with the space organizations of India and Britain, and has conducted exchanges with space-related bodies of Algeria, Chile, Germany, Italy, Japan, Peru and the United States. Here we will look specifically at Chinese cooperation with

Brazil, France, Russia, the ESA, and the UN.

BRAZIL

China's well-established collaboration with Brazil focuses on the Earth resources satellites. Following the successful launch of the Sino-Brazilian Earth Resources Satellite 02 in October 2003, the two governments signed supplementary protocols on further joint research and manufacturing of satellites 02B, 03 and 04, as well as on extended cooperation in a data application system, maintaining the continuity of data of Sino-Brazilian Earth resources satellites and expanding the application of such satellites' data region wide and worldwide (CNSA 2009).

FRANCE

"China and France have developed extensive space exchanges and cooperation. Under the mechanism of the Sino-French Joint Commission on Space Cooperation, the exchanges and cooperation between the two countries have made important progress in space science, Earth science, life science, satellite application, and satellite TT&C." (IOCSC 2006) such as the Double Star, the Galileo Project and the Sino-French Gamma-ray Burst Mission SVOM.

RUSSIA

"The space cooperation between China and Russia has produced marked results. Within the framework of the Space Cooperation Sub-Committee of the Committee for the Regular Sino-Russian Premiers' Meeting, a long-term cooperation plan has been determined. In addition, exchanges and cooperation in the sphere of manned spaceflight have been carried out, including astronaut training." (IOCSC 2006)

According to the summary of the 11th Meeting of the Russian – Chinese Space Cooperation Subcommittee Completed in Beijing in November 2010 (Roscosmos 2010), Russia and China announced that they had finalized arrangements to cooperate to accomplish goals including Mars and Phobos exploration using the Russian spacecraft Phobos-Grunt and Chinese microsatellite Yinghuo-1, which are to be launched in 2013. Both countries stressed that through their united efforts, mutually beneficial goals are expected to be accomplished. Roscosmos gave a wide-ranging Q&A interview to Novosti Kosmonavtiki magazine in February 2010 concerning the current state and future plans of the Russian space program, including the planned Soyuz replacement spacecraft, the new spaceport in the Far East, Russo-Chinese cooperation, and other topics.

Under the Space Cooperation Program for 2010-2012, Russia and

China will work on 53 topics, including exploration of the Moon and other planets of the Solar system. At the 11th meeting of the Russian-Chinese Space Cooperation Subcommittee in Beijing, an agreement was signed whose purpose is to establish reciprocal notification mechanism for planned and executed launches of ballistic missiles and space launch vehicles in Russia and China in order to avoid any unclear and unforeseen situations.

ESA

A major successful international space science cooperation of China is the Sino-ESA Double Star Satellite Exploration of the Earth's Space Environment. In addition, China's relevant departments and ESA have implemented the "Dragon Program," involving cooperation in Earth observation satellites. So far more than 16 remote-sensing application projects have been conducted in the fields of agriculture, forestry, water conservancy, meteorology, oceanography and disasters (ESA 2007; IOCSC 2006).

United Nations

According to the report on China's space activities in 2006 (IOCSC 2006), "in the light of the Charter on Cooperation to Achieve the Coordinated Use of Space Facilities in the Event of Natural or Technological Disasters, China has acceded to a disaster mitigation mechanism consisting of space organizations from different countries." In addition, "in cooperation with UN, China has hosted UN/ESA/China basic space science workshops and a UN/China workshop on tele-health development in Asia and the Pacific. China has also hosted, in collaboration with the Multilateral Cooperation Secretariat of the Asia-Pacific Space Cooperation Organization and the UN Economic and Social Commission for Asia and the Pacific, training courses and symposia on space technology applications, and has provided financial support for these activities. China has also taken part in a program promoting the application of space for sustainable development in Asia and the Pacific organized and implemented by the UN Economic and Social Commission for Asia and the Pacific" (IOCSC 2006).

Model of International Space Cooperation

Two obvious reasons for international space cooperation is that space activities require high technology and high investment. Thus, a primary objective of international space cooperation is to develop space technology and space capability while also addressing the financial requirements through space cooperation or commercialization, and to carry out space science research which benefits the whole of humanity.

Space Science Research Cooperation

During the Sino-European joint mission DSP, a large amount of observation data were collected and many important scientific results have been achieved. For example, by the end of 2008, Chinese scientists had published 71 papers, among them 69 in journals listed in Science Citation Index (SCI) and special issues of Annale Geophysicae, Journal of Geophysical Research, and Science in China (Guo & Wu 2009).

Through this program, scientists from both China and ESA have benefited from the abundance of data collected by DSP and ESA's Cluster program. This program also marked the beginning of more extensive international space science cooperation, especially with ESA and other European countries.

Following DSP, there are several on-going space science research cooperation programs. The Space-based multi-band astronomical Variable Objects Monitor (SVOM) mission is a gamma-ray bursts mission in development between China and France for launch in 2015. The China-Russia Joint Exploration of Mars and Phobos mission will be launched in 2013. The China-Russia Joint World Space Observatory for Ultraviolet (WSO-UV) is also in the development phase, and there are also some international cooperation in the programs of China's manned spacelab, space station, lunar exploration, and deep space exploration projects.

Commercial Cooperation

Here are four examples of commercial cooperation in China's space program, including arrangements with Brazil, Nigeria, Venezuela, and a France-Hong Kong joint effort.

Brazil

By collaborating with developing country, Brazil, China successfully found a South-South cooperation pattern. In the 2002 Protocol between China and Brazil, both parties agreed to share the benefits of the satellite system, and the issue of intellectual property rights protection, trade issues concerning satellite launches, and flight insurance in space cooperation are all addressed in the protocol.

NIGERIA

China launched a communications satellite "APSTAR VI" into orbit in April 2005. In December 2004, China signed a commercial contract for a communications satellite with Nigeria, providing in-orbit delivery service to that country.

VENEZUELA

In November 2005, China signed a commercial contract for a communications satellite with Venezuela, providing in-orbit delivery service and associated ground application facilities.

FRANCE-HONG KONG

In 2010 China announced that it would engage in a commercial satellite launch for a "French-made communications satellite for the Hong Kong-based APT Satellite Holding Limited in the first half of 2012. The satellite, dubbed APTSTAR-7 and made by the Thales Alenia Space, will be sent into space by China's Long March 3B/E carrier rocket at the Xichang Satellite Launch Centre in southwestern China, according to a statement issued by the China Great Wall Industry Corporation (CGWIC), the contractor of the launch service" (XNA 2009).

FUTURE DEVELOPMENTS

China has set a goal for its development as "becoming a generally modernized and moderately developed country by the mid 21st century," and has put space development in very high priority and drawn up a new development plan for China's space industry, defining development targets and major tasks until 2050 in 4 steps: now to 2015 as the first step, by 2020 as the second step, by 2030 or 2035 as the third step, and by 2050 as the fourth.

The following are some of the plans and strategic studies carried out in China:

- Outline of the 11th Five-Year Program for National Economic and Social Development published by the state council in 2006;
- National Guideline for Medium-and Long-term Plans for Science and Technology Development (2006-2020) (CSC 2006a) published by the state council in 2006;
- Eleventh Five-Year Plan (2006–2010) for Space Science Development issued by CNSA in 2007;

- 12th Five-Year Plan for China's Science and Technology Development formulated in March of 2011 (CSC 2006b);
- Space Science & Technology in China: A Roadmap to 2050 (CAS 2009).

According to these plans, China will launch and continue key space projects, including manned spaceflight, lunar exploration, Mars and Venus exploration, space science satellite program, high-resolution Earth observation, and new-generation carrier rockets. It will also strengthen basic research, develop frontier space technology, and accelerate progress and innovation in space science and technology.

SHORT AND MEDIUM-TERM PLAN
SPACE TECHNOLOGY DEVELOPMENT AND APPLICATIONS

On space technology development and applications, China's goals in short and medium-term plan are (IOCSC 2006):

- To significantly improve the capabilities and reliability of carrier rockets in space;
- To build a long-term, stably operated Earth observation system, and a coordinated and complete national satellite remote-sensing application system;
- To set up a complete satellite telecommunications and broadcasting system, and enhance the scale and economic efficiency of the satellite telecommunications and broadcasting industry;
- To establish a satellite navigation and positioning system, and bring into being China's own satellite navigation and positioning application industry;
- To achieve the initial transformation of applied satellites and satellite application from experimental application type to operational service type.
- To enable astronauts to engage in extravehicular activities, and achieve spacecraft rendezvous and docking; to realize the lunar-orbiting probe; and make original achievements in space science research.

Some space programs have been launched and implemented accordingly. For example, the mega-project "High-resolution earth observation system," "Airborne remote sensing system," and "National satellite remote sensing ground (network) system network" are going to be implemented. By 2020, groups of small satellites for monitoring meteorology, ocean, resources, environment and forecasting disaster, three-dimensional mapping satellites and their application systems, are to be built up.

The integrated construction of satellite remote sensing, aerial remote sensing, ground-based (network) systems and application systems is to be carried out, and the "National Spatial Information Infrastructure" is to be greatly strengthened. In addition, the second generation of the Compass (BeiDou) satellite navigation system will be set up to cover the Asia-Pacific region in 2012 and a satellite navigation and positioning system with global coverage will be established by 2020 (People's Daily Online 2010).

According to Yang Jun, director of the National Meteorological Satellite Centre, China expected to launch its first high resolution, stereoscopic mapping satellite for civilian use in the second half of 2011, and will launch 14 meteorological satellites over the following 10 years (Dasgupta 2010).

SHORT AND MEDIUM-TERM PLAN
SPACE SCIENCE

Based on the Outline of the 11th Five-Year Program for National Economic and Social Development and the National Guideline for Medium- and Long-term Plans for Science and Technology Development (2006-2020) (CSC 2006a) the Eleventh Five-Year Plan (2006–2010) for Space Science Development, and 12th Five-Year Plan for China's Science and Technology Development (CSC 2006b), China's space science program goals in short and medium-term are:

To carry out the manned space flight program, covering all fields of space science research in spacelab and space station (Aerospace & Defence News 2010):

- Spacelab: to launch Tiangong-1 and Shenzhou 8 without crew in 2011 and Shenzhou 9 and 10 with crew in 2012, and finally the full spacelab Tianggong-2 in 2014;
- Space station: by 2022, several large modules will be launched with crew service and operation, working in orbits for at least 10 years;
- To carry out the lunar exploration program, with a possibility for a manned lunar landing;
- To carry out a deep-space exploration program, targeting the Sun, Mars, Venus, other planets and objects in the solar system,

with a possibility for a manned Mars mission;
- To launch a series of space science satellites around 2014-2016:
  o The Hard X-ray Modulation Telescope (HXMT) satellite, China's first space astronomy satellite to study black holes;
  o The SJ-10 recoverable scientific research and technological experiment satellite, to conduct research on microgravity and space life science;
  o The KuaFu mission, to study the solar influence on space weather;
  o The quantum science satellite, to study quantum key distribution for secure communication and long distance quantum entanglement;
  o The dark matter detection satellite, to search for dark matter and study cosmic ray acceleration.
- By taking full advantage of international collaboration in space science, to implement the following programs:
  o China-Russia Joint Exploration of Mars and Phobos;
  o World Space Observatory for Ultraviolet (WSO-UV);
  o China-France Space Viable Object (SVOM) satellite for studying gamma-ray bursts and cosmology;
- To carry out technology development for future space science missions.

LONG TERM PLAN

In the first few years of the 21st century, major space powers including the US (NASA), Europe (ESA), Japan (JAXA), and others have launched various space exploration missions, and have made substantial and ambitious plans to expand human activities into deep space.

Here we summarize the major elements of the long term strategic plan of China's space program as expressed by the Chinese Academy of Sciences (CAS) in the document entitled "Roadmap for Strategic Research in Key Fields of Science and Technology in China through 2050" (CAS 2009), in which space science, applications, and technology are included with particular emphasis in the report entitled *Space Science & Technology in China: A Roadmap to 2050* (Guo & Wu 2009).

Space science, applications and technology are all research fields that are essential for China's development. It is further found that science and technology have advanced side by side, and there has always existed a strong and mutually beneficial relationship between them: "science leads technology, and technology promotes science" as shown in Figure 13.

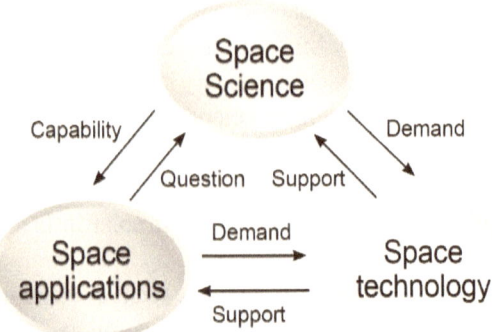

Figure13: Mutual demanding and supporting relationships among space science, space applications and space technology (Source from *Space Science & Technology in China: A Roadmap to 2050*, Guo & Wu 2009)

The roadmap report formulates the strategic goals and vision for space science, applications and technology of China, all while taking into account China's intent to set up the goals, tasks and technical routes for different stages, which would serve as an operationally feasible strategic guide.

Satellite programs with specific scientific goals will constitute the backbone of China's space science. As for space applications, emphasis is on Earth observation and its integrated applications, and the construction of Digital Earth Scientific Platform and Earth System Simulation Network Platform. For the development of space technology, the combination of science and technology is stressed, and those technical supports that can contribute to the realization of goals set for space science and application receive particular attention.

SPACE SCIENCE & TECHNOLOGY IN CHINA: A ROADMAP TO 2050

This roadmap addresses all major fields of space science, space applications, and space technology. Space science includes all the scientific research fields that are related to space. Space applications in this roadmap refer mainly to earth observations from space and other subsequent applications. Space technology deals mainly with all kinds of enabling technologies that provide means of and services to space science carried out through 2050.

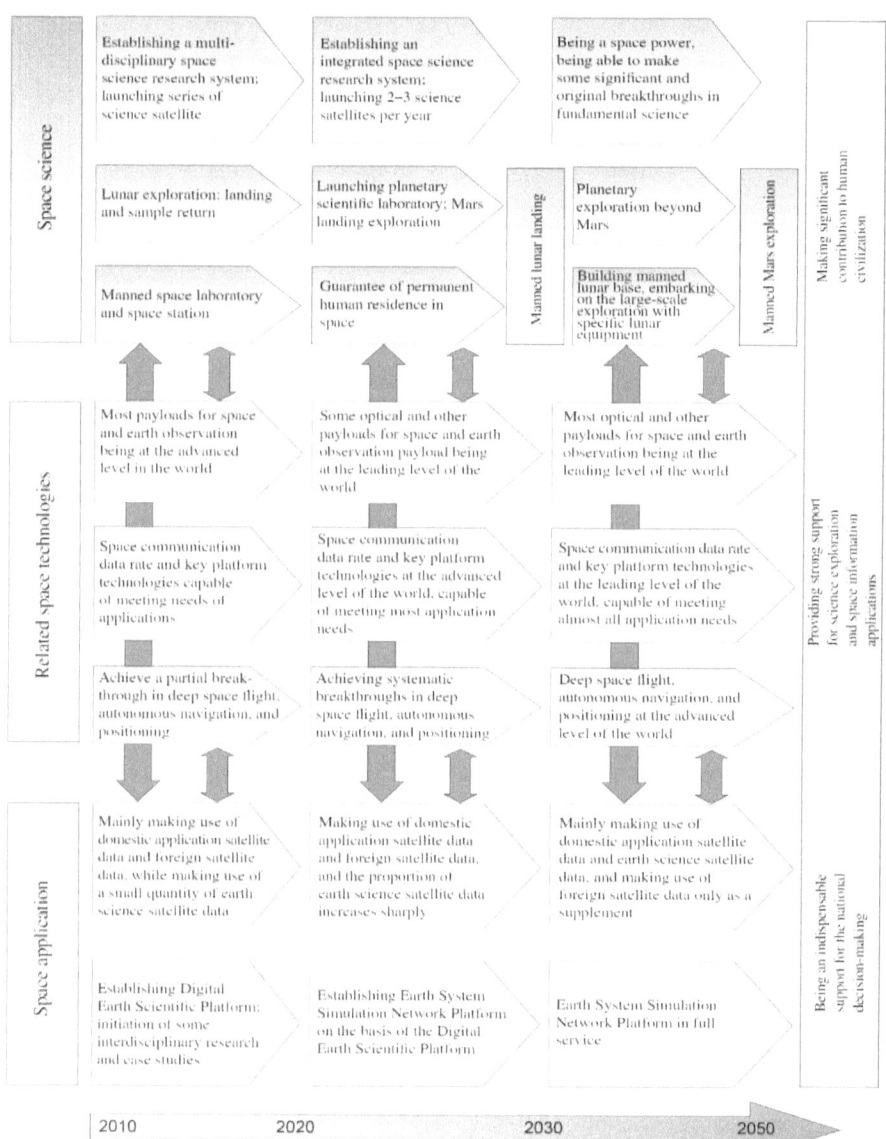

Figure 14: China's Roadmap for space science, applications and technology through 2050
(Source: Space Science & Technology in China: A Roadmap to 2050, Guo & Wu 2009)

Figure 14 illustrates the main goals for and relationships between space science, space applications and space technology in the roadmap through 2050. The strategic goal of space science (Strategic Goal 1 in the roadmap) is set with the anticipated future development of space technology in China, such as an optical telescope with extremely high spatial resolution exceeding that of the Hubble Space Telescope for imaging the deep space and seeking new discoveries. Meanwhile, extremely high spatial resolution and deep observations will produce extremely large amounts of data, requiring fast data transmissions from deep space. In the mean time, it is recognized that new means of propulsion for sailing into deep space are required, in order to arrive at the currently unreachable places by human beings, including beyond the solar system.

The strategic goal of space applications (Strategic Goal 2 in the roadmap) requires a large amount of new observational data, for example, on global climate change and natural disasters. To achieve this goal, new instruments and sensors for monitoring, for example, salinity of the seawater and analyzing water in soil are needed, and also more satellite orbits and more observational parameters for global environmental monitoring are also required. Therefore the report calls for more careful studies and planning on new orbits for satellites and observatories, cluster design, and even possibly an observatory based on the Moon.

The strategic goal of space technology (Strategic Goal 3 in the roadmap) is set to meet the demands of space science and space applications in the roadmap in a coherent way. It is recognized in the report that space technology development should also consider and be driven by the nature of high-tech development and the obstacles and challenges confronting it. Many breakthroughs in space technology are thus inevitably required, which will enable the rapid development of China's overall space program.

SUMMARY OF CHINA ROADMAP OF SPACE PROGRAM TO 2050

If this roadmap is executed even partially, it is clear that space science, applications and technology can be developed fully in China and reach a level compatible with China's national status in the world by 2050, which should be a significant driving force for the development of the country. Looking back at the 50 years' history of China's rapid development of space capability and looking forward to China rapid social and economic development, we foresee that China's future space program should be able to contribute effective solutions to many of the future problems that may face the country, while also placing China in a leading position among the world's space powers, and hopefully contributing many exciting scientific discoveries.

Conclusion

More than 50 years of development of the Chinese space program has resulted in great progress in space science, applications, and the related technologies. Since the releases of the two white papers, many guidelines and strategies, China's space program, including its organization, policies, history, strategies, current status and future outlooks, becomes more visible and better understood by the international community.

The Chinese government has emphasized that China's space program adheres firmly to the national and the international space policy and law. China's space policies are intended to "make innovations independently, leapfrogging development in key areas, shoring up the economy and leading future trends in order to promote independent innovations in space science and technology, make space activities create more economic and social benefits, ensure the orderly, normal and healthy development of space activities."

In addition, China has participated in and led several international cooperation programs on space applications and space science. Many bilateral and multilateral agreements and frameworks have also been setup, allowing mutually beneficial space applications and explorations to be executed effectively.

However, China recognizes that its space science, applications and technology do not match well with its overall status of social and economic development, which in many aspects are still much less advanced compared to that of developed countries.

Therefore, the long term roadmap put forward by strategic study calls for a higher priority to be put on China's space program to provide the much needed support to China's economic, social, scientific and technological development. With the acceleration of social and economic development in China, its space program is anticipated to expand and improve at an ever faster speed, and international cooperation has become and will continue as a vital part of China's space program.

⁂

Le Wang

Le Wang is Associate Professor and Space Project Manager in Xi'an Institute of Optics and Precision Mechanics of the Chinese Academy of Sciences, and a PhD student in Shaanxi Normal University. From Shaanxi Normal University, she obtained her Bachelor's Degree in the Department of Foreign Language from 1991 to 1995, and then her Master's Degree in the School of Foreign Language from 2004-2007. She completed her study in the Space Management Master Program, International Space University, Strasbourg, France, from 2010-2011. Since 2008, she has been studying for her Ph.D. degree in the College of Chinese Language and Literature, Shaanxi Normal University; the subject of her Ph.D. study is *Revivals of the Historical and Cultural City—Study on Comparison between Xi'an and European Cities*.

References

Aerospace & Defence News, 2010, China says manned space station possible around 2020 (online), http://www.asd-network.com/press_detail/31461/China_says_manned_space_station_possible_around_2020.htm [Accessed 5 June 2011]

Center for Space Science and Applied Research (CSSAR), KuaFu Mission (online), http://english.cssar.cas.cn/lc/201008/t20100825_57932.html [Accessed 25 May 2011]

China Aerospace Science and Technology Corporation (CASTC), http://www.spacechina.com/english/ [Accessed 5 June 2011]

China-Defense-Mashup, 2011, *China Developing New Rocket Engines for the Long March 6 and 7*, http://www.china-defense-mashup.com/china-developing-new-rocket-engines-for-the-long-march-6-and-7.html [Accessed 20 November 2011]

China Great Wall Industry Corporation (CGWIC), year unknown, *Launch Records of the Long March Rockets* (online in Chinese), http://cn.cgwic.com/LaunchServices/LaunchRecord/LongMarch.html [Accessed 5 July 2011]

China's Lunar Exploration Program (CLEP), 2008, News release by CNSA on Results and Data Release from China's Lunar Exploration Program (online in Chinese), http://www.clep.org.cn/index.asp?modelname=yxt\index [Accessed 5 June 2011]

China's Lunar Exploration Program (CLEP), 2010, *Chang'E 2 entered into elliptical lunar orbit* (online in Chinese), http://www.clep.org.cn/index.asp?modelname=zt_ceertp_nr&recno=34 [Accessed 5 June 2011]

China's Manned Space Engineering, http://en.cmse.gov.cn/list.php?catid=42 [Accessed 5 June 2011]

China.com, 2011, China's BeiDou Compass System Challenges US's GPS (online in Chinese), http://club.china.com/data/thread/1011/2725/67/20/8_1.html [Accessed 8August 2011]

China National Space Administration, Organization and Function (online), http://www.cnsa.gov.cn/n615709/n620681/n771918/index.html [Accessed 5 June 2011]

China National Space Administration (CNSA), http://www.cnsa.gov.cn [Accessed 5 June 2011]

China National Space Administration (CNSA), 2009, Live broadcast of the delivery ceremony of China-Brazil 02B earth resource satellite, http://www.cnsa.gov.cn/n615708/n620168/n2259528/167288.html [Accessed 5 June 2011]

China's State Council (CSC), 2006a, *The national guideline on medium- and long-term program for science and technology development* (online in Chinese), www.most.gov.cn/yw/t20060209_28601_0.doc [Accessed 5 June 2011]

China's State Council (CSC), 2006b, *The national guideline for science and technology development in the 12th five-year period* (online in Chinese), http://www.most.gov.cn/kjgh/ [Accessed 5 June 2011]

Chinese Academy of Sciences (CAS), http://www.cas.cn/ [Accessed 15 June 2011]

Chinese Academy of Science (CAS), 2009, *Space Science & Technology in China: A Roadmap to 2050*, Science Press Beijing and Springer

CNSA, 2003, *China's Space Activities in 2001* (White Paper) (online), http://www.cnsa.gov.cn/n615709/n620681/n771967/69198.html [Accessed 5 June 2011]

Dasgupta, S., 2010, *China to launch mapping satellite to create network of 35 orbiters* (online), http://articles.timesofindia.indiatimes.com/2010-07-30/china/28288871_1_navigation-satellite-carrier-rocket-china-aerospace-science [Accessed 25 July 2011]

Europe Space Agency (ESA), 2007, Double Star (online), http://sci.esa.int/science-e/www/object/index.cfm?fobjectid=41296 [Accessed 10 June 2011]

Guo, H. & Wu, J., 2010, *Space Science & Technology in China: A Roadmap to 2050*, Science Press Beijing and Spring-Verlag Berlin Heidlberg

International Business Times (IBT), 2010. China Launches GPS satellite into orbit (Online), http://www.thefreelibrary.com/China+launches+GPS+satellite+into+orbit.-a0236012813 [Accessed 23 June 2011]

Information Office of China's State Council (IOCSC), 2006, China's Space Activities in 2006 (online), http://www.china.org.cn/english/features/book/183672.htm [Accessed 5 June 2011]

Information Office of China's State Council (IOCSC), 2011, China's National Defense in 2010 (online in Chinese), http://www.gov.cn/ztzl/zghk50/content_419652.htm

[Accessed 5 June 2011]

Li, D. & Cheng, D., Outlook of the development of the new generation launch vehicles (online in Chinese), http://www.calt.com/xwzx/ztxw/yjybcfs/yzhjdwl/index.html [Accessed 27 July 2011]

Ministry of Education of China (MOE), http://www.moe.gov.cn/ [Accessed 5 June 2011]

Ministry of Industry and Information Technology of China (MIIT), http://www.miit.gov.cn [Accessed 5 June 2011]

Ministry of Science and Technology of China (MOST), Missions of the Ministry of Science and Technology (online), http://www.most.gov.cn/eng/organization/Mission/index.htm [Accessed 5 June 2011]

People's Daily Online, year unknown, Introduction to the Long March Series Rockets (online in Chinese), http://scitech.people.com.cn/GB/25509/55912/83184/index.html [Accessed 5 August 2011]

People's Daily Online, 2010, China to set up global satellite navigation system within 10 years (online), http://english.peopledaily.com.cn/90001/90776/90881/7201448.html [Accessed 17 July 2011]

Russian Federal Space Agency (ROSCOSMOS), 2010, 11-th Meeting of the Russian – Chinese Space Cooperation Subcommittee Completed in Beijing (online), http://www.federalspace.ru/main.php?id=2&nid=10739&lang=en [Accessed 27 July 2011]

sina.com, 2011, Rocket launches this year will break records (online in Chinese), http://news.163.com/11/0304/02/6U94RS0800014AED.html [Accessed 9 June 2011]

Space Foundation. 2008. Space Report 2008. Washington DC: Space Foundation

Space Foundation. 2009. Space Report 2009. Washington DC: Space Foundation

Space Foundation. 2010. Space Report 2010. Washington DC: Space Foundation

Space Foundation. 2011. Space Report 2011. Washington DC: Space Foundation

Space Launch Report, year unknown, www.spacelaunchreport.com [Accessed 5 August 2011]

State-owned Assets Supervision and Administration Commission of the State Council (SASAC), year unknown, Main Functions and Responsibilities of SASAC (online), http://www.sasac.gov.cn/n2963340/n2963393/n2965120.html [Accessed 5 June 2011]

United Nations Office for Outer Space Affairs (UNOOSA). 1967. Doc. A/RES/51/122: ANNEX to Declaration on International Cooperation in the Exploration and Use of Outer Space for the Benefit and in the Interest of All States, Taking into Particular Account the Needs of Developing Countries (online), http://www.unoosa.org/oosa/SpaceLaw/spben.html [Accessed 7 July 2011]

Wen, J., 2011, Report on the Work of the Government in 2011 (online in Chinese), http://www.china.com.cn/policy/txt/2011-03/16/content_22150608.htm [Accessed 18 July 2011]

Wen, J., 2011, Speech at the conference celebrating the 50th anniversary of the country's aerospace industry (online in Chinese), http://www.gov.cn/ztzl/zghk50/content_419235.htm [Accessed 25 June 2011]

Xinhua News Agency (XNA), 2006, China Marks 50th Anniversary Of Aerospace Industry (online), http://www.spacedaily.com/reports/China_Marks_50th_Anniversary_Of_Aerospace_Industry_999.html [Accessed 5 June 2011]

Xinhua News Agency (XNA), 2008, Ministry of Industry and Information Technology inaugurated (online), http://www.china.org.cn/government/news/2008-06/30/content_15906787.htm [Accessed 25 June 2011]

Xinhua News Agency (XNA), 2006, China Marks 50th Anniversary Of Aerospace Industry, http://www.spacedaily.com/reports/China_Marks_50th_Anniversary_Of_Aerospace_Industry_999.html [Accessed 5 June 2011]

Xinhua News Agency (XNA), 2009, China To Launch French-Made CommSat (online), http://www.space-travel.com/reports/China_To_Launch_French_Made_CommSat_999.html [Accessed 27 July 2011]

Yan, Y., 2006, China to remarkably increase investment in science, technology, http://www.gov.cn/english/2006-02/09/content_183777.htm [Accessed 8 June 2011]

Zhu, L., XinHua News Agency, 2006, China to combine military, civilian research organizations (online), http://www.gov.cn/english/2006-02/09/content_183844.htm [Accessed 5 June 2011]

Chapter 11

The Leadership Competition between Japan and China in the East Asian Context

Kazuto Suzuki, Ph.D.
Hokkaido University

© Kazuto Suzuki, 2012. All Rights Reserved.

Abstract

Japan and China, two advanced spacefaring nations, are often referred as rivals in space. Successful manned space program in 2003 and ASAT test in 2007 by China were considered as turning points that led Japan to compete against China, and which potentially introduced the idea of a "space race in Asia." This paper analyzes the objectives, norms and logics of space policy in Japan and China, and discusses the differences between them over what to do in space, and explores the differences of approach. However, Japan and China are also competing for leadership and influence over the region. APRSAF and APSCO, two similar regional space organizations, are the vehicles of this competition. The result of the competition is positive for Asian countries, because it provides them with access to space technology as public goods throughout the region.

Introduction

In Asia, it is often said that there is a space race between Japan and China.[1] Each of the two countries launched satellites for exploring the Moon in 2007, and the similarity of the functions and nature of these two probes gave impression that they saw themselves as competing against one another to be the leading space country in Asia. Similarly, because of the success of the Chinese manned space program, it seems to many that they have ambition to beat the United States, Europe and Russia, and establish superior position in the world's space community.

Many also believe that the competition of Moon probes may extend to a much wider competition among these Asian countries, perhaps even a military one, as it did in the time of the Cold War. In short, some fear that space power has shifted to the East, and that the competition among Asian countries may result in much higher tension and friction among them.

However, this perspective is very much influenced by the Cold War imagery, and I suggest that it may be misleading to employ the Cold War analogy. It is not always true that those who pursue significant objectives in space are also pursuing a hegemonic position in the world. Three countries have their own reasons and logics for sending Moon probes, which need not be associated with the ambition for being hegemonic power.[2] I believe that we have to be open to other possible explanations, which we will examine in this chapter.

Japan

1. Origins of Japanese Space Policy

Since the beginning of its activities in space, Japan has avoided engaging in any security-related uses of space, due largely to the fact that Japan's

---

[1] James Clay Moltz, *Asia's Space Race: National Motivations, Regional Rivalries, and International Risks*, Columbia University Press, 2011.
[2] Kazuto Suzuki, "Is There a Space Race in Asia? Different Perceptions of Space" in N.S. Sisodia and S. Kalyanaraman (eds.), *The Future of War and Peace in Asia*, Magnum Books Pvt. Ltd., 2010, pp.181-200.

Constitution is explicitly pacifist. In addition, in 1969 the Japanese Diet adopted a resolution called "Space Development for Exclusively Peaceful Purposes," which prevents the Japanese defense authority from investing in, owning, or operating space systems. All Japanese space programs therefore exclude any military element, and are conducted exclusively under civilian authority in the name of research and development for new technology.[3]

Although the term "exclusively peaceful purposes" is not unique, as it also appears in the Treaty of Outer Space, or ESA Convention, the interpretation of this clause in Japan is quite unique. During the deliberations in 1969, Diet members argued that this clause should be interpreted similarly to the case of peaceful use of atomic energy. In the Japanese mind, both atomic energy and space are dual-use technologies, that is, technologies that can be used for both civil and military purposes. Also, the newly established Science and Technology Agency (STA) was in charge of both technologies, so Diet members had no doubt that space should be restricted as rigidly as atomic energy. Since Japan has suffered from the trauma of nuclear holocaust in Hiroshima and Nagasaki, there remains strong skepticism about the peaceful use of nuclear technology, and therefore, the Diet made it explicit that the technology should only be used civilian purposes, which means that the defense authority should not be administratively, financially or politically involved in the development and operation of nuclear technology programs. This interpretation relating to "exclusively peaceful purposes" was directly transplanted to space with the same intent.

2. SHOCKS IN POST-COLD WAR PERIOD

For a long time, particularly during the Cold War, the "non-military" nature of space did not present any problems for Japanese leaders. The US-Japan alliance provided the necessary infrastructure for intelligence gathering and telecommunication from space, and Japan's pacifist Constitution prohibited Japanese Self-Defence Force (SDF) from being deployed beyond its own borders.

However, the perception of Japanese people on security matters has been dramatically changed by two events. First, the imminent threat of

---

[3] Sawako Maeda, "Transformation of Japanese Space Policy: From the 'Peaceful Use of space' to 'the Basic Law on Space,'" The Asia-Pacific Journal, Vol. 44-1-09, November 2, 2009.

North Korea became evident when the Taepodong was launched over Japanese territory in 1998, which has caused a big change in the policy paradigm. This incident put the Japanese public as well as the policy community in a panic mode, and there was a strong demand to protect the homeland and prevent North Korea from launching missiles towards Japan. Thus, immediately after the Taepodong launch, the government made the decision to initiate a new satellite program called Information Gathering Satellite (IGS).

The launch of IGS faced serious constraints due to the existing legal interpretation of the 1969 Diet resolution. Although it was clear that the purpose of IGS was to monitor military activities of possible threats such as North Korea, it was presented as a "crisis-management" (note: it was even difficult to mention "dual-use" because it implies the possibility of the participation of JDA) satellite, which would also serve civilian needs in order to comply with the 1969 Resolution. To comply with the previsions of the Resolution, the Section of Intelligence Gathering in the Cabinet Secretariat was designated as the operator of IGS rather than the JDA. This situation led to a wide-ranging understanding among politicians that the legal constraints of the "exclusively peaceful purpose" resolution was too strict to have room to adapt to a new situation, as under the changing security environment in the post-Cold War period, it seemed nonsensical to maintain such a rigid pacifist rule in the face of an imminent threat.[4]

The second major event that has strongly influenced Japanese thinking about its own security was the Japanese Cabinet decision to participate in the Missile Defence (MD) program in 2003, which raised another difficult question for the Japanese space and security community. The issue was understood this way: On the one hand, because of the "exclusively peaceful purpose" resolution, the JDA and SDF would not be able to develop, launch and operate its own early warning or tracking satellite, which gathers crucial information about missile launch from any adversary or potential adversary. Without its own early warning satellite, the JDA would therefore be obliged to depend on the early warning information from the United States. However, if the JDA entirely depends on the US intelligence for initiating the deployment of MD counter-attack missiles, it would touch upon the sensitive issue of "collective defense."

The Japanese government has been taking a unique interpretation of its Constitution Article 9, which is that Japan holds the right of collective defense, but it would not exercise it. The Constitution does not allow the

---

[4] Tsuyoshi Sunohara, *Tanjo Kokusan Spai Eisei (The Birth of National Spy Satellite)*, Nikkei publishers, 2005.

Japanese government to possess military forces for offensive purposes but only for self-defense. Therefore, the Japan-US alliance is based on the understanding that the alliance was actually based on the unilateral exercise of collective defense by the United States, that is, the United States has obligation to support Japan militarily when Japan is under military attack, but Japan cannot do the same because of the unique interpretation of collective defense. So Missile Defense is set and ready for operation, but it would not be able to launch a counter-attack missile unless the command comes from Japan's own early warning satellite, because if a Japanese counter-attack missile is launched by the US command, it would be considered an exercise of collective defense. Currently, the Japanese MD system is designed to launch its counter-attack missile on Japanese command based on intelligence gathered by radars at sea (Aegis fleet) and on the ground, while US satellite early warning signals are used as "reference." Thus, many people in the Liberal Democratic Party (LDP), particularly those who are interested in defense issues, strongly demanded that the government reconsider the "exclusively peaceful purpose" clause of the Diet resolution in 1969.

3. BASIC LAW FOR SPACE ACTIVITIES

Politicians of the ruling LDP launched a study group on legal and political issues of Japanese space activities in 2005, and identified the problems of space policy driven by a bureaucracy. The report was issued in 2006 and urged LDP to propose new legislation concerning regulation of space activities by creating a ministerial post with a portfolio of space, establishing a new government forum for space user ministries, and changing the interpretation of the 1969 Diet resolution. This report was accepted by the politicians not only within LDP but also its coalition partner, Komeito, and largest opposition party, Democratic Party of Japan (DPJ). These three parties submitted a draft bill of Basic Law for Space Activities, and it passed the Diet in May 2008.[5]

The Basic Law defines the direction of a new space policy and a new decision-making structure. First, it will set up a new Minister for Space and Space Development Strategy Headquarters (an intergovernmental coordination body with strong authority). The Minister for Space would be a "specially designated" minister who will not be in charge of the

---

[5] Kazuto Suzuki, "Transforming Japan's Space Policy-making", Space Policy, Volume 23, Issue 2, 2007, pp.73-80.

management of the ministry, who resides in the Cabinet Office for coordinating policies of different ministries. The Headquarters will be composed of all the ministers and some specially appointed members from academia and industry. Although this is an ambitious challenge due to the conservative attitude of the government towards any reform, there are hopes that these new institutions will provide a positive force for more political attention and dynamics in space activities.

The Headquarters is intended to be the final decision-making body for the allocation of budget, which is to occur by bundling all budget requests from various ministries and negotiating with the Ministry of Finance on behalf of those ministries.

Second, the bill states that "Space development of Japan shall follow the Outer Space Treaty and other international agreements, and shall be conducted on the basis of the concept of pacifism in the Constitution," as stated in Articles 1 and 2. This clause of the bill suggests that the traditional interpretation of "exclusively peaceful purpose" as "non-military" would no longer apply. In this regard, the new Basic Law opens up the possibility for the military authority to be involved in development, procurement and operation of space systems.[6]

4. IS JAPAN GOING TO BE A BIG SPACE POWER IN THE NEXT DECADE?

One of the motivations for LDP politicians for promoting the Basic Law for Space Activities was to strengthen Japan's capability for using space in international affairs because they were very concerned about the development of the Chinese space program. Of course, the members were impressed by the successful manned space program, but their concern was not about the competition in the manned space capability nor in the space race for the Moon. Instead, their attention was on the recent development of Chinese action towards other Asian countries.

In 2005, the Chinese government concluded the signing of the establishing agreement for APSCO (Asia-Pacific Space Cooperation Organization) with 7 member states,[7] based on the AP-MCSTA (Asia-Pacific Multilateral Cooperation in Space Technology and Applications)

---

[6] For details, see the Appendix of Saadia M. Pekkanen and Paul Kallender-Umezu, *In Defense of Japan: From the Market to the Military in Space Policy*, Stanford University Press, 2010.

[7] Members are China, Mongolia, Iran, Thailand, Bangladesh, Pakistan, Peru.

with 15 member states.[8]

AP-MCSTA was an organization for developing small satellite technology and user-oriented applications. Both AP-MCSTA and APSCO were initiated by the Chinese government, and they are attracting a lot of attention from developing countries (see the section on China in this paper).

For many years, Japan was the leading country in this region, and JAXA and MEXT were proud to initiate APRSAF (Asia-Pacific Regional Space Agency Forum),[9] which coordinates space programs and enhances the cooperation among the space agencies of various nations in this region. However, this organization focused only on technical aspects of the space programmes of different space agencies, and there was no coordination of strategy or policy. The members are not exclusive, and often, participants expected to gain from what Japan might offer. There was wide dissatisfaction among LDP politicians that Japan was not supporting the needs of developing countries with the transfer of technology and collaborative projects for space hardware.

In response to these demands and fear of losing leadership in the Asian region, JAXA has initiated several projects to go beyond the "talking space." In 2005 at the APRSAF meeting in Fukuoka, JAXA proposed the Sentinel-Asia program. This was inspired by the EU-ESA-sponsored GMES (Global Monitoring for Environment and Security) program to provide regional imagery and data for environment and disaster management. It uses the Japanese Earth observation satellite "Daichi" (ALOS) and NASA's MODIS, as well as software that was developed by Digital Asia Research Center, Japanese IT-ventures. In addition, JAXA initiated the SAFE (Space Application For Environment) program to analyze climate change by monitoring water resources, sea level, forest degradation, and agricultural data.

However, these program were not enough to satisfy other Asian partners, as they demanded further technology transfer. Thus, in 2009 JAXA initiated the STAR (Satellite Technology for the Asia-Pacific Region) program, which mimicked APSCO's SMMS. This program

---

[8] Members are China, Mongolia, Malaysia, Iran, Indonesia, Chile, Ukraine, Thailand, South Korea, Bangladesh, the Philippines, Pakistan, Peru, Argentina, and Russia.

[9] Participants are Australia, Brunei, Canada, Sri Lanka, Germany, France, Pakistan, Italy, Japan, Bhutan, Cambodia, Thailand, Laos, Malaysia, Mongolia, Nepal, New Zealand, Bangladesh, China, Chile, India, Indonesia, Kazakhstan, South Korea, Singapore, Philippines, Turkey, Russia, Vietnam, Israel, Ukraine, Myanmar, UK, USA and Taiwan.

includes Malaysia, Thailand, India, South Korea, Indonesia and Vietnam, and is focused on developing small satellites (Micro-STAR and EO-STAR) together with JAXA.

It was a big step for APRSAF to become more of a technology-oriented forum, but the commitment of JAXA for this program seems to be ambivalent. Because of security concerns, JAXA is not free to transfer all necessary technologies for developing small satellites under various legal frameworks on export control and "peaceful use of space." Although the Basic Law for Space activities allows JAXA to perform intensive international cooperation, other related laws are not sufficient to provide enough room for JAXA to commit to these programs.

To summarize these points, Japanese space policy has changed dramatically in recent years, and it seems that the direction of the policy is going towards "utilization of space" rather than "developing technology." This objective has been supported by politicians, whose interest is in taking leadership in the Asian region and using space as a "tool" for providing benefits to other countries, especially developing countries, including providing security-related services such as disaster monitoring and confidence building measures. At the same time, politicians are concerned about the cost and benefit of space activities, and it would be difficult to promote "big projects" such as manned-space programs. Thus, it can be said that Japanese space efforts may not be glamorous, but will focus on pragmatic and effective programs to positively influence other countries in the Asian region.

5. CHINA

Chinese space activities and its decision-making process are opaque at best. In the decision-making process of military and civilian programs it is not clear which agencies and companies are involved in the program, and when it comes to the military program it is almost impossible to penetrate into the decision-making community at all. Because of this opaqueness, analysis of Chinese space policy involves a lot of guessing and speculation. However, it is possible for us to discern the general trend of the Chinese space program and its normative understanding on how space should be used.[10]

---

[10] Perhaps one of the most important work on Chinese space analysis is Gregory Kulacki and Jeffrey G. Lewis, *A Place for One's Mat: China's Space Program, 1956–2003*, American Academy of Arts and Sciences, 2009, which would help us understand the objectives and intentions of Chinese space activities.

THE SLEEPING DRAGON IS WAKING UP

The emergence of China as a space power started in 2000 when Chinese State Council issued its first White Paper on Space Activities, the first public statement on what China aims for and has achieved in space. It emphasizes on the one hand utilization of space for peaceful purposes and promotion of the benefit to all mankind, but on the other hand protection of China's national interests and strength, and implementation of its national development strategy.[11]

This dichotomy of global/national ambivalence can be seen in many aspects of the Chinese strategy for space. On the one hand, China National Space Agency (CNSA) and China Aerospace Science and Technology Corporation (CASTC), both established in 1993, are institutionally under the Committee on Science and Technology Industry for National Defence (COSTIND). These two major organizations are strongly influenced by national political climate and strategic objectives, but they are relatively autonomous institutions and insulated from the defense community, and there is little communication and exchange of information between the defense authority and CNSA.

But on the other hand, the China Academy of Sciences (CAS) and various technological institutions are generally autonomous from national political objectives and tend to emphasize the importance of Chinese contribution to humanity.

The most remarkable aspect in this White Paper is the emphasis on application programs. China has not invested extensively in application technology for a long time, but rather the Chinese government sought to acquire application satellites from foreign manufacturers. However, the US export control restrictions, known as ITAR (International Traffic in Arms Regulation), put satellite and space technology as controlled items under the Munitions List. This meant that export of US-made satellites, as well as satellites made with any components and parts produced in the US, needs to be approved by the US Department of State. Because of the competitiveness of the US space technology, it would be almost impossible for non-US manufacturers to avoid using US components and parts, and therefore the ITAR restriction was, in effect, a de facto exclusion of China from the international space market. Thus, it became imperative for the Chinese government to invest in application technology to meet the growing demand for space-based infrastructure and services.

---

[11] The Information Office of the State Council, China's Space Activities, a White Paper, November 22, 2000.
<http://www.spaceref.com/china/china.white.paper.nov.22.2000.html>

Following the White Paper of 2000, the State Council published a subsequent White Paper in 2006.[12] The objectives and principles of the second are not substantially different, but this paper expresses a stronger conviction that the Chinese space program is on a steady track, particularly noting a series of successes in its manned-space program. Also, the second White Paper emphasizes the civilian and peaceful nature of the Chinese space program, and is thought to be a response to American National Space Policy, which was issued a few months before. The US space policy document included a certain nuance that its space programs aim at protecting US territory and national interest, and states that the US will use its technical superiority to prevent any country from disrupting American space activities. The Chinese response therefore emphasized China's own national interests and intentions.

The third White Paper,[13] which was published in 2011, was not dramatically different from the previous two White Papers, but there was a growing confidence in its space activities. The major focus was on the manned-space flight with particular emphasis on the successful launch of Tiangong-1, the Chinese space station, and docking maneuver with unmanned Shenzou-8. Also, the third White Paper stressed the success in application satellite programs such as BeiDou/Compass navigation system and various Earth observation and communication satellites. Furthermore, the third White Paper focused on the importance of the space debris issue, which would threaten the safety of manned spacecraft. This is ironic because it was China that increased the amount of space debris by its attempt to exercise Anti-satellite (ASAT) capability in 2007. This White Paper did not reflect on its own action, but emphasized that the Chinese government is monitoring space debris and committed to mitigate it.

6. LEADERSHIP IN ASIA AND DEVELOPING WORLD

The dichotomy between global/national objectives leads the Chinese government to be aware of the need to cooperate with other countries and international organizations, but in fact the Chinese approach to international cooperation seems to be a hegemonic one as well as

---

[12] The Information Office of the State Council, China's Space Activities in 2006, October 12, 2006. http://news.xinhuanet.com/english/2006-10/12/content_5193446.htm

[13] China's space activities in 2011, December 30, 2011. http://www.chinadaily.com.cn/cndy/2011-12/30/content_14354558.htm

cooperative.

China represents itself as a leader of developing countries, and this self-image gives a certain aspect of hegemonic space policy. For example, the Chinese government has initiated the "Declaration on International Cooperation in the Exploration and Use of Outer Space for the Benefit and in the Interest of All States, Taking into Particular Account the Needs of Developing Countries"[14] stressing that international space cooperation should be based on equality, mutual benefit, peaceful utilization and common development. This policy suggests that China would provide its technical know-how and space-based services for developing countries, whereas major Western countries are not providing sufficient support. From China's perspective, developing countries are disadvantaged by their lack of industrial and technological capabilities, and they are not able to participate in the international space community. Thus, China positions itself as a leader of developing countries, helping to improve their technical capabilities and bringing them toward the capability level of the international space community. Malaysia can be seen as one example which benefited from its cooperation with China. Malaysia, as a member of AP-MCSTA and APSCO, has engaged in various programs that China offered for developing countries, and built its capability for space engineering. The most important element for Malaysia with regard to the cooperation with China is to develop manned-space capability because the Malaysian space agency has put high priority on sending Malaysians to space.

This hegemonic approach can also be seen in the Chinese endeavor to establish regional space cooperation organizations. In 1992, China, Pakistan and Thailand signed an MoU for establishing an international organization for space technology and application cooperation. This cooperation created an international space cooperation agency called AP-MCSTA (Asia Pacific Multilateral Cooperation in Space Technology and Applications) and attracted many developing countries in Asia, including South Korea, Iran, Indonesia, Mongolia, and Bangladesh. The objective of AP-MCSTA was to promote multilateral cooperation in space applications, but it was clear that China had the leading expertise, and many

---

[14] United Nations General Assembly, Fourth Committee "Declaration on International Cooperation in the Exploration and Use of Outer Space for the Benefit and in the Interest of All States, Taking into Particular Account the Needs of Developing Countries," A/51/590, 4 February 1997.

participating states became "students" of China[15]. In fact, teaching and training of human resources were the main objectives of AP-MCSTA.

It offered programs for remote sensing data analysis on environmental studies / protection, natural resource exploitation, as well as in disaster monitoring and prevention, which do contribute to the promotion of capacity building for the Asia-Pacific Region. The organization has held seven international conferences, attended mainly by engineers and scientists along with some policy makers, and has focused on exchanging information and holding workshops for various application programs.

Among those workshops, Cooperation in Small Multi-Mission Satellites (SMMS) was perhaps the largest and most important. SMMS was created by China, Iran, Republic of Korea, Mongolia, Pakistan, Thailand and Bangladesh for developing microsatellites for communications and remote sensing. China had previously been developing smallsat technology from the early 1990s by sending students to the University of Surrey, and several universities played incubator for such technologies brought back by those students. The SMMS project was, as a result, functioning as a technology transfer mechanism from China to other members.[16]

This Chinese leadership through AP-MCSTA has further developed by creating APSCO (Asia Pacific Space Cooperation Organization). In 2005, China, Bangladesh, Indonesia, Iran, Mongolia, Pakistan, Peru and Thailand signed the APSCO Convention, and Turkey joined in 2006.[17] Since the APSCO Convention entered into force only in 2008, it would be difficult to judge this organization based on its achievements. However, it is clear that China is exercising its leadership in this region (interestingly, while excluding Japan), and is creating opportunities for the member states of APSCO to develop their autonomous capabilities, which could then be launched on the Chinese Long March Rocket, which is heavily restricted in the international launch service market due to the American ITAR.

Another aspect of hegemonic approach is the bilateral cooperation with resource-rich countries. China has been collaborating with Brazil on the Earth resources satellite program, and following the successful launch of the China-Brazil Earth Resources Satellite 2 (CBERS-2) in October

---

[15] He Qizhi, "Policy and Legal Implications of Asia-Pacific Space Cooperation", in Chia-Jui Cheng (ed.), The Use of Airspace and Outer space for All Mankind in the 21st Century, Kluwer Law International, 1995, pp.49-56.

[16] See http://www.apsco.int/SMMS.aspx

[17] But Indonesia and Turkey have not yet ratified the Convention, thus the members of APSCO are limited to 7 countries.

2003, the Chinese and Brazilian governments signed supplementary protocols on the joint research and manufacturing of follow-on satellites, and on cooperation in a data application system, maintaining the continuity of data of CBERS and expanding the application of such satellites' data region-wide and worldwide. Obviously, this project aimed to penetrate Brazilian resource research, particularly the rich possibilities for offshore oil.[18]

In addition, China signed a commercial contract for a communications satellite with Nigeria, providing in-orbit delivery service in 2004, and subsequently with Venezuela, providing in-orbit delivery service and associated ground application facilities in 2005. These were commercial contracts, but they imply several important issues. First, China uses space technology as a bargaining chip for securing resource supply. There is a growing number of developing countries which need satellite communication capability for improving their national infrastructure, and China, promoting its self-image as the leader among developing countries, took advantage of its superior capabilities in space technology and offered its space system, including launch services, at a very low price. (It was said that the contract for developing, launching and operating satellites altogether costs only USD 250 million, which is considered to be about half a price of major satellite developing countries can offer.) Second, these contracts suggest that Chinese industrial capability has reached the level of the international commercial market. The increasing reliability of Chinese technology, together with the success of its manned-space program, enhanced the confidence of the Chinese space community that they could enter into the commercial market. Third, the contracting partners are not the allies of the United States or Western countries. Both Nigeria and Venezuela are, to some extent, anti-American countries, or countries where Western governments have difficulty penetrating due to their Human Rights records. Some may cast doubt that China is forming an "Anti-American Coalition," but it seems that China is trying to avoid penetrating countries where there are strong ties with the West, while satisfying its own needs for access to additional resources.

---

[18] Yun Zhao, "The 2002 Space Cooperation Protocol between China and Brazil: An Excellent Example of South–South Cooperation", *Space Policy*, Volume 21, Issue 3, August 2005, pp.213-219.

7. SHOOTING DOWN SATELLITES

Finally, we cannot avoid discussing the Chinese experiment at shooting down its own satellite in 2007. This action has little to do with space-based services, but this incident tells us much about how China regards space, and how the Chinese decision-making system works.

First of all, the Chinese decision to shoot down its aging weather satellite, FY-1, was taken by the military authorities without much consultation with the space community or the diplomatic corps, which explains why there were mixed messages from different Chinese authorities. The Ministry of Foreign Affairs, the "window for outside world," was not able to confirm or deny the fact that China took out its own satellite for some time following the act. The Ministry stressed that this was an act of peaceful use of space, since the action was not infringing with any international commitment. The reaction indicates that the Ministry of Foreign Affairs was not informed, and its policy explanation was not well prepared.

Second, the reason and rationale for the military to exercise its capability for shooting down the satellite was apparently a response to the US National Space Policy of 2006. From China's perspective, the US policy can be seen as a clear statement of the intent to maintain space dominance. Since China had already been excluded from the international launch market, it was natural for China to understand that the new US policy would be an extension of its intent to deny China entry into the international space community. Thus, Chinese military authorities considered that demonstration of its capability to shoot down the satellite would force the US to reconsider its position vis-à-vis Chinese space activities.

Third, the Chinese space community, in contrast to its military authority, was well aware of the negative impact of satellite destruction in orbit, and the subsequent spread of space debris. Since China is increasingly dependent on its space-based infrastructure and services, the civilian community clearly recognized that the creation of more debris would undermine the effectiveness of China's own space systems.

This is the reason why Chinese government has fiercely promoted the protection of the space environment and the prohibition of damaging space objects. In fact, China has actively participated in activities organized by the Inter-Agency Space Debris Coordination Committee, started the Space Debris Action Plan, and strengthened international exchanges and cooperation in the field of space debris research. It also proposed together with Russia, establishing a guideline for protection of space environment in 2007. China also strongly promotes prohibition of "weaponization" of space in PAROS (Prevention of an Arms Race in Outer Space) under CD (Conference on Disarmament). It seems contradictory for China to demand such international regulations, but it is understandable if we take into account that the civilian and military space communities in China at times lack coordination.[19] In any case, as the United States is strongly opposed to setting up a legal framework to restrict American activity in space, it would therefore be difficult to foresee any progress in this domain.[20]

8. Conclusion: It's not a Space Race, but Competition of Leadership

As discussed above, it should be considered misleading to suppose that there is a competition between Japan and China for improving their respective military capabilities in space. It is true that Japan has launched Information Gathering Satellites (IGS), and China is developing a lot of application technologies which might be useful for military space systems, but as we have seen, their space policies did not originate due to military demands. The space community, as well as public perception, has not been not associated with defense community from the beginning.

Also we have to be aware that space is only a tool for military activities. By treaty it is prohibited to place weapons of mass destruction in the orbit, and most countries have agreed to avoid the "weaponization" of space. China itself is the sponsor of the treaty to ban space weaponization in PAROS. While many countries use space for collecting military intelligence, communicating with troops, and sending positioning signals for missile navigation, it is typically the military forces on the ground that engage in military actions.

Also, we note that the ASAT test in 2007 made every nation, including China, even more aware of the danger of a space battle. The destruction of additional satellites would inevitably create larger amounts of debris, which would affect not only an enemy, but also their own satellites. Whatever the reason for shooting down a satellite might be, the

---

[19] Bruce W. MacDonald, *China, Space Weapons, and U.S. Security*, Council on Foreign Relations, Council Special Report (CSR) No.38, 2008.

[20] Theresa Hitchens, "The United Nations and Its Efforts to Develop Treaties, Conventions, or Guidelines to Address Key Space Issues Including the De-weaponization of Space and Orbital Debris," in Joseph N. Pelton and Ram S. Jakhu (eds.), *Space Safety Regulations and Standards*, Elsevier, 2010.

consequence of the action would be enormous and unfavorable to any spacefaring nations.

After all, any space activities are determined by the fairest laws, the Laws of Physics, which do not discriminate between civilian and military spacecraft, enemy and friend, or good and bad intentions. Based on this understanding, it is implausible that the Chinese ASAT would invoke overt military competition in space.

However, it is true that there is competition between Japan and China for regional leadership. The rivalry between APRSAF and APSCO is stimulating both Japan and China to use these organizations as vehicles for exercising their leadership. The unintended consequence of this leadership is the enhancement or empowerment of other Asian countries, as they are the beneficiaries of space-based services and technology transfer. Countries which have high ambitions for developing space capabilities now have easier access to high technology and possible international cooperation with experienced partners, and thus it can be said that this rivalry shall not be regarded as a space race, but as a healthy competition for providing public goods for the region.

•••

KAZUTO SUZUKI, PH.D.

Kazuto Suzuki is Professor of International Politics at Graduate School of Law of Hokkaido University, Japan. He graduated from the Department of International Relations, Ritsumeikan University, and received a Ph.D. from Sussex European Institute, University of Sussex, England. As an expert of space policy, he has been working as an advisor for the Space Development Committee of Liberal Democratic Party of Japan, Society of Japanese Aerospace Industry, and Mitsubishi Electric Co., and Senior Policy Researcher for JAXA. He also has been closely involved in the development of the Japanese space decision-making process including the establishment of the Basic Law for Space Activities of 2008 and Mid-term Plan for Space Activities of 2009. He is a member of the International Academy of Astronautics and the Chairman of the Space Security Committee of the International Astronautical Federation. He has worked in the University of Tsukuba from 2000 to 2008, and moved to Hokkaido University. He has conducted research on the International Political Economy perspective in Space Policy together with nuclear energy policy, in which role he contributed to the Independent Investigation Commission of the Fukushima nuclear accident. He has also conducted research on export control policy, science and technology policy, counter-terrorism, and policies on market regulation.

CHAPTER 12

THE EUROPEAN – AFRICAN PARTNERSHIP IN SPACE APPLICATIONS

CHRISTINA GIANNOPAPA
RESIDENT FELLOW, EUROPEAN SPACE POLICY INSTITUTE

© Christina Giannopapa, 2012. All Rights Reserved.

INTRODUCTION

The African continent, with over one billion people, has the fastest growing population in the world. The majority of the population lives in poverty, and the 53 states on the continent are still classified by the United Nations and International Monetary Fund as developing countries. They are facing problems in meeting the basic needs of their people in basics such as food, water, housing, and in healthcare and education.

Sustainable development in Africa requires access to data, information, knowledge and in all of these areas space technologies through the use of satellites can provide tremendous benefit. The transversal nature of space touches a wide range of policy issues, including environment, agriculture, health, security, education, and disaster management, and effective use of space assets can significantly facilitate growth and sustainable development across the continent.

Although there is much activity in the African continent regarding space applications, little is known about how space applications are utilized by African actors, and how cooperation between Africa and Europe is organized and conducted on various levels. Hence, this Chapter attempts to give such an overview.

It begins by discussing the various bodies involved in African-European cooperation, and the various African and European actors with a space interest in particular. It then examines the political, economic, social, technological and legal barriers to making greater use of space applications, before offering conclusions and outlook on how this situation can be improved, and space can be used more widely for sustainable development.

No other continent can benefit more fundamentally from space applications than Africa, and it will be through partnerships such as the one between Europe and Africa that will be instrumental for realizing this great potential and improving the living conditions and prospects of its vast population.

The Africa–EU Partnership

Africa and Europe as geographic neighbors share a long-standing history and traditions of cultural and social exchange. The cooperation at various levels between the European and African countries dates back centuries, but over the last decade it has been significantly strengthened and sealed by the political commitments of their leaders.

The first Africa-EU Summit took place in Cairo in 2000, setting the basis for a constructive dialog and joint action. Since then, with the establishment of the African Union (AU) in 2002, Europe has found an institutional partner, and over the last decade Africa and the European Union have developed common political strategies, policy documents, and action plans as a basis for their cooperation. These include the 2004-2007 Strategic Plan of the African Union [1], the 2005 "EU and Africa: Towards a Strategic Partnership" and the 2008 "The Africa-European Union Strategic Partnership" [2]. Subsequently, the First Action Plan (2008-2010) for the implementation of the Africa-EU strategic partnership was adopted in December 2007. The Third Africa-EU Summit took place in Libya on the 29-30 November and the Second Action Plan (2011-2013) was adopted [3,4].

The joint Africa-EU strategic partnership defines long-term policy orientations, based on a shared vision and common principles, on human rights, freedom, equality, solidarity, justice, the rule of law, and democracy. It intends to bridge the efforts of the Africans and the European Union and its Member States for development in the African continent through coordination of economic cooperation and promotion of sustainable development. The joint strategy provides the basis for a long term approach, and is implemented via short term (three year) Action Plans with concrete and measurable outcomes.

In order to finance the Action Plans, financial instruments are used such as [1]: the European Development Fund (EDF), the Development Cooperation Instrument (DCI), the European Neighbourhood Policy Instrument (ENPI), the Instrument for Stability, as well as the Thematic Programmes and through EU financial institutions, such as the European Investment Bank (EIB). Individual European member states provide also support when needed. From the African side, African financial institutions are involved, as is the African Development Bank (ADB).

In particular, the Strategy's First Action Plan, which was adopted in December 2007, outlines eight areas for strategic partnership with priority action for the period 2008-2010:

1. Peace and Security
2. Democratic Governance and Human Rights,
3. Trade, Regional Integration and Infrastructure,
4. Millennium Developments Goals,
5. Energy,
6. Climate Change,
7. Migration, Mobility and Employment,
8. Science, Information Society and Space Partnership.

The eighth explicitly mentions space, and it is recognized that space applications provide an essential platform for Africa's regional and continental sustainable development and can assist in tackling the problems Africa is facing including monitoring climate change. The priorities within this area are to: support the development of an inclusive information society in Africa; support science and technology capacity building in Africa and implement Africa's science and technology consolidated plan of action; enhance cooperation on space applications and technology..

The aim is to fully integrate space related issues in specific dialogues and cooperation initiatives in areas such as environment and resource management, climate change, peace and security. Various activities have been developed focusing on utilizing the potential of space applications to better manage natural resources, improve living conditions of populations, and promote sustainable development.

Areas of focus have been telecommunications to bridge the digital divide; use of the Global Monitoring for Environment and Security (GMES) for Africa for monitoring of climate change, desertification or fires, and water and food resources; use of the European Geostationary Navigation Overlay Service (EGNOS) and in the future Galileo for navigation applications. Integrated space applications have already been facilitating humanitarian aid operations and improving security of populations through integrated space applications.

Additionally, various projects have been focusing on space technologies and scientific applications as contributors to support creating knowledge based society for Africa. The main actors in for these activities and various projects mentioned in the action plan are the African Union Commission (AUC), EU, European Member States, the European Space Agency, EUMETSAT.

Figure 3 shows schematically the European and African Actors involved in space applications in Africa.

Due to the transverse nature of space there are also other areas which make use and benefit from space based information and satellite applications, such as Partnership 3 for infrastructures and Partnership 6 in climate change. In particular, climate change and space systems can contribute mainly in promoting climate observation for the African continent, enhancing links to global climate observatory systems, and strengthening climate-monitoring and forecasting capacities.

The Climate Information for Development in Africa Programme (CLIMDEV Africa) is one of the major programs implemented under this partnership, and recognizes the usefulness of space systems. It is a joint initiative of the AUC, United Nations Economic Commission for Africa (UNECA) and African Development Bank (ADB), and aims to find ways to overcome the lack of necessary climate information, analysis and options required by policy and decision-making at all levels.

Another project, the African Monitoring of the Environment for Sustainable Development (AMESD), is involved in the establishment of the necessary infrastructure for climate change information systems. This project is implemented by the AUC for Phase I, 2007-2012 and is aiming at strengthening the capacities of African institutions to use satellite-based

Earth Observation information for decision making in various environmental themes that are impacted by climate change (agriculture, land degradation, water management, etc). AMESD builds on the success of the preparation for the use of Meteosat Second Generation in Africa (PUMA) programs and paves the way for the Global Monitoring System for the Environment and Security (GMES) in Africa. The program is being executed in close cooperation with the Regional Economic Communities (RECs), EUMETSAT, UNECA, World Meteorological Organisation (WMO), United Nations Environment Programme (UNEP), European Commission (EC), EU and African Member States.

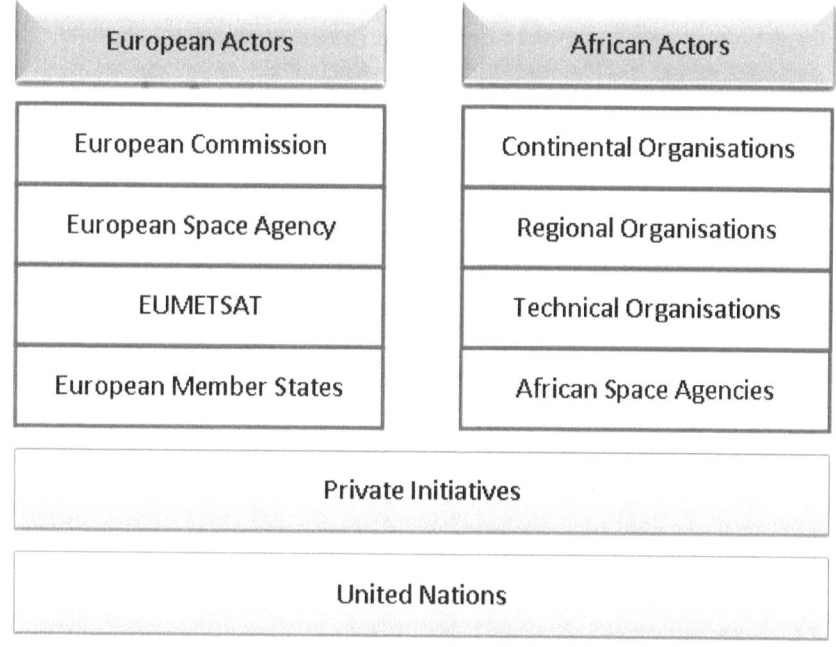

Figure 3: Main African, European and International Actors involved in space applications.

The Strategy's Second Action Plan covering the period 2011 to 2013 was adopted in November 2010 and maintains the eight areas for strategic partnership already identified by the First Action Plan. Under Partnership 8, where explicit reference is made for space, the progress made under the First Action Plan has been assessed, and during the third Africa-EU Summit and it was agreed to launch a high-level Science and Technology policy dialogue between senior officials at Ministerial level. Additionally, the Summit agreed that for the implementation of the future activities it is essential to reinforce cooperation with the RECs, and coordination will take the form of meetings to take place every six months between the AU and the RECs. In Partnership 8 it is recognized that in order to achieve the Millennium Development Goals (MDGs) it is important to raise science and technology capabilities and use of Information and Communication Technologies (ICT) and applications.

AFRICAN ACTORS

The African actors include organizations that operate at the continental level such as the African Union (AU), regional organizations typically the Regional Economic Communities (RECs), technical organizations such as specialized agencies and institutes in different areas and the African space agencies. Figure 4 shows the various policy relevant actors in Africa and the matrix relationship for cooperation between them.

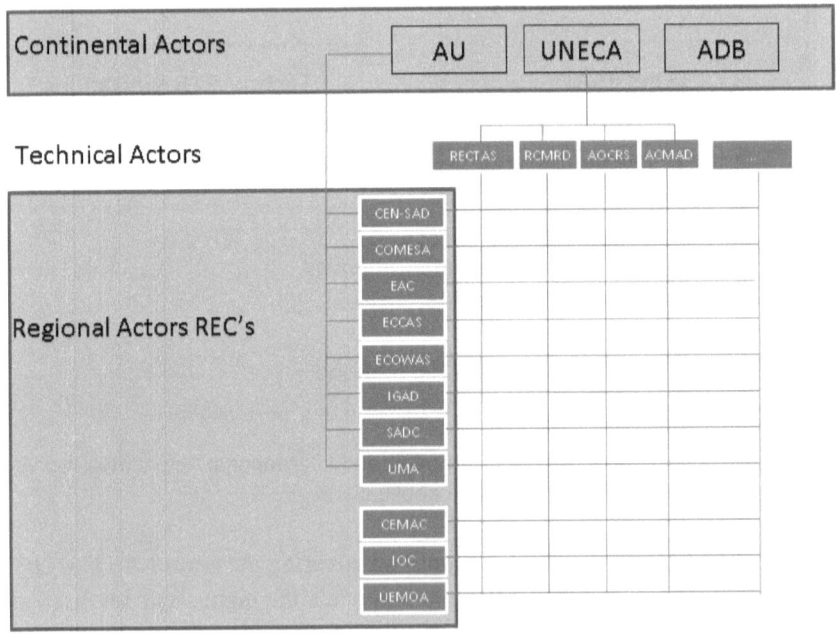

Figure 4: Policy Relevant Actors in Africa

CONTINENTAL ACTORS

At the continental level the African Union (AU), the United Nations

Economic Commission for Africa (UNECA) and the African Development Bank (ADB) have a joint secretariat. These three institutions have distinct roles. The AU has the political mandate, the ADB has the financial mandate and the UNECA is responsible for economic issues. The three institutions largely operate independently, so the role of the secretariat is to assist in coordinating their actions and see how best to advocate for common policy positions.

Historically, the predecessor of the AU is the Organization of African Unity (OAU). Four Summits led to the establishment of the AU on July 9, 2002. In this process there has been the establishment of institutions such as the African Central Bank, the African Monetary Union, the African Court of Justice and especially, the Pan-African Parliament. Additionally, the Regional Economic Communities (REC's) have been strengthened and consolidated as the pillars for achieving the objectives of the African Economic Community and realizing the envisaged Union. The African Union Commission (AUC) is the Secretariat of the Union with executive functions. It represents the Union and protects its interests under the auspices of the Assembly of Heads of State and Government as well as the Executive Committee (See Figure 5). The headquarters of the Commission are located in Addis Ababa, Ethiopia. The role of the AUC is similar to its counterpart in Europe - the European Commission (EC) of the European Union (EU). The AUC consists of chairperson H.E Dr Jean Ping from Gabon, his deputy H.E. Erastus Jarnalese Onkundi Mwencha from Kenya, and a number of commissioners dealing with different policy areas. The space portfolio falls under the Commissioner for Human Resources, Science and Technology H.E. Dr Jean Pierre Onvehoun Ezin, from Benin. The AUC initiated the establishment of a pan-African University, which is a network of African higher education and research institutions, with thematic hubs in each of the five geographic regions of Africa (Eastern, Western, Central, Southern, and Northern). The space discipline has been fostered in South Africa. The first four thematic institutes are expected to be launched in 2011.

Many agencies, departments and programmes of the United Nations are active in Africa. A selected list includes the following:

- UN Secretary General
- UN Security Council
- UN Special Adviser on Africa
- UN Cartographic Africa Renewal
- UN African Missions
- Climate Change-Copenhagen

- FAO, Sustainable Development
- And many others.

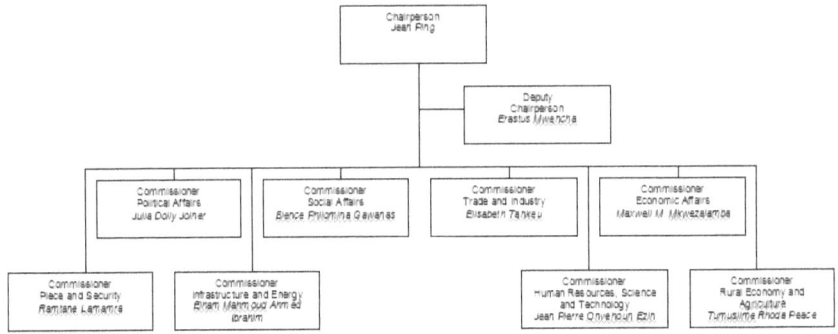

Figure 5: African Union Commission

Within the United Nations family of specialized agencies, the FAO (Food and Agricultural Organization), the ITU (International Telecommunications Union), WMO (World Meteorological Organization), IMO (International Maritime Organization) and UNESCO (United Nations Education, Scientific and Cultural Organization) have been actively involved with satellite remote sensing technology, space communications, satellite meteorology and space science as it pertains to Africa.

The United Nations Economic Commission of Africa (UNECA) is one of the UN's five regional commissions. UNECA plays a very active role in African development, particularly in the area of ICT and Science and Technology. However, UNECA does not have a specific space program; space technologies and their applications are part of other programs. The choice of activities it undertakes is based on needs expressed by Member States and the RECs. In that context, the focus of UNECA service delivery is at the regional and sub-regional levels with its five Sub-Regional Offices (SROs). The SROs serve as vital links between policy-oriented analytical work generated at headquarters and policy making at the sub-regional level. UNECA also provides technical assistance and policy advice to African countries and the RECs. Under its auspices the technical bodies RECTAS, RCMRD, AOCRS and ACMAD were created to achieve the technical implementation of various projects.

Although they do not have specific space programs, some organizations and programs of the UN use satellite-provided information and applications to achieve their goals. The World Food Programme (WFP) is the world's largest humanitarian agency that fights hunger around the world, and is very active in Africa. Satellite and radar imagery provides the WFP and its partners with valuable information regarding climate change, early warning, disaster management and emergency response. It can also provide information for food security. The UN Food and Agriculture Organization (FAO) has the mandate to raise levels of nutrition, improve agricultural productivity, better the lives of rural populations, and contribute to the growth of the world economy. The UN Environment Program (UNEP) has the mandate to coordinate the development of environmental policy consensus by keeping the global environment under review and bringing emerging issues to the attention of governments and the international community. The UN Human Settlements Program (UN-HABITAT) is the UN agency responsible for human settlement, and focuses on promoting socially and environmentally sustainable towns and cities, with the additional goal of providing adequate shelter for all.

The World Meteorological Organization (WMO) coordinates environmental satellite matters and activities, and provides guidance on the potential of remote-sensing techniques in meteorology, hydrology and related disciplines and applications. The United Nations Platform for Space-based Information for Disaster Management and Emergency Response (UN-SPIDER) aims at ensuring that all countries and international and regional organizations have access to and develop the capacity to use all types of space-based information to support the full disaster management cycle. UN-SPIDER has been involved in a number of initiatives that make space technologies available for humanitarian and emergency response.

In order to avoid fragmentation in the UN actions, since January 2007, eight countries have been piloting approaches to delivering as "One" according to the circumstances of the countries. Among these, four are in Africa: Cape Verde, Mozambique, Rwanda, Tanzania.

REGIONAL ACTORS

From the 1960s to the 1990s, African nations created regional economic communities targeting economic cooperation and integration. Today there are numerous regional blocks in Africa, also known as Regional Economic Communities (RECs), which facilitate mutual economic development among various African states. Many functioned for some years, and then after decades were revised and re-established. Africa has the highest number of such cooperation and integration agreements. There is no country in Africa that does not belong to at least one grouping, and most belong to two or more of these groups.

Africa's regional integration was given a boost in 1991 by the adoption of the Abuja Treaty, which established the African Economic Community (AEC) with the objective "to promote economic, social and cultural development and integration of African Economies in order to increase economic self-reliance and promote an endogenous and self-sustained development."[1] The foundation of the Treaty states that the African Economic Community must be established largely through the coordination, harmonization and progressive integration of the activities of the RECs.[2]

TECHNICAL ACTORS

In Africa there are various technical actors including the space agencies of some countries, the technical institutions which have been created under the auspices of UNECA, as well as numerous other local technical centers, institutes, organizations, and associations.

Five African countries have their own space agencies:

1. Algeria (Agence Spatiale Algérienne ASAL),
2. Egypt (National Authority for Space Science and Remote Sensing, NARSS),
3. Morocco (Centre Royal de Télédétection Spatiale, CRTS),
4. Nigeria (National Space Research and Development Agency, NASRDA) and
5. South Africa (South Africa National Space Agency SANSA).

Discussions about the establishment of a continental institution which would work in cooperation with the AU date back to 1991, and the AU project for the establishment of an African Space Agency was finally launched in 2010.

Various technical institutions act as the technical executive for

---

[1] OAU, Abuja Treaty, Article 4.
[2] The REC's that have become pillars of the African Union are: Community of Sahel-Saharan States (CEN-SAD), Common Market for Eastern and Southern Africa (COMESA), East African Community (EAC), Economic Community of Central Africa States (ECCAS), Economic Community of West African States (ECOWAS), Intergovernmental Authority on Development (IGAD), South African Development Community (SADC), and Union du Maghreb Arabe (UMA). The other REC's, often described as sub-regional economic communities (SECs), are: Communauté Économique et Monétaire de l'Afrique Centrale (CEMAC), Indian Ocean Commission (IOC), Union Economique et Monétaire Ouest Africaine (UEMOA), Communauté Economique de Pays des Grands Lacs (CEPGL), Mano River Union (MRU), Southern Africa Customs Union (SACU), West African Monetary Zone (WAMZ).

various projects under the auspices of UNECA. They cooperate in a matrix-like relationship with UNECA, the AU and the RECs. These are:

- Centre for Training in Aerospace Surveys (RECTAS), focusing on geo-informatics, remote sensing, geographic information systems and cartography,
- The Regional Center for Mapping of Resources for Development (RCMRD), focusing on geo-information and information technology,
- The African Organization of Cartography and Remote Sensing (AOCRS), focusing on cartography and remote sensing and
- The Centre of Meteorological Application for Development (ACMAD), focusing on information for weather, climate and environment.

European Actors

The main European Actors highlighted for the African-EU partnership that are mentioned in the first and second Action Plan are the European Commission, the European Space Agency (ESA), EUMETSAT, and the European Member States. Apart from these, there are also many NGOs that are operating in African and various private bodies and companies.

The European Commission (EC) under various directorates is working very closely with the African Union Commission (AUC). With the establishment of the European External Action Services (EEAS) the role of the EU delegations in Africa and RECs capitals has been strengthened. In particular, the permanent delegation of the EU to the AU in Addis Ababa has an important role in coordinating EU policy and action relating to the AU, helping and supporting the AU in the areas outlined in the Africa-EU strategy. There are numerous projects in Africa, including two very important projects in relation to space that are associated to the EU's space Flagship Programs, Galileo together with the European Geostationary Navigation Overlay System (EGNOS), and Global Monitoring for Environment and Security (GMES).

EGNOS is Europe's precise positioning and timing system, augmenting satellite signals currently using the GPS signal. In the future they will use Galileo to combine space, ground and control system components which can potentially cover all of Africa to provide augmented navigation signals supporting aviation, maritime, road, farming, mining, energy, land planning, and search and rescue efforts. EGNOS has already been tested in cooperation with the Agency for Aerial Navigation Safety in Africa and Madagascar (ASECNA) for the aviation sector, demonstrating benefits including safer landings, increased efficiency, more reliable air services for passengers, and better access to remote regions of Africa.

The GMES and Africa initiative was initiated by the Lisbon declaration on GMES and Africa (2007) in response to the Maputo Declaration (2006). It aims at strengthening the capacities and developing infrastructure for the intense and coherent exploitation by African users of Earth Observation (space and in-situ), data, technologies and services in support of the environmental policies for sustainable development in Africa. The GMES and Africa Action Plan was expected to be adopted by the end of 2011.

The European Space Agency (ESA) promotes cooperation in space research, technology and applications, and is broadly involved in international cooperation with other spacefaring nations and also developing regions including Africa. Two important projects strengthen the cooperation between Africa and ESA, which are the GMES and TIGER initiatives. The TIGER initiative started in 2002 in response to the urgent need for action stressed by the Johannesburg World Summit on Sustainable Development (WSSD) regarding water resources. It is an ESA-funded project aimed at helping Africa through the use of Earth observation technologies to overcome problems in collecting, analyzing and disseminating water-related geo-information.

EUMETSAT's main activity to deliver weather and climate-related satellite data provides support in developing countries including those in Africa. Since 1996, EUMETSAT has been active in Africa, cooperating with ASECNA in training activities on satellite meteorology in Africa. With the regional center AGRHYMET, which focuses on food security and increasing agricultural production, it cooperates to foster the use of satellite data for operational and development activities, including rainfall estimates, training and agro-meteorological and hydrological applications. It cooperates also with ACMAD for the use of METEOSAT data and in the PUMA project. In 2008 it signed a cooperation agreement with the African Union Commission regarding the African Monitoring of the Environment for Sustainable Development (AMESD) program, which focuses on extending the operational use of Earth observation technologies and data to environmental and climate monitoring applications. EUMETSAT also participates in the GMES initiative and in a user forum for Africa to reinforce the well-established dialogue with the African user community in order to optimize the use of EUMETSAT satellite data.

Many European member states also have direct cooperation agreements or memoranda, and are funding projects directly with various African countries. This cooperation generally occurs without the involvement of the United Nations, the European Union or any other intergovernmental organization residing in Europe. Here we will provide four examples among the many existing agreements and projects.

An intergovernmental agreement between France and Algeria was signed on February 1, 2006, which was implemented by the French National Centre for Space Studies (CNES) and the Algerian Space Agency (ASAL). It aims to support and promote scientific, technical, industrial and commercial cooperation between the two states in the study and use of outer space for peaceful ends.

In 1995, Germany's Aerospace Centre (DLR) signed a framework agreement with the Council of Scientific and Industrial Research (CSIR) of South Africa, focused on cooperation in research and technology, and has been used since then as a basis for additional forms of cooperation.

Italy has had agreements with the government of Kenya since 1964, and maintains a presence in Malindi, Kenya, where the San Marco Base has been set up. The "Agreement between the Government of the Republic of Kenya and the Government of the Republic of Italy concerning the Satellite Tracking and Launching Station at the San Marco Malindi, Kenya" was signed in Nairobi on March 14, 1995, and was subsequently renewed until June 30, 2012; as of this writing it is currently under re-negotiation. Under Italian Legislative Decree 128/03, as amended by the Statute approved under Legislative Decree 213/09 the responsibility for managing the base has been entrusted to the Italian Space Agency (ASI) since January 1, 2004. Previously, the responsibility resided with the University of Rome "La Sapienza," through the CRSPM (Centro Ricerche Progetto San Marco). The location of the San Marco Base, which has been recently renamed "Luigi Broglio Space Center (BSC)," on the Indian Ocean is ideal for launch and support of equatorial satellites that provide TT&C activities and acquire Earth observation data over Central and Eastern Africa.

In 2006 the United Kingdom entered into a Memorandum of Understanding between the British National Space Centre (BNSC) and the Algerian Space Agency (ASAL). This MOU provides a framework of collaborative activities and review areas of common interest in the civil aspects of space, and facilitates the interchange of information, technology and personnel in areas of mutual interest. The United Kingdom had previously cooperated with Algeria on its national satellite AISAT-1.

THE AFRICAN ENVIRONMENT

To assess the environment under which the African-EU partnership in space applications is realized, we examine the issues that facilitate or impede progress in political, economic, social, technological and legal factors.

POLITICAL FACTORS

Although the political commitment of African leaders to work together and with other international partners to solve problems for African citizens, and to achieve societal, economic and political integration, is being realized step by step through the African Union and various Regional Economic Communities (RECs), the political commitment for space and satellite applications remains lacking, which seems largely to be due to the fact that there is insufficient understanding of the benefits, and a lack of communication of the success stories.

Consequently, space policy remains uncoordinated at the continental level, as well as at the regional level. Further, there are insufficient links generally between the space initiatives and policies and other policy areas, including agriculture, environment, health, transport, and security, and a broad understanding of how space can assist in achieving policy objectives is missing. At the continental level the appropriate institutional mechanisms exist for realizing the benefits of space, there is often insufficient coordination between various operational institutions.

ECONOMIC FACTORS

During the past 10 years there have been significant efforts by the RECs to integrate the various markets, but many of them are still fragmented. The African space market is mostly dominated by government activities largely because low income levels and limited economies of scale make Africa unattractive for Foreign Direct Investment. License fees for satellite systems are generally high, and coupled with the high cost of satellites, this makes it difficult for companies to survive in a limited market environment. Hence, there is a need for low cost systems to facilitate improved access to expand the market for African entrepreneurs and non-African companies.

SOCIAL FACTORS

Space applications can assist in providing solutions to the African population addressing basic needs such as food and water security, health care, education, early warning, disaster management and emergency response, etc. Nevertheless, the benefits of space applications are not sufficiently understood by decision makers as well as the wider population, and there are few people educated in the management and operation of space-based assets. Various space projects in different areas have been developed for Africa but very few are sustainable beyond the pilot phase because often the local community of the end users is not involved from the beginning and does not have the feeling of ownership.

In many cases, appropriate government bodies have not been identified to take responsibility for running and maintaining a project and its maintenance, and as the projects developed in Africa are often conceived by developed countries without full awareness of the infrastructure restrictions of the under-developed countries.

TECHNOLOGICAL FACTORS

Most African countries focus their space activities in three main categories, infrastructure development, use of satellite applications, and capacity building. The majority of the projects focus on space applications, but there is clearly a need for more capacity building in order to improve local technical skills. A Pan-African University is currently being established, and space will be covered by the South Africa unit.

LEGAL FACTORS

African markets for telecommunications and satellite imaging are fragmented, and license requirements and regulations vary from country to country. The current regulatory environment hinders private initiatives by African and foreign investors, and thus delays the benefits that space systems could offer in support of sustainable development. Furthermore, as noted, there is little coordination between various policy and regulatory frameworks in most African countries regarding space and the space component of other sectors, and it often happens that different ministries purchase the same information assets from satellites, a duplication that increases administrative and economic overhead. Current taxation systems on imports can affect the development of satellite applications since nearly all equipment, devices and services need to be imported. It also appears that taxation is duplicated for the same goods and services at various levels resulting in accumulative excessive burden for the importer.

Conclusions and outlook

Our analysis of the key factors leads to broad recommendations for five key areas.

1. The discussion of political factors indicates that the needs of African states have to be translated into concrete topics and goals where space can assist in meeting them, and then it's necessary to identify the responsible actors, and establish links with the appropriate governments.

2. The national needs of the African countries that do have existing capabilities in space competence need to be better reflected in other national policies in areas such as agriculture, environment, health, transport, and security in a coherent manner. Coordination should be achieved at three levels, national, regional and continental. This can be facilitated by coordination meetings, workshops and conferences.

3. The AU, ECA and REC's are important actors in Africa, and coordination among them regarding space based information and applications should be strengthened and formalized.

4. In addition, the development of a pan-African space policy would streamline the current fragmented policy environment and support all African countries in working together at the regional and continental levels to tackle their common goals. One area in which there is a vital need is disaster management and emergency response, and this would make a suitable first topic for a pan-African space policy. This topic is affecting regionally a number of countries and joining efforts would significantly contribute to sustainable development and poverty eradication.

5. For the implementation of policies, formalized institutional relations for space based information and applications are needed. The first point of coordination should be access, use, and sharing of space based information. In particular, maps and GIS would be the first step to formalize inter-actions and institutionalize use of these assets on the regional level. In this area an early focus should be in early warning and emergency response for disaster management.

Over the long term this should lead to institutionalization at the continental level. The establishment of pan-Africa institutions could take the form of a Space Agency that would be beneficial for centralized purchase of space-generated data including maps and satellite images. This should be a long term plan starting first at national and regional levels, as different countries have different needs in the short term, as well as differing levels of development and capacity to use this information.

Success stories concerning the use of satellite applications for the benefit of African citizens and for sustainable development need to be communicated in way that political figures are able to build support. Relevant information needs to be prepared by consortia that would include representatives from technical experts, end users, entrepreneurs, administrators, societal and political groups, etc.

The current institutional framework in Africa is appropriate but concrete links of cooperation need to be established between the main African actors like the AU, UNECA and the REC's. Common strategic partnerships between them are essential to support sustainable development. Adequate resources in the form of funding, personnel and technical expertise are essential, and funds provided from the United Nations, European member states, and various private groups should be better aligned in terms of funding, time scale, and expected targets in order to better utilize the available resources, avoid duplication, and create complementarities. The funding mechanisms should also be aligned such that they cover the entire cycle of a satellite applications project, from pilot project to hand-over, operation, capacity building, and ongoing maintenance.

The improvement of the economic regulatory frameworks to support foreign direct investment will also make the African countries far more open, permitting profit repatriation and providing economic measures and other incentives to attract investment in the space sector and other sectors as well.

In order to enhance awareness of the social benefits that space applications can bring it is necessary to increase communication. Appropriate information mechanisms need to be set up to communicate the benefits at the local level, and education regarding space and its applications should be enhanced in schools and universities, supported by learning materials on the benefits of space and its contribution to development should be provided. Educational programs are of course long-term measures with many social benefits. The pan-African University, currently under establishment, is expected to play an important role in the development of sustainable development of Africa.

From a legal perspective it's important to implement coordination with policies, harmonize licensing procedures and improve taxation laws. Regarding the implementation of appropriate regulations, it is necessary to foster initiatives to build institutional capacity in making treaties, setting standards, formulating policy, drafting regional integration protocols and alignment of regulations at a pan-Africa level. African governments should coordinate their policies and regulations regarding commercial and non-commercial space applications (e.g., data policy). They should foster policy and regulatory harmonization to create larger common markets based on regional economic communities while increasing private sector participation. This would facilitate exchange of assets and reduced duplication in purchasing the same asset. With regard to the purchase of information from satellites, it is important to facilitate cross border exchange and harmonize information infrastructure, to minimize cost and maximize the relevant benefits. Initiatives should be promoted for investment in access and cross-border information infrastructure through effective partnerships between public, private and not-for-profit sectors, in order to achieve universal access and full inter-country exchange of information and data.

Aligning licensing procedures would remove administrative bottlenecks, which suggests that common licensing and authorization procedures is preferable to the current situation of differing rules country by country. An internet-based license application and dissemination of regulatory information would significantly assist potential investors.

In addition, a more coherent taxation system would enhance growth, so African countries are being encouraged to accelerate the conclusion of double taxation treaties (DTTs), which can make it more attractive for foreign investors by helping them to avoid paying taxes twice on the same transaction.[3] Taxation is a leverage to support demand for satellite services in terms of tax-emption or negative taxation – financial support - in order to improve citizens' rights as members of society (freedom of self-expression, right to be informed, freedom of press, etc).

---

[3] The majority of African countries have signed multilateral agreements dealing with the protection of FDI, such as the Convention establishing the Multilateral Investment Guarantee Agency (MIGA) and the Convention on the Settlement of Investment Disputes between States and Nationals of Other States.

Actor	Proposed Actions
European Union (European Commission and other European institutions)	<ul><li>Strengthen its neighbor policy towards Africa focusing on Space as a strategic leverage crossing different thematic issues, such as extension of infrastructure, extra-EU transport, energy, industrial development and trade, etc</li><li>Enhance thematic coordination of European activities of European Actors</li><li>Enhance coordination of EU financial instruments, European Investment Bank and Member States' contributions</li><li>Establish a closer dialogue with the UN agencies present and active in Africa and coordinate funding</li><li>Establish direct dialogue with Regional Economic Communities in Africa, the eight pillars of the AU and the others</li><li>Promote the use of space in other areas of the strategic partnership for the 2011-2013 Action Plan</li><li>Promote the benefits of space for Africa's sustainable development to African politicians through the High Representative of the European Union for Foreign Affairs and Security Policy and External Action Service representatives in African countries</li><li>Regularly maintain an overview mapping of funded projects, bilateral agreements, best practices, etc.</li><li>Assist the African Union with obtaining adequate resources for its targets</li><li>Move towards a co-funding (Africa-European) system with African states and the Africa Development Bank to increase sustainability</li><li>Perform assessments of public funding activities and share best practices with European Member States</li><li>Implement better mechanisms for follow-up impact assessment of pilot projects</li></ul>

EU Member States	• Improve coordination between national activities in Africa and EU activities • Maintain an overview of national activities regarding space based information and applications for Africa • Perform assessments of public funding activities and share best practices with other European Member States and the EU
European Space Agency	• Ensure that the actors involved in ESA projects properly deal with the transition from pilot projects to operational projects and assure the commitment of African government relevant bodies to continuation and local acquisition in terms of knowledge end technology transfer • Promote the coordination of space efforts through African space agencies for the space faring nations of Africa and through other governmental and technical bodies where space is relevant for those that are users of space and those that are not • Participate in education regarding the use of space at different levels
EUMETSAT	• Promote coordination of space efforts on a regional level through its links with African government bodies and technical agencies • Increase support for capacity building efforts in Africa • Enhance the role of Africa to acquire data in-situ in order to merge with satellite data
United Nations	• Promote the use of space at the level of decision makers and through UNECA for the sustainable development of Africa • Coordinate the efforts between the different UN agencies and programs regarding space use as "one UN" • Establish in UNECA a specific department for space and coordinate and support the RECs in this area • Establish coordination mechanisms with the EU regarding space activities in Africa • Perform cost benefit analysis of the use of space as part of awareness campaigns in order to underline shadow-prices and cross-sectional benefits • Increase coordination of activities through the joint

	secretariat between the AU, UNECA and AfDB
African Union	• Enhance its international relationships in order to show local needs that could be satisfied through space technology
• Enhance its role as a catalyzer of Africa's Vision for Space. The main objectives of space policy for Africa would be to focus on implementing those policy aspects that would allow successful use of space based information and applications for sustainable development
• Increase the resources of the AUC in relation to space
• Establish a position responsible for space as a shared competence among African Member States which deals with space topics such as telecommunications, navigation, remote sensing, etc and is responsible for maintaining a database of African capabilities and projects and promoting the use of space for sustainable development
• Encourage Africa entrepreneurship
• Promote a bottom-up approach to projects
• Establish closer coordination with the RECs and establish policies that cover regional needs and identify how space can assist in development
• Develop a capacity-mapping database at the continental level (experts, institutions, data, information, infrastructure, etc.)
• Map the different ongoing activities conducted by regional and international agencies and identify end users
• Collect in a systematic way and disseminate case studies and lessons learned, promoting the benefits of space with arguments to inform politicians and assist them in supporting a case for space
• Promote cost benefit analysis of projects that use space assets
• Increase coordination of activities through the joint secretariat between the AU, UNECA and AfDB
• Develop mechanisms for increasing co-funding activities with African capital
• Promote education on space at all levels of education: primary, secondary, tertiary and continuous learning

	(e.g., school, university, on the job training, etc.) • Perform studies on the economic benefits of space for African development • Promote the implementation of space policy and space components in other policies for capacity building at various institutional levels and different actors (end-users, decision makers and trainers) • Ensure sustainability by continuous training to mitigate "brain drain" • Prepare and conduct workshops for key decision makers
Regional Economic Communities	• Promote a fair investment climate in order to attract FDI for development and deployment of space • Take space up as a topic for sustainable regional development • Create a map of existing mechanisms and capacities in the region (experts, institutions, data, information, infrastructure, etc.) • Conduct an evaluation of the use of space from different mechanisms in the region • Promote space based information and applications as a key development component for sustainable development • Promote the development and adoption at the regional level of a space policy reflecting the needs of the regions • Open intra-regional dialogues for continental integration • Perform cost benefit analysis of the use of space for development and comparison with no use of space. A particular case would be disaster management and disaster forecasting • Establish regional centers for single point acquiring of data and maintaining a common database • Promote regional development in download capabilities, information processing and capacity building • Facilitate the establishment of a critical mass of providers and users at the regional level • Organize regionally early warning and emergency response in disaster management and coordinate with UN efforts • Facilitate cooperation between universities, schools, technical centers and the local community for training

	on space assets and sustainability • Coordinate with member states to harmonize appropriate rules and regulations, licensing procedures, taxation etc. regarding space • Facilitate horizontal cooperation to establish institutional groups • Strengthen multi-institutional mechanisms

African Member States	• Develop a map of national needs and space relevance
	• Develop a map of national capabilities (experts, institutions, data, information, infrastructure, etc.)
	• Facilitate local activities and commitment to build up capacity and avoid reliance on foreign actors
	• Conduct an evaluation of the use of space from different mechanisms in the country
	• Explore ways to promote and benefit from public-private partnerships
	• Facilitate courses in space based information and applications and recognize such professionals in civil service positions
	• Coordinate with other member states to harmonize appropriate rules and regulations, licensing procedures, taxation etc. regarding space
	• Strengthen multi-institutional mechanisms
	• Coordinate through the RECs to develop a common strategy on needs and how space can provide solutions and share costs for data acquisition, processing and capacity building
	• For passive users and active users[4]
	   o Identify other policy areas where space could be beneficial
	   o Promote the development and adoption at the regional level of a space policy or space policy component in other policy areas reflecting the needs of the nation
	   o Establish a centralized national approach and

---

[4] *Passive users* are African countries that do not have any space capabilities. They only receive passively information already processed by others.
Active users are Africa countries that have the capacity to process information offered.
Active developers: are those African countries that themselves have the capacity in space activities and typically have a space agency and more advanced space policy components either self-standing or as parts of other policies.

	creation of centers for acquiring, distributing and maintaining databases of space information to avoid duplication of acquisition ○ Promote the implementation of space policy and space components in other policies for capacity building at various institutional levels and ensure sustainability by continuous training to mitigate "brain drain" ○ Promote on-the-job training ○ Facilitate communication between information providers and users in the country to bridge the gap between the two communities. • For active developers ○ Promote cases of success to other countries and assist them in development ○ Develop mechanisms to attract FDI
African Space Agencies	• Coordinate their efforts on a regional level • Establish links with other technical bodies in Africa that could benefit from space • Establish links with academic networks • Coordinate their efforts on a continental level • Strategically inform and assist in making politicians and decision makers aware of the benefits of space and provide them with arguments and cases • Identify a strategic plan for international relationships with European entities such as EU, ESA and EUMETSAT
African Technical Organizations	• Strengthen existing networks and create new networks for cooperation and coordination of activities between the technical actors • Strengthen the existing institutions to be able to take up new technologies • Map and monitor the current local degree of technical feasibility for space • Focus on GIS and mapping activities • Strengthen capacity building in radar and image processing technologies • Explore opportunities to acquire information via African receiving stations • Promote and lobby professionals in the fields of space

- based information and applications
- Increase the involvement of the local community
- Strategically inform and assist in making politicians and decision makers aware of the benefits of space and provide them with arguments and cases
- Facilitate data sharing among institutions

This review of African actors and activities regarding space utilization and satellite applications may come as a surprise to those who see Africa as a continent that is not engaged in technology development. Africa, to the contrary, is a continent on the rise, and while it will be still some time in the future before Africa will see full-fledged space powers (either nations or regions), it will steadily increase the use of space applications for economic and societal development. The first "African" International Astronautical Congress, which took place in late 2011 in South Africa also represents a positive trend.

Long term success will depend on a strong and sustained effort from all actors to reach agreed upon and still more ambitious goals. Space applications as an instrument for development have a particularly strong potential in Africa, and in partnership with existing spacefaring nations and alliances this will become reality.

•••

CHRISTINA GIANNAPAPA, PH.D.

Christina Giannopapa has been Resident Fellow at the European Space Policy Institute (ESPI) since January 2010. Prior to joining ESPI, she had ten years experience in engineering. From 2007–2009, she served as Technical Officer for the European Space Agency (ESA), where she was responsible for overseeing projects in the field of life and physical science instrumentation. Previously, she held positions in academia in Eindhoven University of Technology, the Netherlands, where she currently holds an Assistant Professor position.

She has worked as a consultant to various high-tech industries in research and technology development. In policy she worked briefly in DG Research, European Commission. In her academic years she has received various academic scholarships and has numerous publications in peer reviewed journals. She holds a Ph.D. in Engineering and Applied Mathematics from the University of London, UK and an MEng in Manufacturing Systems Engineering and Mechatronics, University of London, UK. Additionally, she has attended professional education in Law and International Management.

REFERENCES

This Chapter is based on research undertaken by the author for ESPI Report 26 "European-African Partnership in Satellite Applications for Sustainable Development. A Comprehensive Mapping of European-African Actors and Activities," September 2010, download at www.espi.or.at, and the article "Improving Africa's benefit from space applications: The European-African parternship", Space Policy 27 (2011), 99-106. Updates have been made, in particular in view of the Third Africa-EU Summit, which took place in November 2010.

[1] African Union, "Strategic Plan of the Commission of the African Union, Volume 3: 2004-2007 Plan of Action. Programmes to speed up integration of the continent", Addis Ababa, Ethiopia, May 2004.

[2] Council of the European Union, General Secretariat, "The Africa-European Union Strategic Partnership", June 2008, Brussels, ISBN 978-92-824-23752.

[3] Council of the European Union, "14th Africa- EU Ministerial Meeting", Luxembourg, 26th April, 9041/10 (Press 92).

[4] African Union, "Joint Africa EU Strategy Action Plan 2011-2013", Tripoli, Libya, November 2010.

[5] "Strategic Plan of the African Union Commission", Volume 1: Vision and Mission of the African Union, May 2004, African Union.

CHAPTER 13

European SatCom Policy:
A tool of international cooperation between Europe and Africa

VERONICA LA REGINA
SENIOR RESEARCHER, INTERNATIONAL INSTITUTE OF SPACE COMMERCE,
DOUGLAS – ISLE OF MAN

© Veronica La Regina, 2012. All Rights Reserved.

INTRODUCTION

The purpose of this chapter is to explore how European Satellite Communications (SatCom) policy can accelerate African development. One of the current European policy priorities is the Digital Agenda, with the ambition to bring broadband to every European citizen by 2013. This policy initiative promotes European growth as smart, sustainable and inclusive. The targets of the Digital Agenda can be pursued by different technologies, including SatCom, the advantage of which is to assure a uniform quality of service over a huge area of coverage. Due to their geographical proximity, an appropriate SatCom solution that assures broadband for Europe can also serve a large part of Africa.

The last European Space Council in 2010 mentions the partnership on space with Africa as a way to support sustainable development of African continent and also in support of the Millennium Development Goals (MDGs). The expansion of broadband is a critical factor in development, and in meeting the MDGs. Broadband can deliver health services and education, as well as supporting cultural diversity and the generation of economic activity and management of climate change, natural disasters and other global crises. Highlighting the need for governments to raise broadband to the top of the development agenda and speed up its rollout, the importance of providing affordable broadband in the least developed countries is recommended.

It should be stressed that SatCom is often the best tool in the provision of broadband services in rural and remote areas, and in the case of natural disasters when land-based infrastructure is not available. In addition, this policy approach is consistent with the international space law principle that the exploration and use of outer space shall be carried out for the benefit of all people.

This chapter has two sections, the first describing the setting of broadband issues for Europe and Africa in terms of policy and technology, and the second presenting the current and planned SatCom capacities serving Europe and Africa where partnership with Africa can be enhanced.

THE SETTING: THE DIGITAL AGENDA FOR BROADBAND PROVISION IN EUROPE

Broadband as network and connectivity provision has numerous positive externalities[1] to enhance the welfare of every member of global society. This is also the rationale behind the European policy initiative, the Digital Agenda[2] to deploy broadband throughout Europe. The motivations behind

---

[1] For a comprehensive overview of the main positive impacts coming from broadband provisions ITU-Broadband Commission for Digital Development, see *Broadband: a platform for progress*, ©ITU/UNESCO, June 2011. It is available at: http://www.broadbandcommission.org/report2/full-report.pdf

[2] European Commission, Communication from the Commission to the European Parliament, the Council, The European Economic and Social Committee and the Committee of the Regions, A Digital Agenda for Europe, COM (2010) 245 final/2,

the Digital Agenda came from various external factors, including socio-economic and technological issues. Previously distinct communication networks and services are today converging onto one network thanks to the digitalization of content, the emergence of IP, and the adoption of high-speed broadband. But convergence has other aspects as well, including network convergence,[3] service convergence,[4] industry/market convergence,[5] legislative, institutional and regulatory convergence and also co-operation,[6] device convergence,[7] and converged user experience.[8] The movement towards convergence has been based on the evolution of technologies and business models, and this process has led to entry of new players into the market, increasing competition among players operating in different markets, and the necessity for traditional operators to co-operate with companies that were previously in other fields.

As a result, convergence touches not only the communication sector, but involves a wider range of activities at different levels across all market dimensions,[9] including the manufacturers of terminal equipment, software developers, media content providers, ISPs, etc.

What broadband requires is:

- *High speed* communication systems for transferring of complex data-packages;

Brussels 2010. Available at http://ec.europa.eu/information_society/digital-agenda/documents/digital-agenda-communication-en.pdf

[3] It is driven by the shift towards IP-based broadband networks. It includes fixed-mobile convergence and 'three screen convergence' (mobile, TV and computer).

[4] It stems from network convergence and innovative handsets, which allows access to web-based applications and the provision of traditional and new value-added services from a multiplicity of devices.

[5] It brings together in the same field industries such as information technology, telecommunication, and media, formerly operating in separate markets.

[6] It is taking place between broadcasting and telecommunication regulation. Policy makers are considering converged regulation to address content or services independently from the networks over which they are provided (technology neutral regulation).

[7] Most devices include a microprocessor, a screen, storage, input device and some kind of network connection. Increasingly they provide multiple communication functions and applications.

[8] There is unique interface between end-users and telecommunications, new media, and computer technologies.

[9] To get a comprehensive view of broadband the concept of business ecosystem see ESPI Perspective n. 59, Veronica La Regina and Chris Wilkins, *The Appropriateness of Public-Private Partnerships for SatCom in Delivering the Digital Agenda*, April 2012, http://www.espi.or.at/images/stories/dokumente/Perspectives/ESPI_Perspectives_59.pdf

- *Always-on systems* where there is no limit in terms of time access and location;
- *Two-way capability* to guarantee interaction.

The EC has described its goals for the Digital Agenda in the following terms: "The objective is to bring *basic broadband* to all Europeans by 2013 and seeks to ensure that, by 2020, (i) all Europeans have access to much higher internet speeds of above 30 Mbps (*fast broadband*) and (ii) 50% or more of European households subscribe to internet connections above 100 Mbps (*ultra-fast broadband*)." [10]

Achieving these targets requires two types of technological action: extending the network reach, and upgrading the network to higher capacity. Three groups of technologies are required including wired, terrestrial wireless technologies, and SatCom.

Technology Action		Variable	Fixed	Terrestrial Wireless	SatCom
Extending the network		Investment	High	Medium	Low
	System Requirement	Backhaul	High	High	Low
		Last - Km	High	Medium	Low
		Number of People served	High	Medium	Low
		Land Coverage	High	High	Low
Upgrading the network		Investment	High	Medium	High
	System Requirement	Backhaul	High	High	Low
		Last - Km	High	Medium	Low
		Number of People served	High	Medium	Low
		Land Coverage	High	High	Low

Table 1: Matrix of Technology Actions and technologies for broadband implementation

---

[10] European policy initiative, Europe 2020, pursues and establishes the concept of smart, sustainable and inclusive growth. See, EC, Communication from the Commission to the European Parliament, the Council, The European Economic and Social Committee and the Committee of the Regions, Europe 2020 Flagship Initiative - Innovation Union, SEC (2010) 1161, Brussels, 6.10.2010, COM(2010) 546 final.

At present, satellite capacity already in orbit can easily provide the basic broadband where no alternative network exists[11] and where terrestrial networks are not economically affordable as shown in Table 1.

There is significant scientific literature[12] following the seminal work of Haavelmo[13] which shows that investment in infrastructure increases GDP. There have also been several studies demonstrating that investment in broadband also enhances growth of GDP, the rate of job growth, cost saving and efficiency, serving as a communication and transaction platform for the entire economy that can improve productivity across all sectors. Broadband networks are increasingly being recognized as fundamental for economic and social development, and advanced communication networks are a key component of innovation ecosystems.

Broadband networks also increase the impact and efficiency of public and private investments that depend on high-speed communications as a complementary investment to other infrastructure such as buildings, roads, transportation systems, health and electricity grids, allowing them to be "smart" and save energy, assist care for the aging, and improve safety.

As noted above, the main challenge of SatCom is upgrading capacity, which is fixed from the beginning of its development. The development of a SatCom system does not have an exit strategy.

Currently, fixed technology has generally been adopted, but as it does not exist in remote locations, the extension of the network is costly. The cost should be lower with terrestrial wireless, but it still requires ground infrastructure such as radio bridges. This often includes a satellite backhaul connection through Very Small Aperture Terminals, usually coupled with wireless technologies such as Wi-Fi. This combination allows access to telecommunication and data services even to more remote areas, albeit with limited and expensive bandwidth.

Backhauling by satellite is affordable with low effort, as the addressable footprint is usually bigger than a single country.

Government investment in broadband networks will likely include a mixture of extending access to unserved/underserved areas and upgrading infrastructure in areas that already have connectivity.

There are a number of technological choices to be considered, each

---

[11] It is digital divide due to lack of infrastructure.
[12] See inter alia W. T. Stanbury, *Perspectives on the New Economics and Regulation of Telecommunications*, IRPP, 1996 ; OECD, *The role of communication infrastructure investment in economic recovery*, 2009.
[13] Haavelmo, Trygve, *A Study in the Theory of Economic Evolution*, Amsterdam: North-Holland, 1954.

of which has benefits and drawbacks. When policy makers focus on connectivity the variables to consider include the number of impacted users, the marginal improvements users will receive, the capacity of the network, longevity and upgradability of the system, and the strategic value of the projects.

Governments also consider the number of users who will benefit from any network investment, and in the case of the Digital Agenda, the social objective is to not exclude anyone. Delivering low-speed broadband to areas previously without connectivity will likely have a different impact than upgrading network capacity for existing users from 10 Mbitps to 100 Mbitps.

The long term impacts of network investment on productivity are linked to their practical life span. Networks with long predicted life spans will produce higher aggregate impacts on GDP and growth than those which may need to be upgraded or rebuilt after only a few years. SatCom offers a nearly limitless upgrade path, while other technologies may not be able to accommodate higher speeds without significant investment in new network infrastructure.

In addition to these technology-related issues, the geography and demographics of Africa are also important factors in examining the relationship between Europe and its neighbor to the south.

THE GEOGRAPHIC RELATIONSHIP BETWEEN EUROPE AND AFRICA

The geographic position of Europe is generally presented from a European-centric point of view,[14] as shown on the left side of Figure 1. The image to the right shows the same geography from a different perspective that is highly relevant to the SatCom conversation.

The European continent was originally sandwiched between several future continents, and was ripped from Africa when North America and Greenland pulled away. Then the block struck Asia, and all of Asia and Europe pivoted on Iran. As Europe was driven back into the Mediterranean, compression raised the mountains along Southern Europe and North Africa, while friction formed mountains in Scandinavia. The final stop built the Ural Mountains.[15] The plate tectonics are shown in Figure 2.

---

[14] Probably the first maps of Europe are dated around 1100 with the Crusade routes.
[15] "The Geological and Tectonic Framework of Europe," by J.A. Plant, A. Whittaker, A. Demetriades, B. De Vivo, and J. Lexa, in *Geochemical Atlas of Europe*. Part 1 -

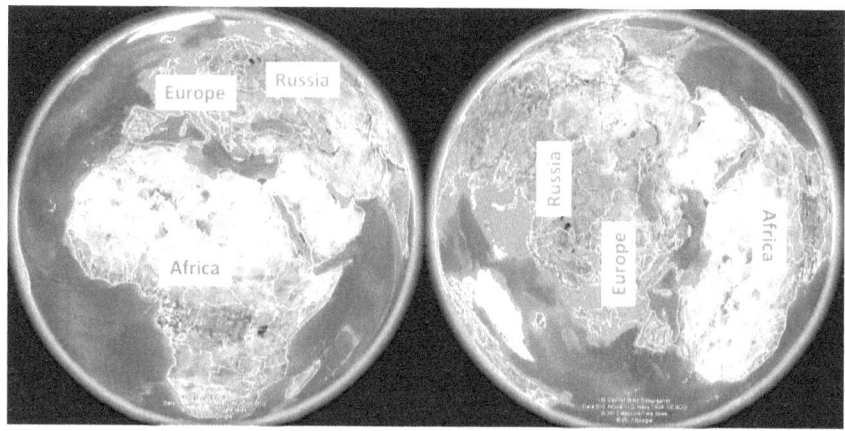

Figure 1 : Geographical position of Europe, Russia and Africa on the Globe
Source : ©Google Earth – Elaboration by the author

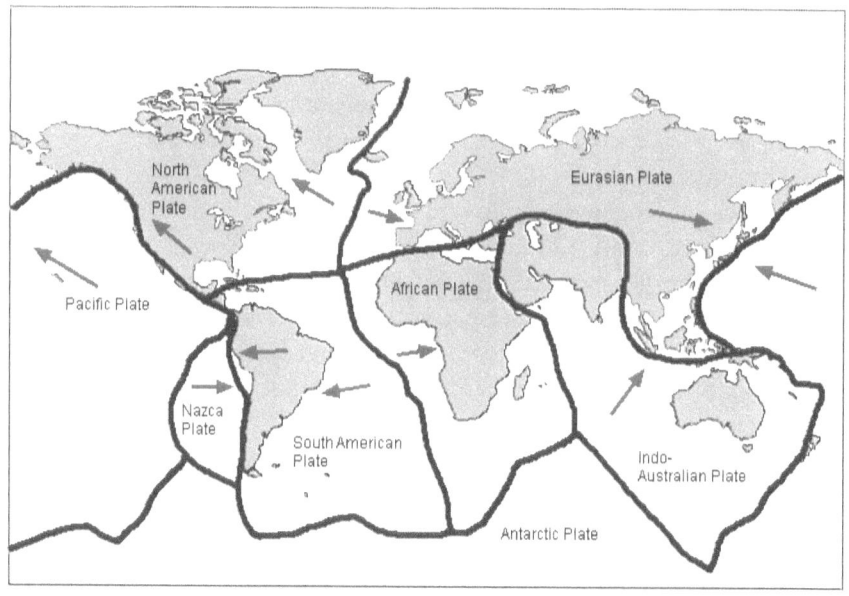

Figure 2 : Global map of tectonic plates
Source : http://www.geography-site.co.uk/pages/physical/earth/images/plates.gif

The Eurasian plate is 85% land mass, with a highly undefined plate

---

Background Information, Methodology, and Maps, Salminen R. (Chief-editor), 2005.

boundary on its southern border connecting to the African, Indian and Australian plates. The continents are converging, and for many millions of years the northern edge of the African tectonic plate has descended under Europe's. The main geologic features of the present-day Mediterranean result from two major processes, the tectonic displacement caused by the sub-conduction of the African plate underneath the Eurasian plate, and the progressive closure of the Mediterranean Sea involving a series of submarine-insular sills.

This combination of factors results in earthquakes, volcanic eruptions, droughts, tsunamis and similar phenomena that often have as consequences critical conditions with huge impact on critical infrastructures such as power plants and distribution networks, transportation networks, and communications systems. Further, the complex relationships among various critical infrastructure elements result in infrastructure interdependencies.[16]

Disasters often directly impact telecommunication infrastructures, while a disaster in itself increases the need for communications, so high data traffic occurs and increased throughput and bandwidth are needed. Further, high traffic volumes can adversely impact on communications among response teams, which increases the reaction time for rescue and relief operations.

Satellite communications providers support recovery efforts following such disaster events by prioritizing the allocation of satellite transponders to the ministries of Defense, National Police Agencies and other government and public sector institutions that conduct disaster relief activities. People access the Internet through satellite connections as mobile and fixed common carriers use satellite communications for replacement of their damaged networks.

Public utilities such as railroad companies and electric power companies also become dependent on satellites. Railway operators employ satellite connections to minimize the impact of land-based disasters by transmitting early warnings and other information from disasters detectors to moving trains.

DEMOGRAPHIC OVERVIEW

Demographically, Africa is the second most populated area of the world,

---

[16] Saifur Rahman, *Impact of Natural Disasters on Critical Infrastructures*, 2005, http://www.emisa.vt.edu/2DOC/EAPC_Panel-2_06.pdf

accounting for some 15% of the world's population, but only 2.5% of world GDP. Europe's population is 11% of the world's, and 7% for the EU-27.[17]

Africa's population growth rate is double the world's, but its population density is lower, at 39 per km^2, while the world average is 51 people per km^2. This geographic dispersion requires significant resources and complex systems to assure welfare, safety and security across vast territories.[18]

Africa has high socio-economic inequality, with a few rich persons and many poor. With a Gini coefficient of 51%, Africa has the worst income distribution in the World.[19] This causes Africa to have great exposure to external economic shocks, and thus African inflation ran at a damaging rate of 8% per year between 2000 and 2010.[20]

Africa is crucial to Europe's energy supply, as it exports 120M tons of crude oil annually, while the EU exports back to Africa 23M tons of refined oil products.

With these factors in mind, let us now examine the policies that link Europe and Africa.

Policy Initiatives Affecting the Relationship Between EU and Africa

The policy relationship between Europe and Africa encompasses a number of initiatives, including bilateral agreements and partnerships between EU and the African Union, and between European member states and various African states.

Here is a list of some of the pertinent agreements.

Year	Policy Initiative	Remarks
2000, 3-4 April	Cairo Declaration and Cairo Action Plan	Initial Africa – EU Summit
2001, 11 October	1st Africa-Europe Ministerial Meeting	Joint declaration on terrorism
2002, 28 November	Africa-Europe Ministerial Meeting	Initial policy development from human rights to trade and markets

---

[17] Eurostast, *EU-Africa Summit - Revival of EU27 trade in goods with Africa in the first nine months of 2010*, News release, STAT/10/178, 26 November 2010.
[18] Niall Ferguson, *Civilization: The West and the Rest*, The Penguin Press HC, 2011.
[19] UNECA, *African economies at the start of the 21st Century*, 2008.
[20] OECD, Statistic Reports, 2011.

2003, 10 November	EU-Africa Dialogue - Ministerial Troïkas Meeting	Enhancement of the dialogue; establishment of the governance models
2004, 1 April	2nd EU-Africa Ministerial meeting	Peace and security are the key-topics
2004, 4 December	Africa – Europe Dialogue 3rd Meeting of the Troïkas	Ministers recognized the widening digital gap between Africa and the rest of the world as a serious setback in the efforts to bring about sustainable development, poverty eradication and integration of the continent in the global information society.
2005, 13 April	4th EU-Africa Ministerial meeting	Initial plan on capacity building on peace and security
2005, 2 December	5th EU-Africa Ministerial meeting	Enhancing and monitoring the dialogue
2005, 19 December	The EU and Africa: Towards a strategic partnership	Initial proposal for EU and Africa partnership
2006, 8 May	6th EU – Africa Ministerial Troika Meeting	The Partnership is intended to both have a wide scope and promote interconnectivity in regional infrastructure networks (roads, rails, ports, energy, water and ICT).
2006, 10 October	7th EU - Africa Ministerial Troika Meeting	The European Commission informed the meeting about the launch in July 2006 of the EU – Africa Infrastructure Partnership, which constitutes the EU's response to the NEPAD short-term Infrastructure Action Plan. Ministers welcomed the Partnership, which will support programs in the sectors of transport, energy, water and sanitation and ICTs that facilitate interconnectivity at continental and regional level.
2007, 15 May	8th EU – Africa Ministerial Troika Meeting	The launch in September 2007 of the EU Africa Infrastructure Partnership and Steering Committee to support AU/NEPAD infrastructure Short Term Action Plan (i-STAP) and Medium to Long Term Strategic Framework (MLTSF) programs in the sectors of transport, energy, water and sanitation and Information and Communication

		Technologies to facilitate interconnectivity and access to services at continental, regional and national levels in Africa.
2007	Commission/Council Secretariat Joint Paper, Beyond Lisbon Making the EU-Africa Strategic Partnership work	In the area of Information and Communications Technology (ICT) the partnership aims at bridging the digital divide that limits access to modern telephony and internet services. It should address the harmonization of policy and regulatory frameworks, the investment in broadband infrastructure and support non-commercial e-services.
2007, 7-8 December	Lisbon Declaration - EU Africa Summit	Initial step of Africa Forum and further EU – Africa Partnership
2008, 16 September	10th Africa – EU Ministerial Troika Meeting	Enhancement of the governance model
2008, 10 November	Council Conclusions on trilateral dialogue and cooperation between the European Union, China and Africa	The Council urges Africa and China to cooperate with the EU in establishing trilateral dialogue and cooperation.
2008, 20 - 21 November	11th Africa - EU Ministerial Troika meeting	Establishment of "Africa-EU Partnership on Science, Information Society and Space"
2009, 28 April	12th Africa - EU Ministerial Troika meeting	Development of the people-centred dimension of the Partnership
2009, 14 October	13th Africa - EU Ministerial Troika meeting	Implementation of the Joint Africa-EU Strategy
2010, 26 April	14th Africa – EU Ministerial Meeting	Implementation of the Joint Strategy
2010, 19 November	15th Africa – EU Ministerial Meeting	Preparations for the 3rd EU Africa Summit
2010, 29 – 30 November	Tripoli Declaration 3rd Africa EU Summit	Enhancement and strength of the cooperation
2010, 30 November	New Joint Africa – EU Strategy (JAES)	Promotion for an inclusive and sustainable growth with the involvement of private actors and entrepreneurship

Table 2: The main key-policy documents of Africa – EU International Relations

The Lisbon Declaration in 2007 led to further partnerships and the

Millennium Development Goals, a robust peace and security architecture in Africa, the strengthening of investment, growth and prosperity through regional integration and closer economic ties, the promotion of good governance and human rights, and the creation of opportunities for shaping global governance in an open and multilateral framework. Here is a clear reminder of strategic partnership as enabler of the Joint Strategy and the Action Plan.

The first joint Africa–EU strategy[21] identified four objectives: the Africa-EU political partnership; the promotion of peace, security, democratic governance and human rights, fundamental freedoms, gender equality, sustainable economic development and Millennium Development Goals (MDGs); the sustenance of a multilateral system for addressing global challenges; and support for people-centered partnership including non-state actors. The same policy initiative establishes a strategy framework of four areas: peace and security; governance and human rights; trade and regional integration; and key development issues. This last cluster foresees policy as a tool for accelerating progress towards the MDGs. For the purposes of this chapter, the relevant points are n. 59 as an element of "Human and social development;"[22] n. 66 under "Environmental Sustainability and Climate Change"[23] and n. 84 under "Development of Knowledge-based Societies."[24]

This partnership required a further action plan, named "First Action Plan 2008 – 2010 for the implementation of the Africa – EU Strategic

---

[21] It is linked with the previous African Union Constitutive Act and Strategic Framework 2004-2007 and the EU Africa Strategy of 2005.

[22] From the 1st Joint Africa-EU strategy: (…) *Furthermore, building upon the 2007 Addis Ababa Declaration on Science Technology and Scientific Research for Development, Africa and the EU shall strengthen their cooperation in these areas. In this context, attention will also be paid to space-based technology, applications and sciences.(…)*

[23] From the 1st Joint Africa-EU strategy: (…) *Africa and the EU should strengthen existing cooperation mechanisms and programs relating to the use of space technologies and space-based systems.(…)*

[24] From the 1st Joint Africa-EU strategy: (…) *Africa and the EU will strengthen their cooperation in building knowledge-based societies and economies. Both sides recognize that the development of S&T and innovation is one of the essential engines of socio-economic growth and sustainable development in Africa; that competitiveness in the global economy is increasingly dependent on knowledge and innovative ways of applying modern technology, especially Information and Communication Technology (ICT); and that meeting the MDGs requires a special global effort to build scientific and technological capacities in Africa. Thus partnerships and investments advancing access to ICT infrastructure, access to quality education, and the development of science and technology and innovation systems in Africa are crucial for attaining all other development goals. (…)*

Partnership." This document proposed a list of priority actions, and among them there are two specific actions of relevance here, number 8, "Africa – EU Partnership on Science, Information Science and Space" established the objective, the expected outcome, the related activities, the involved actors and sources of finance. Telecommunication is an element of all the priority actions of this section, as they are a crucial element to bridging the digital divide and scientific divide, and to enhancing cooperation in the use of suitable and affordable space applications to support African development.

This plan brought deployment of EASSy,[25] a 10.000Km submarine fiber optic multi-point cable system along the East coast of Africa, which became operational in 2010. The project was funded by Infrastructure Trust Fund of 3.6 million Euros.

Ancillary actions, more focused on applications, have been the EU project of AfricaConnect for 12m Euros, supporting the deployment of regional research and education networks in Africa, and their interconnection with the European GEANT network. The objective is to provide the African scientific community with better access to research and education resources, and higher capacity access the internet. The AXIS project mobilized by Luxembourg supports the deployment of local internet exchange points through the Infrastructure Trust fund, with initial contribution of US$4m.[26]

The 3rd Africa–EU Summit took place in Tripoli on November 29-30, 2010 resulting in the Tripoli Declaration. The partnership was confirmed and a new Joint Africa–EU Strategy (JAES) 2011-2013 was adopted, which includes the involvement of private actors and entrepreneurship.

---

[25] The Eastern African Submarine Cable System (EASSy), an undersea fiber optic cable system connecting countries of Eastern Africa to the rest of world was switched in 2011. Fiber is coming into Africa by the bucket load.

[26] From Luxembourg's Department of International Cooperation : *En 2010, la Coopération luxembourgeoise a ainsi répondu favorablement à une demande du Fonds fiduciaire l'UE-Afrique pour les infrastructures auquel elle a contribué deux millions d'euros au démarrage en 2007 et a décidé d'appuyer deux projets financés par ce fonds : le projet « Satellite-enhanced Telemedicine and eHealth for Sub-Saharan Africa » et le projet « African Internet Exchange System (AXIS) ». Pour ces deux projets, Lux-Development joue pour le compte de l'Etat luxembourgeois le rôle de « lead financier ». L'agence est responsable pour la mobilisation et la gestion de l'assistance techniqueLe projet de télémédecine est portée par une « Telemedicine Task force » au sein de laquelle figurent la Commission de l'Union africaine, le NEPAD (Nouveau partenariat pour le développement de l'Afrique), la Banque africaine de développement, les organisations africaines régionales, l'OMS (Organisation mondiale de la santé), la Commission européenne et l'ESA (Agence spatiale européenne). Cette dernière est le principal porteur du projet qui consiste dans une première phase (4 ans) en une série d'études avec un budget de 4,17 millions d'euros.*

These two peculiar attributes of growth come from "Europe 2020," a 10-year strategy proposed by the EC in March 2010 for reviving the economy of the EU that aims at "smart, sustainable, inclusive growth" with greater coordination of national and European policy.[27]

The Tripoli Declaration emphasized issues related to technology transfer to promote a knowledge-based society, which requires a solid and consistent telecommunication network and related infrastructure to support the integration of Africa into the global economy. The EU committed itself to collect 0.7% of its Gross National Income by 2015 for aid spending in Africa with a substantial fund of 50 billion Euros of Oversea Development Aid (ODA) to support the overall Africa–EU Partnership.

This declaration is linked with the Action Plan 2011 - 2013, issued and adopted by the 3rd Summit EU–Africa. The Action Plan identifies 8 thematic partnerships,[28] of which telecommunications are an element of partnership number 3, Regional Integration, Trade and Infrastructure, and number 8, Science, Information Society and Space.

The Action Plan also has a section describing cross-cutting issues in terms of political architecture, financial structure, and policy tools. Advocacy for backbone infrastructure will be done as a policy of regional economic integration, trade and infrastructure. The partnership for Science, Information Society and Space includes telecommunication as a cross-cutting element supporting the deployment of the other development

---

[27] Europe 2020 identifies five headline targets: 1. Employment: 75% of 20-64 year-olds to be employed; 2. R&D and Innovation: 3% of the EU's GDP to be invested in public and private R&D for innovation; 3. Climate Change & Energy: reducing greenhouse gas emissions 20% (or even 30%, if a satisfactory international agreement can be achieved to follow Kyoto) lower than in 1990, providing 20% of energy from renewable sources, saving of emissions through at least 20% increase in energy efficiency; 4. Education: Reducing school drop-out rates below 10% at least 40% of 30 - 34 year-olds completing third level education (or equivalent); 5. Poverty & Social Exclusion: at least 20 million fewer people in, or at risk of, poverty and social exclusion. These targets will be implemented through national targets in each European country, reflecting different situations and circumstances. They are highlighted through 12 Flagship Initiatives which contain three segments of Growth: smart, sustainable and inclusive.

[28] The eight thematic partnerships are :
1. Peace and Security;
2. Democratic Governance and Human Rights;
3. Regional Integration, Trade and infrastructure;
4. Millennium Development Goals;
5. Energy;
6. Climate Change and Environment;
7. Migration, Mobility and Employment;
8. Science, Information Society and Space.

objectives of the Action Plan.

The role of network and connectivity is described mainly as support for the development of an inclusive information society in Africa, in line with the guidelines of the Europe 2020 Digital Agenda and the Africa Union ICT development framework. Implementation is expected to be done by multiple stakeholder groups such as the Africa Union Commission, the EC, national entities dealing with ICT policy, research communities, NGOs, private companies, and international organizations to enhance cooperation at all levels to address poverty reduction, economic growth, social development, and regional integration.

THE ROLE OF SATCOM IN THE AFRICA – EU PARTNERSHIP

The core issue inhibiting the African continent from achieving intensive

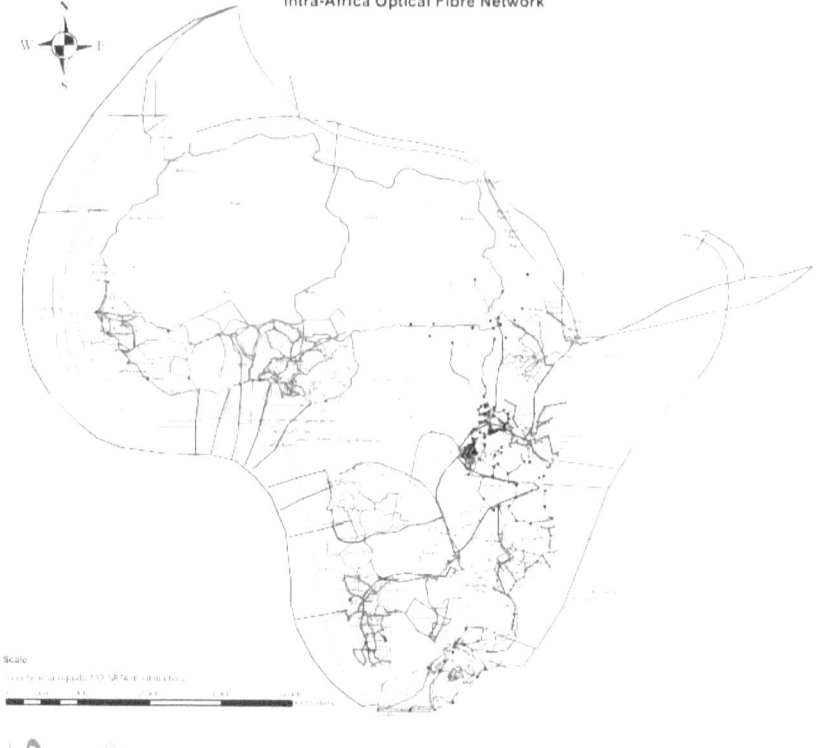

and wide broadband access is the lack of infrastructure, including

Figure 3: Intra-Africa Optical Fibre Network
Source: Mon, 07/12/2010 - 17:03 — ©UbuntuNet

telecommunication networks and electricity systems.

The coastlines benefit from the backbone and cable infrastructure developed to connect Africa with the other regions, but huge areas lack any telecommunication infrastructure, as shown in Figure 3. This prevents Africa from achieving its proper role at the global level, and its interaction with other regions is actually declining. African internet traffic is about 0.7% of overall world traffic.

If broadband brings several positive externalities, its absence implies numerous negative externalities. Global traffic in terms of billions of minutes is shown below in Table 3.

Interregional Flow	Traffic (B.ns of minutes)	Interregional Flow	Traffic (B.ns of minutes)
US Canada – Europe	12,2	Europe – US Canada	8,8
Europe – US Canada	8,8	US Canada – Europe	12,2
US Canada – Asia	21,7	Europe – Asia	12,7
Asia – US Canada	6,8	Asia – Europe	5,4
US Canada – South America	30.3	Europe – South America	3,5
South America – US Canada	5,5	South America – Europe	1,3
US Canada – Africa	2,7	Europe – Africa	6,4
Africa – US Canada	0,5	Africa – Europe	0,9
IN- Flow Total	21, 6	IN- Flow Total	19,8
OUT – Flow Total	66, 9	OUT – Flow Total	31,4
Asia – Europe	5,4	South America – US Canada	5,5
Europe – Asia	12,7	US Canada – South America	30,3
Asia – South America	0,3	South America – Europe	1,3
South America – Asia	-	Europe – South America	3,5
Asia – Africa	2,3	South America – Africa	-
Africa – Asia	0,5	Africa – South America	-
Asia – US Canada	6,8	South America – Asia	-
US Canada – Asia	21,7	Asia – South	0,3

		America	
IN- Flow Total	34,9	IN- Flow Total	34,1
OUT – Flow Total	14,8	OUT – Flow Total	6,8
Africa – Europe	0,9		
Europe – Africa	6,4		
Africa – South America	-		
South America – Africa	-		
Africa – Asia	0,5		
Asia – Africa	2,3		
Africa – US Canada	0,5		
US Canada – Africa	2,7		
IN- Flow Total	11,4		
OUT – Flow Total	1,9		

Table 3: Interregional Traffic Flow, 2012
Source: © Telegeography, with elaboration by the author

An assessment of the internet traffic within and between each region, and shows that Africa is quite isolated from the rest of the world. To bring African infrastructure towards world standards, it needs upgrading and extension of national networks; development of communication capabilities and rural communications; rapid expansion of communications skills and manpower development institutions; appropriate use of new communication technologies for national and interregional communications; continuing expansion of the terrestrial communication network project (Pan-African Telecommunications Network, PANAFTEL); and establishment of telecommunication, broadcasting and related equipment manufacturing facilities.

Despite the lack of capability, demand for telecommunications in the developing economies of Africa does exist. There is also the abundant supply capacity of the developed countries willing to serve this market. However, the mutually beneficial arrangements acceptable to both sides have still to be created.

Each African country has its own specific problems, but some common constraints are evident. Most African countries have established basic national telecommunications networks that are small and concentrated in the urban areas. International and some of the intra-African traffic is provided by satellite Earth stations. Rural areas in Africa remain especially poorly served, and with some exceptions, are given low priority by national telecommunication administrations. The low initial revenue potential from rural services is in part responsible. In operation, these services are sometimes inefficient, slow, and very expensive.

Alternatively, international services operating within the same network are faster and often of better quality. In spite of the progress made in the development of the Pan-African Telecommunications Networks (PANAFTEL), it is still limited in scope. Utilization of the network by its potential users for radio broadcast, television services, aeronautical and meteorological services and press and news agencies, etc. is very low, in part due to operations difficulties, poor maintenance, in some cases high tariffs, absence of operational agreements, shortage of trained manpower and spare parts, and poor management, as well as the need for upgrading some of the earlier installations, utilization of new technologies and interconnection to other sub-regional section of the PANAFTEL Network.

Satellite broadband access can be a fast and feasible solution to complement terrestrial broadband services and ensure coverage to the most remote areas. The technology has improved enormously in terms of reduced transmission delay and throughputs of capacity per time unit.

Early in 2013 the West Africa Cable System (WACS), an ultra-high capacity, fibre optic, submarine cable system that will link countries in Southern Africa, Western Africa and Europe, will be switched on.

With one stroke a huge of amount of a new capacity will come to Africa. The system will play a key role in reducing digital divides across the region and connecting Africa with the rest of the world, but this is by no means the only new network. Figure 5 shows current and planned telecommunication networks by terrestrial technologies which are along the coastline. This again has the weakness of not serving peoples living in the internal areas and in locations remote from the main cities. The main owners of these infrastructures are multinational companies[29] of developed nations, and their goal is to seek business profit.

---

[29] *See* http://manypossibilities.net/african-undersea-cables/

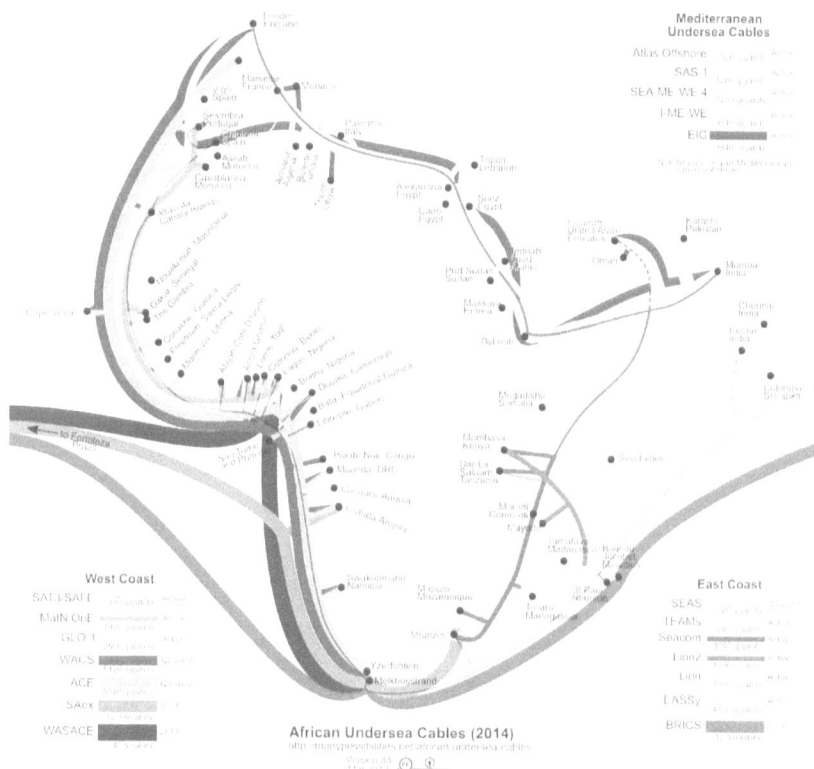

Figure 4: Planned Telecommunication Networks for Africa
Source: http://farm6.staticflickr.com/5120/7087121729_9de109f3b9_b.jpg

The implementation of a terrestrial telecommunication network covering the entire African land is huge expense and also requires long lead time. Thus, an obvious alternative is to consider SatCom, already present in the area guaranteeing connectivity through C-Ban and Ku-Band, and also with the potential of Ka-Band.

Figures 5 and 6 show the current availability of KU and C band SatCom over Africa. In addition, several commercial ventures are operational for Ka-Band over Africa,[30] and there is also an innovative SatCom system, O3B,[31] with the specific mission of providing broadband to Africa.

---

[30] ESPI Report 32 – Veronica La Regina, SatCom Policy in Europe, May 2011; http://www.espi.or.at/images/stories/dokumente/studies/ESPI_Report_32_web.pdf

[31] *See* http://www.o3bnetworks.com/

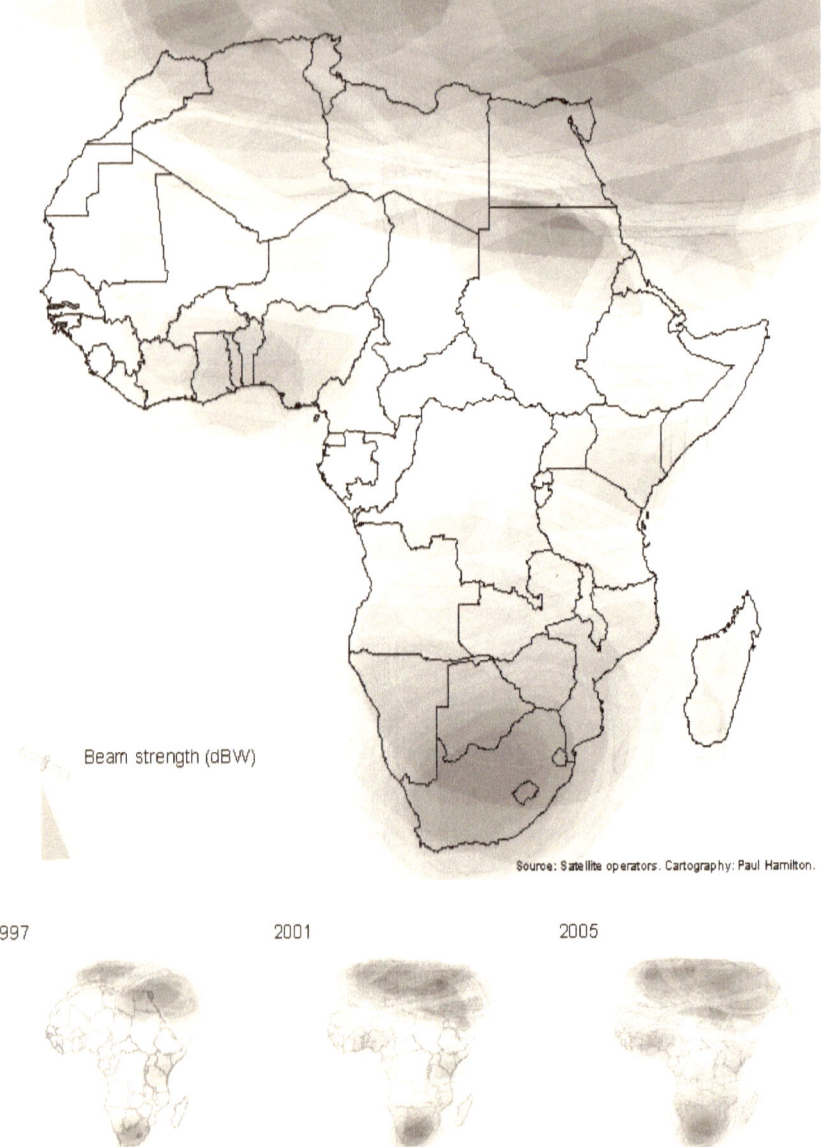

Figure 5: Ku Band Coverage Over Africa
Source: Paul Hamilton

Figure 6: C Band Coverage Over Africa
Source: Paul Hamilton

The critical point of SatCom provision in Africa is the lack of a fair and equitable distribution of the benefits. SatCom is used as an element of the overall business ecosystem of telecommunication as a provider of backhaul and back-up services, and thus the satellite operators including

Intelsat, Eutelsat, SES, Arabsat, Rascom, etc. are dealing with local telecom operators and not directly with end users. This business model provides higher profit margin[32] to satellite operators, but the overall economic ecosystem does not benefit from the entire value-added that could be created.

Conclusions and Policy Recommendations

In conclusion, SatCom's particular strength in cost-effectively providing communication over large areas, and its easy deployment for end user solutions, makes the sector of special interest in the present era of globalization feeding the phenomenon of convergence.

Two key social applications in which satellite communication contributes are telemedicine and tele-education. Offering quality medical care and education is a challenge in many developing countries, particularly in rural areas. Telemedicine helps connect local health care workers with support from information and personnel in other areas.

Through tele-education, students connect with teachers, curricula and course assignments from a distance. In addition, SatCom can play a crucial role in disaster management.

Europe has made significant achievements[33] in the sector in terms of technology, regulatory frameworks and market share. At the same time, new efforts are required to enhance these achievements and to address new challenges in a strategic way.

The main challenge as reflected in the UN Summit on the Millennium Development Goals, concluded in September 2010 with the adoption of a global action plan to achieve the eight anti-poverty goals by their 2015 target date, and the announcement of major new commitments for women's and children's health and other initiatives against poverty, hunger and disease. Europe has already started dialog on the eight goals that, in a broader sense, establish the concept of smart, sustainable and inclusive growth. Of particular relevance is goal number 8, "Develop a global partnership for development," and sub-target 8F which suggests establishing partnerships making available the benefits of new technologies, especially information and communications. Here, the role of the EU should be to provide broadband in a sustainable way for and with

---

[32] ESPI Report 32 – Veronica La Regina, *SatCom Policy in Europe*, May 2011; OECD, New Space Economy, Febraury 2012.
[33] ESPI Report 32, *idem supra*.

Africa, which means sustaining partnerships with local actors and stakeholders using existing capacities, and transferring the necessary competences that are lacking locally.

To achieve this goal it is worth noting the contribution that SatCom can make to the EU Digital Agenda for the EU itself and, using the same effort in terms of infrastructure, for EU goals for its neighbours, Africa and the Middle East. This involves a set of considerations from the industrial and regulatory points of view. It is crucial to enhance the synergies between the traditional terrestrial-based telecommunication sector and SatCom through an integrated industrial policy, as proposed by the EC in 2010. Implementing the dialog in an integrated way also helps in reducing the time and costs required for adoption of policy initiatives.

The importance of the Digital Agenda presents the opportunity to start new dialogues among European stakeholders, as both public and private players. This process will bring benefits in several areas, such as a more equitable international and European communication regulatory framework, an industrial policy that enhances the competitiveness of the EU, a market more satisfying for consumers, a space policy more integrated to deliver economic, societal and strategic aims.

All these issues are supported by a set of synergetic policy recommendations, including:

- Enhance the existing framework of EU–African partnership, which should involve also SatCom as an available and prompt key enabling technology for broadband provision;
- Enhance the role of private actors in potential innovative public-private partnerships with innovative business models to get higher distributions of benefits coming from broadband commercialization;
- Strengthen the international dimension of SatCom to overcome lack of national and local telecommunication policies favorable to broadband;
- Establish a synergetic view of the role of SatCom as a cross-sectorial enabling element of policy initiatives in telecommunications, industrial policy and technology transfer, international relations for trade, peace and security, and socio-economic growth.

In conclusion, the objective for the EU should be an integrated vision for SatCom that would deliver the Digital Agenda as a key component with an appropriate space policy that supports the concept of European growth and competitiveness to enhance the European role as global actor. A coherent and appropriate SatCom system for broadband, delivering services as one of the Millennium Goals for the EU and for Africa is an opportunity to give substance to the fundamental principle of international space law that the exploration and use of outer space shall be carried out for the benefit and interest of all.

•••

VERONICA LA REGINA

Veronica La Regina is visiting Research Professor at International Space University for Space Business and Management, and Senior Research Fellow of International Institute of Space Commerce on the Isle of Mann. She has also been a Resident Fellow of the European Space Policy Institute in Vienna.

Previously she was employed at Telespazio SpA in the department of business strategies and marketing, and has also held a position at Wave Energy Centre in Lisbon, and as economics researcher at Osservatorio Filas in Rome.

She holds a Masters Degree in Institutions and Space Policy (2009) at Italian Society for International Organizations (SIOI) in Rome, Ph.D. Studies in Economic Sciences (2004) at State University of Milan, Graduate Studies in Math and Statistics (2001) at University of Rome Sapienza, and a Graduation in Law (1999) at LUISS G. Carli in Rome. She also has been an invited lecturer for energy economics and space issues, and is an active researcher in the field of Satellite Communications.

CHAPTER 14

INTERNATIONAL COOPERATION IN SPACE, FROM USSR TO RUSSIA

OLGA ZHDANOVICH, MSc
SECRETARIAT OF THE EUROPEAN COOPERATION FOR SPACE STANDARDISATION

© Olga Zhdanovich, 2012. All Rights Reserved.

This chapter presents a historical overview of international cooperation in space during the period of the Union of Soviet Socialist Republics (USSR) and to the present days of the Russian Federation. The post-Soviet era and current international cooperation in space are described through interviews with people who were or still are involved in the joint space programs. I asked Dr. Carlo Mirra from EADS/Astrium, who is responsible for the ISS Increment and Mission Integration from European side, Dr. Antonio Verga from ESA who was involved for a number of years in the Foton and Bion cooperative ESA-Roscosmos projects, and Dr. Mark Belakovskiy from the Institute of Medical and Biological Problems (IMPB), which recently completed a successful international experiment for a flight stimulation of mission to Mars – MARS 500, to share their opinions on the cooperation with Russia.

Introduction

During the Cold War from 1946 to 1991, the world was divided into two political camps: the Soviet Union and its partners (socialist countries, East), and the USA with its allies (capitalist countries, West).

Cooperation in space was conducted in the USSR with socialist countries through the Program Intercosmos. The USSR also cooperated with the Western world, mainly with France, Sweden and Austria. Despite severe competition in space, the Soviet Union and USA permanently spoke about possible cooperation, and that resulted in the Apollo-Soyuz Test Project (ASTP) in 1975, and was followed by the Mir-Shuttle Program, the Mir-NASA Program, and then the International Space Station Program (ISS).

In the period of the USSR, many projects were done on an exchange basis, not for profit. With the end of the Cold War and the decline in priority of space and the increase in importance of financial matters, the Russian space program has become more commercially oriented. Today Russia engages in vast international cooperation in the area of human space flight, especially long duration space flight, an area where the Soviet Union accumulated significant experience, more than any other nation.

Organization Of International Cooperation In Space, From The USSR To Russia: Institutions And Activities

This section presents a brief history of the birth of the organizational infrastructure for the national space program and international cooperation in the USSR, and in the Russian Federation after 1991.

During decades of the Cold War there was a saying that the USSR was closed by an "iron curtain," and Western countries were unclear as to what was happening behind this "curtain." For partners of the Soviet Union from the socialist countries, the situation was quite the opposite, as many had a close working partnership with the Soviet Union on matters pertaining to space.

In 1992 the Russian Space Agency (Roscosmos) was founded. The main difference from the USSR space program organization is that during Soviet times there was no single entity or structure doing space activities. Instead, there were many organizations defining and implementing different aspects of the Soviet Space Program, including the Ministry of General Machine-building, the Soviet Academy of Sciences, and the Inter-ministerial Scientific and Technical Council for Space Research under the umbrella of the USSR Academy of Sciences. Other Ministries also contributed to the development of the national space program, mainly through the Inter-ministerial Council and the Commission for the Military and Industrial issues under the Central Committee of the Soviet Communist Party.

Permanent international cooperation in space officially started in 1958. After successful activities for the International Geophysical Year, the United Nations (UN) General Assembly adopted its resolution 1348 (XIII) on December 13, 1958. This resolution formed the basis for the international cooperation in space by creating the ad hoc Committee on the Peaceful Uses of Outer Space (COPUOS). Eighteen countries joined this Committee, including Argentina, Australia, Belgium, Brazil, Canada, Czechoslovakia, France, India, Iran, Italy, Japan, Mexico, Poland, Sweden, the Union of Soviet Socialist Republics, the United Arab Republic, the United Kingdom of Great Britain and Northern Ireland, and the United States of America. One of the issues that the Committee was requested to report on to the 14th UN General Assembly was international cooperation in space:

> "(b) The area of international cooperation and programs in the peaceful uses of outer space which could appropriately be undertaken under United Nations auspices to the benefit of States irrespective of the state of their economic or scientific development, taking into account the following proposals, inter alia:
> (i) Continuation on a permanent basis of the outer space research now being carried on within the framework of the International Geophysical Year;
> (ii) Organization of the mutual exchange and dissemination of information on outer space research;
> (iii) Co-ordination of national research programmers for the study of outer space, and the rendering of all possible assistance and help towards their realization."

The 13th UN General Assembly requested the Secretary-General "to recommend any other steps that might be taken within the existing United Nations framework to encourage the fullest international cooperation for the peaceful uses of outer space."

As a UN permanent body, COPUOS was endorsed by the General Assembly in 1959, resolution 1472 (XIV) on December 12, 1959 with the mandate to promote international cooperation and peaceful use of outer space. In 1959 this Committee included twenty four countries, while today there are seventy-one countries which are members.[1]

The Soviet Space Program was a closed program, in that before 1991 almost all governmental documents were classified. Recent publication of USSR space program documents from 1946 to 1967 (Baturin, 2008) presents what happened from a government perspective. In 1965 the start of international cooperation in space occurred when the Government of the USSR officially approved the policy.

On December 10, 1959 by the Order of the Soviet Government N1388-618, "About development of research for outer space" the Inter-ministerial Scientific and Technical Council for Space Research, under the umbrella of the USSR Academy of Sciences, was formed (Baturin, 2008). For decades this Inter-ministerial Council was the central "think-tank" for the development of the Soviet Space Program. The Council was headed by Academician M. Keldysh, who was also the President of the Soviet Academy of Sciences. All chief designers of the Soviet space program, including S. Korolev, V. Chelomei, M. Yangel, and others, were members of this Council.

In response to the UN General Assembly resolution 1472 (XIV), the Soviet Government on March 10, 1960 issued the Order N300-188 "About USSR participation in the international organization for the peaceful use of outer space." This Order obliged the Inter-ministerial Scientific and Technological Council for Space Research, which was under the USSR Academy of Sciences, to prepare agenda items for consideration in the UN COPUOS by the request of the USSR Ministry of Foreign Affairs. Other Soviet ministries, including the State Committee for Defense Technology, State Committee for Radio-electronics, and State Committee for shipbuilding were instructed to provide necessary materials related to peaceful use of outer space to support the development of position of the USSR representatives in UN COPUOS. (Baturin, 2008).

In October 1963 the Institute of Space Biology and Medicine was

---

[1] Website of the UN Office for Outer Space Affairs, 2012.

founded in the Soviet Union under the USSR Ministry of Health. In 1965 the Institute was renamed as the Institute of Medical and Biological Problems (IMPB). Since then, the institute has played the key role in the area of space biology and medicine in Russia. In 1994 IMBP received status of the State Research Centre and in 2001 the Institute became a scientific center of the Russian Academy of Sciences.

The year 1965 marked the start of USSR international cooperation in space. In April the Soviet Government issued Order N302-103 "About cooperation of the USSR and socialist countries in the research and use of outer space." The Order proposed to organize a meeting to discuss fields of cooperation with countries that were considered friends of the Soviet Union.

On May 15, 1965 the Institute of Space Research of the USSR Academy of Sciences was founded.

On May 30, 1966, by the Order of the USSR Council of Ministers and Central Committee of the Communist Party N421-130 the Council for the International Cooperation in Research and Use of Outer Space under USSR Academy of Sciences was formed. Lately it received the name Intercosmos Council. The Council coordinated the work of other ministries and committees involved in international cooperation projects and programs (Russian Academy of Sciences, 2012). This was done mainly through the Inter-ministerial Scientific and Technical Council for Space Research.

During the Soviet Union era the International Council Intercosmos of the Soviet Academy of Sciences, Institute of Space Research and Institute of Medical and Biological Problems were the known faces of the space program, and were playing the key role for the international cooperation in space.

In February 1985 the General Department for the Development and Use of Space Technology was created inside the Ministry of General Machine Building, known as USSR Glavkosmos. This Department has become the point of contact for international activities of the Soviet Space Program until 1992. Glavkosmos' goal was to commercialize Soviet space technology internationally, as well as promote applications of space technology inside the Soviet Union. At that time it appeared as a Soviet Space Agency to the outside world. For the very first time the whole civil part of the Soviet Space Programs was suggested for cooperative projects with the actual names of the Soviet space technology items that were used inside the Soviet Union by Soviet engineers. Before 1985, for example, the majority of the Soviet Union satellites were known to the rest of the world simply as "Cosmos" satellites.

The booklet entitled "Commercial space services rendered by Glavkosmos of the USSR" states that Glavkosmos' responsibility was the "organization and coordination of work to develop and use space technology," as well as "enforcement of international agreements in the field of development and use of space technology to implement programs of international cooperation." Glavkosmos offered the following space technology available for international partners:

- Launching a wide variety of payloads into various orbits utilizing Soyuz, Vostok, Molniya, Tsyklon, Cosmos, Energiya-Buran, Zenit and Proton launchers;
- Performing scientific and technological experiments in microgravity through spacecraft Foton and Bion and/or using specialized hardware as Splav-2 and Kashtan (for material science and crystal growth), Pion-MA (for biological research) etc.
- Flights to Mir-Soyuz-Tm-Progress-Kvant space station complex;
- Data from remote sensing satellites Okean and Resurs-O;
- Leasing communication channels of the Soviet communications spacecraft, etc. (Glavkosmos, USSR, 1990).

Glavkosmos signed agreements with the European Space Agency, and research onboard the Foton and Bion satellites was part of the cooperation agreement. In 1990-1991 joint space flights of a Japanese journalist, British and Austrian cosmonauts were performed.

After the collapse of the Soviet Union, the Russian Space Agency was founded in 1992 using Glavkosmos and the Ministry of General Machine building as its base.

USSR – FRANCE COOPERATION IN SPACE

France was the pioneer of international cooperation with the Soviet Union, and 2011 marked the 45th anniversary of cooperation with Russia in space. Cooperation began in June 1965 with the Order of the USSR Council of the Ministers N510-195 "About cooperation between USSR and France in the field of research and use of outer space," which requested the USSR Ministry of Foreign Affairs to inform the Ambassador of France in Moscow of the readiness of the USSR to start negotiations with France (Baturin, 2008). The Order defined the main fields of cooperation as:

- Data exchange from satellites launched by the USSR and France
- Joint research of upper atmosphere with meteorological rockets launched from India
- Development of the joint program for a launch of joint meteorological and geophysical rockets with data exchange follow-up
- Development of experiments for the radiotelephone communications and transmission of black and white and colored TV programs from the satellite Molniya-1
- Research in space communications, space meteorology and space studies for agreed programs
- Development of the joint scientific instruments by Soviet and French engineers and scientists
- Launch of French satellites by the Soviet boosters and placements of various scientific equipment developed by the French scientists on the Soviet satellites (Baturin, 2008)

The Order also notes that this is not an exhaustive list. The agreement was signed in 1966, and on the occasion of the signing French President General De Gaulle became the first foreigner to visit the USSR cosmodrome at Baikonur.

During 1967-1969 the CNES-Intercosmos Council bilateral commission met twice a year, and since 1970 once a year, and it was referred to as the "Great Commission." Five working groups were created: geophysics, astronomy, space biology and medicine, telecommunications, and meteorology. Some of the many accomplishments achieved during this long lasting bilateral international cooperation are listed below.

- High atmosphere studies where done in the period 1967-1979.
- Study of ionosphere and magnetosphere with balloons was performed from 1968 to 1974. For example, in 1974 six French and ten Soviet balloons were launched from Kiruna, Sweden. French spectrometers, to study electrons and protons of low energy throwing down to Earth atmosphere during aurora borealis, were put in 1971 and 1973 on Russian satellites Oreol-1 and Oreol-2.
- French satellites SRET-1 and SRET-2 were launched in 1971 and 1975 by the USSR piggybacked on Molniya satellites.
- Space geodesy experiments on Luna-17 in November 1970, at Lunakhod-17 in 1973 and Luna-21 in January 1973 with

- practical implementation of laser-reflectors for space triangulation.
- Lunar samples brought back by Luna-16, 20 and 24 were given to French laboratories between 1971 and 1977.
- Space biology experiments were done on 11 spacecraft titled Cosmos in a period from 1973 to 1996.
- Since Cosmos-1514, monkeys were equipped with flying suits produced by French CERMA.
- Elaboration of materials (ELMA) started in 1978. French experiments were performed onboard of Salyut-6 and Salyut-7 Soviet space stations, as well as on 13 Foton satellites from 1985 to 2002.
- Earth observation experiments were done on satellites Meteor-3, Resurs-O, as well as onboard space station Mir (Lardier, 2001).

In 1979 the Union of the Soviet Socialist Republics signed an agreement with France for human space flight. J.-L. Chrétien made his flights in summer 1982 and later in autumn 1988.

In 1989 a new agreement between the Soviet Union and France was signed for a further 10 years in addition to the First Agreement of 1966. The new Agreement covered periodic flights of French cosmonauts, such as the flight of M. Tognini in August 1992.

This was followed by a global agreement for four human spaceflights that included J.-P. Haigneré in 1993, C. André-Deshays in 1996, Leopold Eyharts in 1997, and J.-P. Haigneré in 1999 (Lardier, 2001).

Program Intercosmos

In April 1965 the Soviet Government issued Order N302-103 "About cooperation of the USSR and socialist countries in the research and use of outer space." The Order called for a meeting of representatives of partners of the Soviet Union, and Soviet Premier A. Kosygin sent an invitation to the leaders of those countries.

The meeting was held in Moscow November 15-20, 1965 with the representatives of Bulgaria, Hungary, Cuba, North Korea, Mongolia, Poland, Romania, and Czechoslovakia. China and Albania refused to participate in the meeting, and Yugoslavia did not reply to the invitation at all.

At that time Vietnam also declined to participate due to its ongoing war and the absence of space research activities. Towards the end of the 1970s, Vietnam did send a cosmonaut onboard the Soviet space station. Order N302-103 also defined the following fields of cooperation: visits of representatives of the socialist countries to the Institute of Medical and Biological Problems (IMBP), laboratories of the Academy of Sciences and Medical Academy, Central Observatory of the Hydro-meteorological Service and space data acquisition stations in the Moscow region and Yevpatoria (today, Ukraine). The Order also permitted the use of a number of boosters and geophysical and meteorological rockets for international cooperation, and allowed members of the socialist countries at a defined list of launching pads (Baturin, 2008). At that time the basics of proposed cooperation looked as follows:

- Development of the hardware for scientific research on the Soviet boosters and geophysical and meteorological rockets by the specialists from the socialist countries, with the support of the Soviet scientists;
- Participation of socialist countries in the development of an international system for long range radio telecommunications, TV, and meteorological services, as well as development of the ground segment;
- Data exchange received from Soviet satellites and/or by Soviet scientists;
- Development of radio technical and optical monitoring ground stations;
- Scientific research in space physics, space meteorology, communications, space biology and medicine due to programs approved by the Inter-Ministerial Council Scientific and Technical Council for space research under the USSR Academy of Sciences (Baturin, 2008).

As mentioned, in May 1966 the Council for the International Cooperation in Research and Use of Outer Space under USSR Academy of Sciences was organized.

During a meeting of April 5-13, 1967 in Moscow, the first program of cooperation among the USSR with Bulgaria, Hungary, East Germany, Cuba, Mongolia, Poland, Romania and Czechoslovakia was developed, which lately became the Intercosmos Program. On April 13, 1970 at the meeting in Poland this cooperation program was officially named Intercosmos Program (from two Russian words International Cosmos). The Program was run by the Intercosmos Council and for two decades Soviet

Academician B. Petrov was the Head of the Council, followed later by Academician V. Kotelnikov. The Intercosmos Council was active until 1991, and performed a significant number of space research experiments in areas that include space meteorology, communications, space biology and medicine, space physics, Earth observation, and human spaceflight.

Each country participating in Intercosmos Program had its own coordinating committee, and joint projects were done through Joint Working Groups. No special budget was defined for these projects, and while the Soviet Union offered Soviet space technology for free, each participating country developed their own national space programs in space science and engineering according to their needs and their own budgets. High level Intercosmos Council meetings were conducted regularly.

The first satellite for the Intercosmos research program, Cosmos-261, was launched in December 1968. It mission was to study the upper atmosphere of Earth. On October 14, 1969 the first real "baby" of the Intercosmos cooperation program was launched, a satellite officially named Intercosmos-1.

During the period from 1969 to 1979, 20 satellites of the Intercosmos series were launched, as well as eight meteorological vertical-launch rockets for joint research. In 1976 an automatic universal orbital platform was launched as a new generation of Intercosmos satellites. The newly developed Intercosmos spacecraft was able to carry 2 to 3 times more scientific payload than previous Intercosmos satellites. The spacecraft had also a unified telemetric system that allowed all participant countries to acquire scientific information directly by ground stations on their own territories (Kozyrev, 1979).

A new round of the Intercosmos Program started in 1976 when the Soviet Union proposed to participate in human space flight. Nations that sent cosmonauts to space onboard Soviet space stations Salyut-6 and Salyut-7 include Czechoslovakia, Bulgaria, Hungary, Cuba, Eastern Germany, Mongolia, Poland, and Romania. Later this program was joined by Vietnam and Afghanistan.

The first human spaceflight of the Intercosmos Program was on March 2, 1979 by the joint USSR - Czechoslovakian crew A. Gubarev and V. Remek. In the post-Soviet Union era when Europe started to look into space history, apparently V. Remek was the first European man in space (Kozyrev, 1979).

In 1972 the International Communication Union Intersputnik was created by the USSR and its partners. Intersputnik and Intercosmos had long standing cooperation in space telecommunications. The Intercosmos

Program stopped in 1991 after the split of the Soviet Union.

Soviet Union – United States Cooperation in Space

The story of Soviet–US cooperation in space requires special consideration and a separate history book; in this chapter the major steps of bilateral cooperation are outlined.

During the Cold War, political rhetoric and activities between the USSR and the USA were far from positive. American U-2 spy planes flew over USSR territory, the Caribbean Crisis, the US war in Vietnam, and Soviet troops in Afghanistan all created very heavy and difficult environment for cooperation between the two countries. However, behind the diplomatic and mass media scenes, space engineers and scientists were constantly trying to build cooperation between the two superpowers. The period of political détente in the early 1970s was also quite helpful.

In the article, "United States – Soviet Space cooperation during Cold War," published on NASA's website, former Director of the Institute of Space Research of the USSR Academy of Sciences Roald Sagdeev recalls that in 1957-1958 US President D. Eisenhower wrote a number of letters to the General Secretary of the USSR Communist Party N. Khruschev suggesting cooperation in space. Khruschev did not reply.

However, after the flight of Yuri Gagarin into space in 1961 and flight of John Glenn in 1962, Khruschev wrote to President Kennedy proposing cooperation in space (Eisenhower, 2008). Based on these exchanges, a three-part bilateral agreement between the USSR and the USA was prepared by USSR Academician A. Blagonravov and Dr. H. Dryden, Deputy Administrator of NASA. The 1962 bilateral agreement allowed for:

- Coordinated US and Soviet launches of experimental meteorological satellites with follow up of data exchange;
- Launch of both countries' spacecraft with absolute magnetometers on board with subsequent data exchange to produce Earth's magnetic field in space;
- Joint communications experiments using Echo 2, US passive satellite (NASA, 1975).

The Blagonravov-Dryden negotiations ended with a second agreement in October 1965 with the decision to make the joint publication outlining a Soviet-American research in space biology and medicine. The

Institute of Medical and Biological Problems was the key in this cooperation from the USSR side. The bilateral scientific research was published in three volumes in 1975 in Russian and in English.

In 1969, NASA Administrator Dr. T Paine proposed to the Head of the USSR President Academy of Sciences M. Keldysh that cooperation in rendezvous and docking of a manned spacecraft as well as in space research (NASA, 1975) should be developed. In October 1970 negotiations for the possibility of the US and the USSR to each design a manned spacecraft with compatible docking mechanisms were held in Moscow.

Vladimir Syromyatnikov, father of the USSR docking mechanism recalls in his book *100 Stories about Docking* that when the NASA delegation arrived in Moscow they were quite pessimistic about the possible cooperation. Twenty years later Glenn Lunney, an American Technical Director of the ASTP commented that, "Apollo-Soyuz was a miracle." Three working groups were organized since 1970 and these groups formed the basis for the development of ASTP (Apollo-Soyuz Test Project) project until its docking in orbit. Working Group #1 was responsible for integration design, means of rendezvous and control were the responsibility of the Working Group number 2; and the third Working Group was responsible for docking (Syromiatnikov, 2005).

The next NASA-the Soviet Union Academy of Sciences Agreement of 1971 was focused on:

- Exchange of lunar samples obtained from the Apollo and Luna programs;
- Exchange of weather satellite data between the United States' National Ocean and Atmospheric Administration (NOAA) and the Soviet Hydro-meteorological Service;
- Coordination of network or meteorological sounding rockets along selected meridian lines;
- Exchange of detailed medical information of human body reactions to the space environment (NASA, 1975).

On May 24, 1972 the US President Richard Nixon and the Soviet Union Prime Minister Alexei Kosygin signed the Agreement concerning cooperation in the "Exploration and Use of Outer Space for Peaceful Purposes." The Joint Soviet-American mission was officially announced as the Apollo-Soyuz Test Project. This agreement also required both countries to fulfill a NASA-Soviet Academy of Science agreement of January 1971 (NASA, 1975).

In July 1975, Soyuz 19 and Apollo docked in space. The archive of Jaap Terweij includes the records from the onboard press conference of Soyuz 19 and Apollo spaceships held on July 18. Questions were sent separately from the USA and the USSR, but everything that the cosmonauts and astronauts told the press was strictly controlled and was verified extensively by both sides.

The press conference began with the official statements by Apollo Commander T. Stafford and Soyuz 19 Commander A. Leonov. The US Commander Stafford stated from the orbit: "We are happy to work today on the ASTP program. The success of the flight which is now followed by America, the USSR, and the rest of the world, is a result of the will, cooperation and efforts of the governments of our two countries, the mission commanders and the engineering and technical personnel, and other specialists. Yesterday as I opened the hatch for the first time and said "Hello" to Valery and Alexei, I felt that when we opened this hatch in space we were opening back on Earth a new era in the history of man. How this era develops further will depend on the will, efforts and faith of our two nations and other nations of the world. I am sure this era has a good future ahead of it. It is real pleasure for me to take part in this flight and to work together with the Soviet cosmonauts."

The Soviet Commander Leonov stated that, "We representatives of the two countries are performing this joint flight because our nations and Governments want to work together in the spirit of cooperation, because many specialists in the USA and in the USSR put tremendous effort into making this project a reality. The flight has become possible in the climate of international détente and developing cooperation between our two countries. This flight is an important step along the endless road for space exploration by the joint efforts of all mankind" (Press-kit, 1975).

This question to Vance Brand was transmitted from the US press center: "Now that the Americans have united with the Russians in space for the first time in an international space mission, what do you think manned spacecraft can do in the joint exploration of other planets?" He answered, "I think that we now stand good chances of that. But it is not likely to happen in the near future, to my mind. It will probably be another 20 or 30 years before we shall be ready to start the exploration of other planets. But modern progress is truly impetuous, as evidenced by the development of aviation. And it is quite possible, therefore, that within the next 20 or 30 years the time will come when we shall think of exploring other planets by joint effort. It seems to me it ought to take place. It would save time, effort and financial outlays. It would be the most interesting project that would benefit the whole world" (Press-kit, 1975).

In 1978, American engineers from Johnson Space Center prepared a proposal for the US Space Shuttle to dock with the Russian Salyut station in space including the SPACELAB module. In 1991, US President George Bush and Soviet President Mikhail Gorbachev signed an agreement that included visits of US astronaut aboard Soyuz-TM, and a fight by the Russian cosmonaut on a Space Shuttle mission. They also discussed the possibility of a US module to be docked to the Mir space station (NASA, Shuttle-Mir History/Background/Cooperation Timeline 1962-1993, 2004).

The Joint Cooperation Agreement in space with a Shuttle mission to visit Mir space station (Mir-Shuttle program), a US astronaut onboard Mir, and the US participation in Mars-94 Russian exploration project was signed in 1992. The next year this Agreement was expanded to include 10 Shuttle flights to the Russian Mir space station (Mir-NASA Program). In 1993 Russia was invited to join construction of the International Space Station.

Foton-Bion:
A brief history of European cooperation from the Soviet Union era to modern Russia (1987-2007)

The following Foton-Bion story about the European Space Agency-Russian cooperation on joint microgravity projects for technological experiments onboard Foton and for biological research onboard Bion satellites is written by Dr. Antonio Verga, who participated in a number of years in cooperation with Russia as an ESA project manager. This presents a personal viewpoint, as his story goes far beyond engineering, as he witnessed significant changes in working culture in space, real ice-breaking from the Cold War era, but at the same time he also witnessed dramatic events that my country went through in the post-Soviet Union era.

I am simply jealous knowing that Antonio was privileged to know one of the most classified persons in the USSR, a colleague of Sergey Pavlovich Korolev, Dmitry Ilyich Kozlov. Dmitry Kozlov was the Head of the space enterprise in Samara from the end of 1950s until his death in 2009, and he was the Chief Designer of R-7 booster and Soviet reconnaissance satellites.

"The history of ESA's cooperation with the Russian aerospace world, actually with the Soviet one, started in May 1986 when, in the aftermath of the US-Space Shuttle Challenger tragic explosion, the national delegates at ESA's microgravity program board "urged ESA to look for new flight opportunities for all areas relevant to microgravity, and particularly for opportunities with the Soviet Union." What had until then been strictly forbidden, in line with the European politics about the Cold War, suddenly became keenly recommended. It was the time when Mikhail Gorbachev was planning the reconstruction of the Soviet Union, not knowing that his revolutionary ideas would finally lead to its political and economic fall. From May 1986, ESA flew its microgravity experiments three times with the Soviet Cosmos program, starting in September 1987 with Bion-8 (still called Cosmos 1887 at that time), then followed by Bion-9 in 1989 and Foton-7 in October 1991, just before the USSR's political and geographical entity began to fall into pieces.

I had no chance to participate in any of those pioneering missions, but I collected the personal experiences of many of my colleagues. Those were the times when commercial exchanges were unthinkable between the Western economies and the Soviet Union. Smart enough, the managers of the space business on both sides found a way to cooperate that was efficient and productive at the same time.

The Soviet Union in those days had the capability to launch satellites and human crews into space at a terrific pace and with unimaginable success rates. Conversely, the European space industries were still struggling to find their roles and their identities within a baby ESA that was setting its race for space well behind the American and the Soviet giants. However, the '80s saw tremendous technological leaps in Europe in all fields and disciplines, and the European applied science programs of those days was not equaled by any other academic community in the world. Hence, the cooperation between the young ESA and an old and almost finished Soviet Union was a natural partnership, and its sustaining vehicle was an exchange of services: ESA acquired the flight tickets to low-orbit Soviet platforms in return of technology spin off and support infrastructures. This nature of cooperation "in kind" lasted throughout the transition from USSR to Russia and continued, on a different basis until the most recent Foton flights in 2005 and 2007.

I had the impression from studying the testimonials recorded about those early days that beyond the initial steel-cold approach that the political etiquette imposed at both sides, there was a strong will and stubborn intent to break the ice and let the best ideas of those two very different worlds flow, meld and blend in search of a joint experience, venturing into a new "space romance." While this was certainly less glamorous than the Apollo-Soyuz rendezvous of the '70s, it was made to endure technical and political difficulties and to maintain a steady and constant role in the space stage worldwide. The most significant example of this cooperation is the pre-fab cabin of MOSLAB, which still stands on the IMBP grounds next to Moscow. Its outdated ESA and RKA logos still greet the European delegations on their way from the Sheremetievo International Airport to the city.

MOSLAB has a special chapter within this history. A 70 square meter laboratory with advanced installation for the preparation of biological specimens, it was completed by ESA by 1992 and traded in 1995 to the Russian Space Agency for opportunities given to ESA on three Bion flights and two Foton flights. The last of that series, BION-10, flew in January 1993 and although the Soviet Union did not exist any more by then, Bion-10 still carried the hammer-and-sickle logo on a red field, an homage to the flag that had waved for more than 30 years of journeys into space that mankind started in 1957. MOSLAB was used until the Foton-11 mission in October 1997, and served as ESA's "vanguard," or forefront, in Russian soil, a home base of many experiments duly readied in its labs before their transport to the launch site, the Plesetsk Cosmodrome, 18 hours away by train.

My very first journey to Russia in 1993 had some dramatic moments. It was a cold and rainy evening in September when we arrived in Samara after an endless stop-over and airport commuting stage in Moscow. Samara was pitch dark on that evening, and with no vehicles along any of its deserted streets, except for white and red, noisy trams running in the middle of wide, empty boulevards. At each of their passings, the hotel room where I tried to sleep trembled from floor to ceiling, and the nightmare of an earthquake prevailed my waking and my emotions of those nights.

Samara had just been "opened" in February that year and renamed from the Soviet appellation after Kuybishev, but many of the Soviet marks survived until the late '90s, including bronze effigies of Stalin decorating some of its street corners. In a room of the Samara Aerospace State University I had my first meeting with the project team of TsSKB, called only KB or Konstruktorskoe Bureau (Design Office) at that time. The small TsSKB delegation of four equated exactly in number to me and my colleagues, and all communication went through an interpreter. Out of that group, only Valery Ivanovich and I had the chance to discuss in front of the blue prints and schematics of Bion-M2, on another cold and wet September morning in 2010, exactly 17 years after our first encounter.

TsSKB, nowadays complemented by a glorious suffix to become TsSKB-Progress, has been the largest and probably the most important space factory in Russia or former Soviet Union. Chosen by Josef Stalin himself in the city of Samara, at a safe distance from the risks of foreign military attacks, there the father of the Soviet Union had a special bunker dug for his own safety, 50 meters below one of the most central and most beautiful of Samara's squares. I had to be patient for a few more years, five indeed, before being allowed in May 1998 to admire the powerful capacity of TsSKB's integration halls, where no less than four 43 meters-tall, 3-stage Soyuz rockets were being assembled at the same time. But from that very first and rather dry meeting, I understood that a huge potential was ready to be deployed for our common intents of scientific research in space.

The drama of that first trip of mine culminated on our way back to Moscow the first day of October 1993, where we were to meet with managers of the newly founded Russian Space Agency. We ended up in the intense turmoil between President Yeltsin and the Russian Parliament of that time, supporters to the last tried to seize power and control of the Russian capital. We left one day earlier than planned, just before a curfew state was declared and all international flights cancelled. General Grachev and Boris Yeltsin swept all opposition away by mid-October 1993, but by then I was meditating at home on whether to accept or decline the challenge that I had ahead of me. Something suggested that I embrace that new course.

One year later we were back in Samara and, every year since then, I have been cruising back and forth for meetings, conferences, reviews, integration and test campaigns for five or more Foton missions, starting from Foton-11 in 1997. Meanwhile, Foton-9 and Foton-10 were completed in 1994 and 1995 respectively, yet with opposite fortune, with me being barely an observer or just at the edges of the data telemetry chains.

In 1995 ESA set an office in the heart of Moscow, and with MOSLAB on the brink of being disowned and forsaken, we sought alternative options to the logistic challenges that our project had to face and, sometime, to be pestered by. That was probably one of the most difficult periods, the transition from the stiff but well-structured Soviet organization to the more flexible but often confused Russian administrative regulations left us, more than once, mesmerized in between deceptive and unclear procedures. The help of ESA's office in Moscow became more and more essential, although our center of gravity was in TsSKB and in Samara. Samara was to me and many of us the true mirror of the Russian heartbeat of economy, culture, development, social and popular life. We saw this town falling apart in the late '90s and being reconstructed throughout the very beginning of the 2000s, an architectural revolution that managed to maintain the most striking and charming allure of the old times, and to respect the surrounding natural environment in a manner which has been dynamic and conservative alike. The local chocolate factory was sold to a Swiss competitor, but the hammer and sickle that decorated the gable of its gate resisted for a few more years before disappearing, while downtown the first McDonald's location was created in the space of a fortnight.

After years of frustrating hesitations, TsSKB finally opened their gates to us in 1998, just when the economic crisis hit Russia and put its population down on their knees. Empty food shops and queues at delivery points in street markets were a sharp contrast to our hi-tech instruments, our tool boxes, our satellite phones. But in those two years that closed the 20th century and marked the lift-off of Foton-12 on an unforgettable date, 9/9/99, our friendship with the project team at TsSKB tightened, our reciprocal respect grew, our mutual confidence and trust reached a level that was well beyond the professional boundaries. Little by little, all possible communication channels were surveyed and opened, the exchange of information widened, engineering teams started working together in an unprecedented way, technical solutions were studied and adopted by joint team of specialists, and the interest in our joint activities escalated to upper level managers. In October of that year, Dmitri Ilich Kozlov, General Director of TsSKB and number one of the enterprise invited me, ESA's number 4061, to share a drink for his 80th birthday; I was busy de-installing ESA's hardware from the scorched Foton-12 capsule just returned to Samara, but I dropped everything and ran.

The old barriers of secrecy and suspicion, meaningless remnants of the cold-war era, started crumbling down. Foton became the platform where brilliant technical solutions met, the test bed of our engineering skills, the product of our shared professional experiences, the battlefield where innovative scientific research found its path towards ever-ambitious objectives. To mark all these facets of the program and to hold a tangible example to be studied and further developed, the Foton-12 re-entry capsule was brought to ESTEC and proudly stands in our showcase in the Erasmus building as unique a piece of spacecraft ever flown to and returned from space.

In late November 1999 I flew back from the biting freeze of Moscow after a meeting in Roscosmos. On the flight to Amsterdam, unbelievable but true, I was seated next to Mikhail Gorbachev, who had recently lost his lovely wife Raissa and was heading to Munster to open a new ward of the local hospital, where she received her last attentive care. My command of the Russian language was still very poor and I managed to put together only a few limping sentences. Mr. Gorbachev is a man of immense culture, far visions and great social and popular consideration. He had been the leader of a country that was among the most prominent space powers of the world, yet during the four hour flight over those lands that once dealt with the balance between the NATO countries and the Soviet Union block alliance, he never dropped his attention from my words and always followed with interest my stories and explanations about my fresh and amazing experience with Russian space fellows. I could not, and still I can't, believe that my short and fragmented sentences about the new Russia captured the concern and empathy of a former president of the Soviet Union. I was very embarrassed for not being able to answer several of his questions about the ancient Roman history, one of his favorite studies of a lifetime.

Samara, in those years after Foton-12, became almost my second home town, not knowing actually which one is the first, and I began driving around in rental cars, partying at wedding celebrations and, at some very sad times, mourning deceased colleagues or members of their families. Shop assistants and waiters greeted and waved at me in the central pedestrian area calling by my first name. Police officers stopped me often for offences to the traffic rules, which I had to negotiate hard to clear. The hotel where we used to stay changed its name, its breakfast room, its decoration and its furniture, but always kept the very same room booked for me whenever I had been there. One of the nearby restaurants invited me once to cook some Italian dishes for its customers; we gathered public curiosity among the locals for swimming competitions, skiing trail excursions, basketball games, and even singing and dancing contests. At any achieved milestone of our joint projects, TsSKB never hesitated to host us to celebrate on their cozy boat while cruising the Volga river. That boat, named after the program we have been pursuing and where delicious food and inviting drinks were never at short, was the only ever manned Foton ship.

After the enthusiastic and highly content-rich first international scientific conference that the Samara's academic and industrial institutions organized in summer 2000, the opening of the 21st century was signed by the darkest point in the history of Foton: the launch accident in Plesetsk on October 15, 2002. Foton-M1 was totally destroyed in the crashing explosion of the Soyuz rocket and the arson of its 250 tons of kerosene propellant. The accident, taking two human lives and severe injuries, could have meant the end of that program, but the pride of TsSKB-Progress' freshly nominated director and the convinced stern position of ESA, ready to restart and to continue, made two more missions possible anew with yet a larger complement of scientific experiments and technologically advanced instruments. Since the very first stages of work, the failure investigation board, which I had the honor of being a member of, had been clearly searching the root causes of the failure, a shared and common objective. In no one's mind ever comes the shadow of shame or the ploy of hiding evidence. The inquiry board reached the conclusion of its work before the year was over, with unbiased judgment and sticking to the forensic facts.

Having kissed good-bye to Plesetsk after 10 launch campaigns, ESA travelled along with the Foton machinery moving to Baikonur and invited new partners on board, such as the Canadian and the Italian Space Agencies, and US scientists. The joint work with our Russian partners, TsSKB-Progress, Roscosmos, IMBP, Design Bureau of General Machine Building (KBOM), scrolled out through weeks and months, with a very pleasant and high team spirit, renovated integration halls both in Samara and in Baikonur, nimbler and simpler procedures for transport and Customs clearance, new and more efficient ways to exchange engineering data, and a rather expectedly optimized set of operations both for the ground testing phase and for the in-orbit conduct. When Foton-M2 was about to lift off from Baikonur on May 31, 2005, I bribed the Tu-74 pilot and crew with some colorful stickers of the mission, and all the passengers were allowed to disembark the aircraft and watch the launch from the tarmac, a far but still convenient viewing point. We took off only after the Soyuz rocket completed its orbital injection task; tears were rolling down my cheeks while, standing in an airplane packed with a crowd of Kazakh and Uzbek peasants, their livestock of hens, eggs and any sort of vegetables, fruit, spices, sausages and cheese, I was trying to reach my dear father agonizing in hospital, thousands of miles away. I did not make it on time.

The quality of work and professional relations improved even further for the Foton-M3 mission, despite harsh negotiations for some financial issues risked, at times, undermining the contract. With that mission Foton reached its apex of success: 18 pieces of payload equipment, 40 scientific experiments, 4 technology prototypes, 2 application projects, and 2 educational student experiments, a total of almost 400kg of mass, found their places on board. The mission, assisted from three ground stations and two operational centers in Moscow and in Northern Sweden, was a bursting accomplishment and sent echoes of enthusiasm and recognition in Russia and Western Europe alike. With Foton-M3 completed, the 12th of such projects, ESA scored 3700 cumulative hours in orbit, completed 170 scientific experiments, uploaded and downloaded 1600kg of payload mass.

All this would have been impossible without the support, help, cooperation and friendship of our Russian, former Soviet, colleagues. Twenty years (1987-2007) have passed marked by intense work, highly emotional moments, gratifying achievements, attracting perspectives, and, above all, an everlasting friendship.

Since the Bion and Foton programs lacked the glamour and the show of manned missions, they never hit the front pages of press releases, and both in Russia and in Western Europe, the crew working with these projects were viewed as sons and daughters of a lesser God. Nevertheless, our motivation has always been undefeated and high in the hours spent in shabby and inadequate labs where only our inventive approaches and the unabridged support of our Russian colleagues sustained our struggles to carry on, to move any next step ahead, to try yet once more serving our sole goal: the progress of scientific research that the space environment could alone foster and nourish.

I gathered souvenirs, millions of pictures, celebration pins, books signed by the pupils of the great S. Korolev the true pioneer of the space era, the advice of humble heroes of meticulous tests and repeated trials, the simple but poignant gifts of those who served the big shows from behind the scene, the handshakes of great men and women who made those projects the scope of their entire lives, the moving tails of cleaners, street sweepers, bar keepers and taxi drivers. During those unforgettable years, I learned how and when to drink vodka to celebrate with style, how to toast with acknowledging and encouraging speech, how to wait with patience that time models and shapes our faults into new perspectives. I was taught how to face adverse circumstances, how to dare with calculated risks, how to adapt and focus my view on the essential things, on bare necessities, and how to accept the compromises that often bring about the only possible way forward. One gold medal in the name of Servey Korolev was handed out to me in 2008 by the Samara's Lord Mayor and the Dean of the State University; it hangs proudly next to my office, the only reward I felt I really deserved, proof that I would never give in.

In November 2010, on the way to enliven again ESA's love affair with Bion, idled for 15 years. Bion-M1 and Bion-M2 were just waiting for our consent to join in and proceed with new challenging science goals, but our upper management decided to stop and bring ESA's involvement with those projects to a "controlled end" after 20 years seeded by fruitful results.

I never understood this short-sighted strategy, or better lack of strategic plan, and the only explanation I was ever given was that this kind of endeavor "did not fit any longer in the political landscape of Europe." As a simple-minded engineer, I do not see what such a decision was worth and, as I seldom do, I have kept weeping for the shame and the pity of these missed opportunities. Bion-M1 will take off in September 2012: recalling and leveraging on the practical common sense and well-rehearsed habits of our Russian colleagues, I made sure that a little imprint of ours will be illegally on board." (Verga, 2012)

International Cooperation Of Russia In Space Medcine And Biology

The USSR Institute of Medical and Biological Problems (IMBP) was organized in 1963 and is the leading research establishment for the space biology and medicine in Russia. In response to my questions, Dr. M. Belakovskiy provided a brief report on international cooperation performed by the Institute.

The Institute began international cooperation in 1965 with the Soviet-American project to publish research in space medicine and biology. During flights of the Soviet Space Stations Salyut-6, Salyut-7 and space station Mir from 1977 to 1999, significant experience in space biology and medicine was accumulated in international cooperative projects and programs of the Intercosmos Program as well as others. In the post-Soviet Union era the major cornerstones of international cooperation were the Mir-Shuttle and Mir-NASA programs. Experience received in these programs formed a basis for the international cooperation concerning management, assembly, operation and research for the International Space Station.

Today the Institute of Medical and Biological Problems have joint research and cooperation projects with space agencies, research institutions, industrial companies from more than 40 countries including the USA, ESA, France, Germany, Japan, Canada, China, Italy, South Korea, and Malaysia.

The main fields of international cooperation performed by the IMBP today are as follows:
- Sharing knowledge, results of previous research done by the IMBP;
- Ground simulation and research of influence of conditions of a

space flight on the human body;
- Development and testing of special Space Station equipment for biological and medical research, mostly for the Russian segment of the ISS;
- Pre-flight and post-flight medical check-ups of cosmonauts and astronauts;
- Organization of medical and biological experiments onboard unmanned spacecraft;
- Spin-off of space biology and medicine space experiments and its results, technology transfer;
- Educational programs for schoolchildren, Bachelor level and Master students, PhD and postdoctoral research.

A major part of the international cooperation projects which are performed by the IMBP are connected with the International Space Station. The IMPB participates in the work of the following ISS multi-national medical bodies: Multilateral Medical Policy Board, Multilateral Space Medicine Board, Multilateral Medical Operations Panel, Multilateral Crew Operations Panel (ISS) Integrated Medical Group, Mission Management Team, Stafford-Anfimov Joint Expert Commission, Multilateral Radiation Health Working Group, and Human Research Multilateral Review Board.

That includes development of the regulation and operation documentation for various issues related to the health of the ISS crews, and development of a set of procedures and its practical implementation for the medical, biological, sanitary and hygienic support of the main crews, visiting crews for the ISS, Soyuz-TM and Space Shuttle. The activities of the IMBP also include organization of works for scientific and technical preparation and operation of the onboard equipment for the medical and biological support of the ISS crews.

The IMBP has played an active role in the preparation of space flights of citizens from the USA, Europe, Brazil, Malaysia, South Korea, Canada, and other countries. Their flight programs included medical and biological experiments based on special techniques developed by the IMBP scientists, including:

- A flight program called Centario of the first Brazilian astronaut Pontes Marcos Cesar, which included one biological experiment Gosum (growth of plants from the seeds in the weightlessness conditions) and three biotechnological experiments;
- The first ESA long-duration program, Astrolab, performed by Thomas Rheiter, which included six medical and biological

experiments (for example, ETD, CULT, NOA1,2, IMMUNO, ALTCRISS, CARDIOCOG);
- Program ANGASA, made by Sheikh Muszaphar Shukor, including four experiments on space biology and medicine such as ETD-M, MOP-M, MUSLE-M, FIS and three biotechnological experiments (CIS, MIS and PCS).

There are number of successful cooperative research projects between Roscosms and the European Space Industry which are already completed or still ongoing onboard the International Space Station including:

- MATREOSHKA experiment is looking into development of the onboard means to study consistent patterns of forming of space radiation in major organs of a human body with the use of scientific equipment Pille, LYULIN and dosimeters from Canada. The project is done together with ESA, DLR, CSA and scientific institutions from Hungary, Bulgaria and Japan;
- Experiment IMMUNO is research of immune reaction of a human body before and after space flight. The experiment is performed together with ESA;
- CARDIOCOG is the joint research of adaptation of specific features of a heart and circulatory systems to the conditions of the long duration space flight. The experiment is done together with ESA.

One of the fields of international cooperation that is suggested today by the Institute of Medical and Biological Problems is development of onboard equipment and separate devices for space biology and medicine research. This is done on the basis of bilateral or multilateral cooperation projects. For example:

- The EUROLAB experiment, is done by IMBP jointly with DLR and Koralevski Industry, developing special equipment to study the psychological and physiological conditions of a cosmonaut in a long duration flight;
- CARDIOME is performed together with CNES, development of hardware for a medical checkup of cosmonauts on the Russian segments of the ISS;
- LADA, together with Space Dynamics Laboratory of the Utah State University, USA, is developing space green houses for the

ISS.

Finally, but not least, the Institute of Medical and Biological Problems has deep experience in the ground simulation of long duration space flights with broad international cooperation. The Institute has made a number of ground simulations with long duration isolation, dry immersion, and hypokinesy with the participation of foreign partners. In November 2011 it successfully completed the Russian project Mars-500, an imitation of the human space flight for Mars. Crew members of Mars-500 spent five hundred days in an isolation chamber. The institutions that participated in the project include the European Space Agency as general partner, NSBRI from the USA, DLR, Chinese Cosmonaut Training Center, Italian Space Agency, ANGKASA Malaysian Space Agency, and research establishments from the South Korea, Italy, Germany, Austria, Spain and other countries.

Looking forward in the near term and far term futures, Russian experts in space biology and medicine are interested in the further cooperation on the following themes:

- Adaptation of the human body to the conditions of the short and long term space flights, including interplanetary flights;
- Means and techniques of directed counter measures for the non-positive changes in a human body during space fight;
- Life support during space flight, including EVAs, and development of next generation biological system for life support;
- Search for extra-terrestrial intelligence.

Some of themes listed above are already the subject of cooperative projects, while others wait for further examination and preparation for international cooperation in a near future. (Belakovskiy, 2012)

COOPERATION WITH RUSSIA FOR THE INTERNATIONAL SPACE STATION: A EUROPEAN INDUSTRY PROSPECTIVE

Dr. Carlo Mirra, Senior Manager of EADS/ASTRIUM, is involved in the Increment and Mission Integration for the International Space Station, and has worked in cooperation with Russia for many years. Below are his answers to my questions on European-Russian cooperation for the International Space Station.

1. How long you are involved in the cooperation with Russia? In what projects have you participated?

I have been involved in cooperation with Russia since the first Dutch Soyuz Mission – the Delta Mission in 2004. After several years of experience with western space programs, mainly Shuttle based including Spacelab and SpaceHab, the discovery of the Russian space program was a very attractive and professionally rewarding experience.

Since then I have been involved in the preparation and management of the Eneride Soyuz Mission and the Astrolab mission, performed in direct relationship with Roscosmos. After the ESA Columbus module launch, we have participated in joint research activities and to upload and sometimes operate experiments on Soyuz and Russian segments. In total I have spent nine years in continuous collaboration that have disclosed a lot on the Russian space heritage, their working environment, and relationships.

2. What are the Lessons learned from joint Russian-European cooperative projects?

We have learned a lot from our Russian colleagues in our baby steps toward manned space flight, such as the important steps and milestones to be taken care of during the mission preparation, the attention to crew training and cosmonaut well-being. The completeness of the operational products accompanying the scientific program, the proper design of the launch packages were additional examples that the long lasting experience of the Russian colleagues helped to shape our thinking in preparation for the Columbus operation period.

I particularly remember the working relationship with several colleagues at Russian Space Corporation (RSC) Energia, IMBP and Gagarin Cosmonaut Training Centre (GCTC). It took some time to build some mutual trust, but communication was always open and never teacher-pupil like. I recall the extreme professionalism of the RSC Energia management. They were very sensitive to customer needs, engaged in problem resolution, and responsive to last minute issues. This is what today we would call a "customer-focused" mentality that I did not expect to find in an environment that was not really used to a commercial relationship.

3. What are the benefits to both Europe and Russia from international cooperation in space?

Europe learned a lot from the cooperation with Russia in the manned space field. Not only on how to design and conduct long duration space missions, which we were not accustomed to until Astrolab in 2006, but also in the mission integration process. I am also convinced that these projects have fostered scientific collaboration with Russian research institutions, and today some of these relationships still exist as legacy of this long collaboration.

I believe these experiences were also eye-openers for the Russian colleagues. Only few of them had had contact with European institutions before, and some did not speak English. It was very important for them to develop their working mode with ours.

4. What are the prospects for future further cooperation in space?
The legacy of the first bilateral collaborations in the manned space field exists today: joint research projects, exploitation of know-how on Russian vehicle design and mission definition are helping the operation of Columbus and ATV on a regular basis. Scientific research experiments have just finished (MARS 500), are ongoing (e.g. Immuno) or are going to be re-launched (e.g. PK4).

New initiatives are developing in the field of planetary exploration, long term medical research and robotic technologies. The path is open; it is up to the people who can exploit it to make good use of it in the years to come.

Concluding Remarks

In the summer of 1996 I was a participant at the Summer Session Program of the International Space University in Vienna, Austria. There was an interdisciplinary essay exam question to answer, asking about legal, engineering, and medical issues concerning an experiment done by an astronaut from a non-spacefaring nation onboard the space station, working on Russian hardware, and going down with US Space Shuttle.

For me and for my ISU SSP'96 classmates, this exam question was not from real life, it was more of an inspirational science fiction written by Arthur C. Clarke (who, by the way, was the ISU Chancellor at that time). ISU stands for more than two decades for teaching in international, intercultural, interdisciplinary and peaceful use of outer space, with inspiration and vision for multi-national cooperation in space.

In 1996, the International Space Station was expected to be launched soon; eleven years later, in 2007, at ISU SSP in Beijing, I organized an inter-agency panel on International Space Station utilization. I did share with ISU SSP'07 class my thoughts about this particular exam question at

ISU in Vienna in 1996 when I was SSP participant, but the difference is that by 2007 such cooperation was the reality: I was at that time responsible for the integration of an ESA medical experiment that would be done by a Malaysian astronaut, on the Russian research hardware, onboard the ISS.

So today's ISU exam questions should be dealing with flight of the international crew to Mars. As mentioned above, Soyuz-Apollo crew member Vance Brand predicted that joint flight to other planets would be a reality 20-30 years from middle of 1970s. Now we are 40 years after the ASTP event and we are already halfway through with the International Space Station in orbit, but the question as to when the flight to Mars can come true, in twenty or thirty years still from now, or less, remains open.

I think it is really good that in the era in which private space tourism is beginning, and in which broad commercialization of space is occurring, and quite severe competition in Earth the observation, telecommunications and other space integrated applications areas, there are still many opportunities left for international cooperation in space. Of course, current economic and financial challenges do reflect significantly on funding of national space programs worldwide, that subsequently influence or even sometimes threaten international cooperation projects and programs. From my point of view, it is still quite symbolic that nations of Earth cooperate and learn how to go together from the cradle, Planet Earth, to explore the far ends of the Universe, as the Russian space philosopher and pioneer Konstantin Tsiolkovsky dreamed in the 1900s.

The basis for that was formed by international cooperation started in the USSR with the Intercosmos Program, the Soyuz-Apollo Test Project, USSR-France collaboration, USSR-ESA cooperation, as well as others, all of which have shown that people from different nations and cultures share many of the same goals and aspirations, and can effectively meld their technical expertise to accomplish demanding mission goals while developing mutual respect, friendship, and tremendous new learnings for the benefit of all.

•••

Acknowledgments

I would like to thank Dr. Mark Belakovskiy (IMBP, Russia), Dr. Carlo Mirra (EADS/Astrium, Netherlands), Dr. Antonio Verga (ESA, Netherlands), Mr. Jaap Terweij (SpaceView, Netherlands) for the materials that they provided, and Dr. Noel Siemon, Australia for his support.

Olga Zhdanovich, MSc

Olga Zhdanovich works in the secretariat of the European Cooperation for Space Standardization at ESTEC/ESA in the Netherlands. In 2006-2010 she worked as Payload Integration Manager for the International Space Station via RheaTech, responsible for the integration of various European research payloads for the ISS, as well as University level educational projects and programs.

She graduated with honors from the Moscow Institute of Engineers in Geodesy, Aerial Surveying and Cartography, Russia as engineer of cartography in 1990. Seven years later she received an MSc in Environmental Science and Policy from Central European University, Hungary/University of Manchester, UK. Although her principal area is remote sensing, during twenty years of her professional career she worked as a consultant on commercial application of space technologies in Earth observation, satellite navigation and telecommunication, as well as educational projects. Ms. Zhdanovich is Faculty of the International Space University and Co-Chair of IAF Sub-Committee on Global Workforce Development. She was the coordinator of the Forum on Space Activities in the 21st Century at UNISPACE III in 1999 in Vienna, and was the member of the Russian Federation Delegation to UN COPUOS in 2004-2005. Olga is recipient of national and international awards and scholarships that include a scholarship from the Royal Dutch National Academy of Sciences in 1995 for the International Institute of Applied Systems Analysis Young Scientist Program, and an award from the European Space Agency.

Olga was one of the three founding editors of the *Novosti Kosmonavtiki* magazine in Russia in 1991. She has authored a number of publications as chapters in books and conference papers on various applications of space technology as well as the Russian space program.

CONTRIBUTOR
ANTONIO VERGA, PH.D.

Antonio Verga has worked since 1988 for the European Space Agency at its establishment in Noordwijk, The Netherlands. He holds a doctor degree in nuclear engineering but he has been dealing for the last 25 years with physical science experiments in weightlessness. In this context, ESA deployed him in the early 90s to exploit the possibilities offered by Foton and Bion as platforms for science experiments in Microgravity.

He is currently the Head of the Unmanned Microgravity Platform Office at ESTEC (European Space Research and Technology Centre), Noordwijk, as part of the Directorate of Human Space Flights Programmes, Utilisation Department, Payloads and Platform Division. As the cooperation with unmanned Russian space project is standing still, he is nowadays managing the Sounding Rocket programs at ESA. During his professional carrier, Antonio Verga, participated in 2 SpaceLab missions with NASA, 6 Foton missions, 2 Bion projects, 6 European Sounding Rocket launch campaigns and 1 parabolic flight campaign with the Canadian Space Agency, published more than 30 scientific/technical papers, attended several international conferences, contributed to various issues of space journals and magazines, and lectured at seminars and topical schools for students. He is referee for SPIE and AIAA proceedings and papers. He is father of three children and fills his free time with sport activities and organizing events close to his Italian origin and traditions.

CONTRIBUTOR
MARK SAMUILOVICH BELAKOVSKIY, PH.D., M.D.

Mark Samuilovich Belakovskiy, Ph.D., M.D., graduated from the First Moscow Institute of Medicine (1972) and finished postgraduate courses at the same institute in 1975. He has been working as the Head of Department at the Institute of Medical and Biological Problems since 1988.

Dr. Belakovskiy is a specialist in the field of research in nutrition and human body metabolism in different extreme conditions. He participated in a biomedical support of various activities such as: a) manned space flights on the Soyuz spacecraft series, orbital stations Salyut, Mir and the ISS; b) research on biological satellites; c) preparation and implementation of high-latitude scientific polar expeditions; d) the first expedition of Soviet mountain climbers to Mount Everest; e) expeditions to deserts; f) biomedical support of sportsmen for the participation at Olympic games, World and European championships, etc.

In IMBP Dr. Belakovskiy is responsible for the external relations and international cooperation, technology transfer of space medicine and biology. Since 1988 his Department has signed more than 350 contracts with organizations and companies including ESA, USA, Canada, Japan, Austria, Germany, France, Switzerland, Italy, and Chile. Dr. Belakovskiy took an active part in the preparation, organization and implementation of the international experiments HUBES-94, ECOPSY-95, ANOG-96, SFINCSS-99, MARS-500, performed at the Institute's facilities in Moscow, Russia.

The Department Headed by Dr. Belakovskiy also successfully performs innovative activities both in Russia and abroad. These activities are looking into transfer of space medicine and biology technologies into the daily practice of national public health service and industry.

Dr. Belakovskiy is Laureate of the Russian Federation State Prize Award, he is the Full Member of International Academy of Astronautics and Russian Academy of Cosmonautics named after Konstantin E. Tsiolkovsky.

The results of research carried out by Dr. Belakovskiy are presented in more than 180 scientific publications.

References

Official website of the UN Office for Outer Space Affairs. (2012, March 15). Retrieved March 15 2012.

Baturin. (2008). *Soviet Space Initiative in state documents*. Moscow: RT-Soft.

Belakovskiy, M. (2012). *International Cooperation in 2012*. Moscow: IMBP.

Eisenhower, S. (2008, May 28). *NASA - United States - Soviet Space Cooperation during the Cold War*. Retrieved March 5, 2012, from NASA: http://www.nasa.gov/50th/50th_magazine/coldWarCoOp.html

Glavkosmos, USSR. (1990). Commercial Space Service Rendered by Glavkosmos of the USSR. Moscow: Glavkosmos.

Kozyrev, N. (1979). *Flights for Intercosmos Program*. Moscow: Znanie.

Lardier, C. (2001). 52nd International Astronautical Congress. 35 years of French-Russian space cooperation, IAA-01-IAA-2-2-07. Tolouse.

NASA. (1975). Apollo-Soyuz Test Project, USA-USSR Press kit. Houston: NASA.

NASA. (2004, April 4). Shuttle-Mir History/Background/Cooperation Timeline 1962-1993 . Retrieved February 21, 2012, from NASA: http://spaceflight.nasa.gov/history/shuttle-mir/history/h-b-cooperation.htm

Press-kit. (1975). The Onboard press conference of Soyuz 19 and Apollo crews. Houston.

Russian Academy of Sciences. (2012, January 15). Intercosmos Council in Russian. Retrieved March 25, 2012, from Archives of Russian Academy of Sciences: http://isaran.ru

Syromiatnikov, V. (2005). *100 stories about docking*. Moscow: Universitetskaya kniga.

Verga, A. (2012).

Foton-Bion: a brief-long history from the Soviet era to modern Russia. Noordwijk.

CHAPTER 15

THE CAPABILITY CRITERION:
INTERNATIONAL COOPERATION AND NATIONAL PRIORITIES IN SPACE DEVELOPMENT

CAPTAIN CHRISTOPHER M. STONE
US AIR FORCE

AND

CAPTAIN BRENT ZIARNICK
US AIR FORCE

© Captain Christopher M. Stone and Captain Brent Ziarnick, 2012. All Rights Reserved.

INTRODUCTION

The choices between international cooperation and competition, and finding the right balance between the two, has been a major concern of space diplomats and leaders in Asia, Europe, and North America for much of the last few decades. Debate in policy circles has revolved around questions of cooperation, and whether collective assurance or independence is the best course of action.

For many decades there has also been an air of utopianism in the space enthusiast community which suggests that by moving into space, humanity could eliminate poverty, unlock the secrets of creation, and make everyone free, wealthy, happy, and wise, and in many ways space development does indeed have the potential to increase wealth and also lead to major advances in physics, biology, philosophy, and other fields. But regardless of the great perceived potential of space activity to mitigate many human ills, the same utopian view of humanity's potential in space also brings baggage that may cause space enthusiasts to unwittingly hamper the activities that may bring the very benefits they most desire.

Calls for international cooperation in space are far more prevalent than similar calls for international cooperation in the air, on land, or sea. International cooperation is complicated, expensive, and inefficient in terms of leadership roles and expectations. Globalist-minded space enthusiasts still advocate the idea that global space cooperation offers the best way to heal many of the material and moral problems of today, and a planet working together is the ideal way to accomplish a prosperous future for everyone. David Lasser, the first president of the American Interplanetary Society, summed up this worldview well in his annual report, April 13, 1931:

> *I anticipate, naturally, that in dealing with a question such as the rocket which may be a weapon in future warfare, that full cooperation [among nations] will not be possible, but a beginning can be made. Such international cooperation should be begun and pursued energetically, for it seems to me that the solution of the interplanetary problem is too large to be localized in any group or nation. I can foresee the building of the first space ship only as a joint effort of a united Earth.*[1]

Although Lasser's "first space ship" was envisioned as a complicated manned voyage to the moon, neither the launch of humanity's first satellite, Sputnik I on October 4, 1957, nor Apollo 11's first manned lunar landing on July 20, 1969 was in fact a joint effort of a united Earth. Though Lasser erred in his prediction, the globalist fervor remains.

---

[1] "Annual Report to the American Interplanetary Society, April 13, 1931" reprinted in David Lasser *The Conquest of Space* (1931), Apogee Books reprint 2002. Burlington, Ontario, Canada. Pp 181.

Donald Cox and Michael Stoiko argued for international controls of space missiles in their 1958 book *Spacepower: What it Means to You* with the Cold War in full swing, going so far as to recommend a United Nations Space Force to manage human space efforts. Arguing from many perspectives, they claimed potentially, a foolproof mutual international setup for the control of long-range missiles and the satellites, which have both military and peaceful uses, will do more toward ensuring the survival of civilization as we know it than any single act.[2]

Almost thirty years after Lasser argued that it would take the resources of a united Earth to solve the interplanetary problem, Cox and Stoiko continued the economic argument for international cooperation in a different way:

> *The tremendous cost of the present missile program in terms of natural resources and manpower have forced the United States into a top-level national control over production and future employment of long-range missiles. The Deputy Assistant Secretary of Defense... was given additional responsibilities in October 1957... by the Defense Secretary and the President to oversee the entire Army, Navy, and Air Force missile program... This is the first time in the nation's history that so many complications have arisen over the control of a single weapon of war... Duplication of missile production efforts on an international level can be just as costly, in the long run, as on a national level...*[3]

Eighty years after Lasser and fifty years after Cox and Stoiko, we are nowhere near a united Earth space program, and projects such as the International Space Station seem to offer evidence that international cooperation dramatically increases costs rather than provides savings.[4]

The recently drafted Global Exploration Strategy[5] acknowledges the benefits of both national programs in organized and coordinated actions among spacefaring countries, and potential international cooperation. The document does not promote a united Earth space program, but one that acknowledges that international cooperation may or may not be the best option for all programs.

Thus the space agencies and space advocates are faced with a

---

[2] Donald Cox and Michael Stoiko. *Spacepower: What it Means to You.* John C. Winston Company. Philadelphia, Pennsylvania. 1958. p 161.
[3] *Ibid*, p 163.
[4] Jones, Tom, *Sky Walking: An Astronaut's Memoir.* HarperCollins, 2006.
[5] Global Exploration Strategy of 2007. http://www.globalspaceexploration.org/

dilemma. Some advocates view international cooperation in space as a mostly unrestricted good and self evident requirement for space exploration and development, while others take a strictly nationalist viewpoint, and believe that their nation must go it alone.

So how should nations and commercial space entities address the dilemma? The answer must be to find a method to realistically examine the potential costs and benefits of international cooperation vs. national efforts or a combination of the two in space projects. This chapter argues for the "Capability Criterion," a method for individual actors (state or private) to decide to engage in international cooperation on a particular project or not based on a logical framework.

What is the Capability Criterion?

The Capability Criterion is meant to be a benchmark or framework that any individual agent, whether organization, company, or nation, can decide whether or not they should engage in a particular international cooperation endeavor. The logic of the framework is that if the criterion is not met, the agent should presumably not cooperate. Conversely, if the criterion is met, then the engagement in the proposed cooperation makes sense.

The criterion is also a method by which the likely benefits of cooperation can be measured. It does not consider international cooperation in any form as a good in and of itself, but solely measures benefit to the individual state or commercial entity. In this way, the Capability Criterion can dismiss utopian globalist sentiment and cut to the heart of the matter: Is the proposed cooperation good for the individual nation or company that is considering cooperating? The criterion uses space capability as the measuring stick to determine whether cooperation benefits the individual entity. "Space capability" is defined here as the ability to do something (anything) in space, and "additional capability" is simply the ability to do something in space that one could not do before. The metric should be very easy to apply even in the proposal stage of any international cooperation project.

The Capability Criterion is this:

> *An actor should engage in international cooperation if and only if completion of the cooperative effort's objective will result in the actor achieving a higher level of enhanced space capability than it would be able to achieve by acting alone while exerting a reasonable but strenuous level of political will and resources.*

The Capability Criterion has three major elements

First, the sole measure of whether an agent should engage in cooperation is whether or not it will personally gain in space capability from the deal. If the agent can do more in space than it could prior to the cooperation, then it is worth pursuing. If the cooperation does not produce more space capability for the actor, it is not in its interest to participate, good feeling or unity of effort notwithstanding. Enhancing space capability, not goodwill, is everything.

The second element of the criterion is an essential modification of the first. It is readily apparent that if only the first part mattered, the Capability Criterion would be a blank check for any type of cooperation. Indeed, cooperative activity in space would result in more capability in space if comparison was made *vis a vis* no space activity at all. Therefore, the first concept must have a graded modification, in that when the agent compares its anticipated space capability after the completion of the cooperative effort, it does not compare it against lack of space activity at all, but rather against an independent hypothetical effort aimed at a similar objective.

By way of a non-space private sector example, a guiding principle of the Hilton Head Island Chamber of Commerce reads:

"Partner with organizations only in areas where we cannot accomplish the objective of the partnership with our own organization alone."[6]

This guiding principle identifies the fact that sometimes partnerships exist for their own sake, without any real, analyzed benefit to one or more partners, and that partnerships are, by their very nature, more complex organizations than their originators in terms of oversight and command structure. Partnerships simply lead to more difficult (and expensive) projects than independent projects. Knowing this, the Hilton Head Chamber does not partner unless the worthy objective cannot be met alone. The default setting of the organization is independent work.

To bring this concept in a space context, a guiding principle of the European Union Space Policy communication reads:

> *International cooperation should also serve as a market opener* **for the promotion of European technology and services** *in the space field and so help* **strengthen this strategic industrial sector [of Europe]**... *The EU, in close collaboration with the*

---

[6] Hilton Head Island, South Carolina Chamber of Commerce "Guiding Principles" July 2011.

> ESA, will continue to maintain and strengthen its "space dialogues" with its strategic partners – i.e. the United States and Russia – with a view to increasing cooperation. These dialogues seek to identify areas where there is **mutual benefit** in cooperation... [Authors' emphasis][7]

In both the local business sector and supranational organizations, partnerships are meant to be a tool to be used only when necessary, not a goal unto themselves.

The second element of the Capability Criterion sets the agent's default position to independent work because it forces the agent to consider working towards the goal independently. If it can get more space capability in a independent effort (implicitly accomplishing the cooperative's goal as well, since space capability must match or exceed in a independent effort or the agent would choose to cooperate per the Global Exploration Strategy), then no cooperative effort is needed because the agent can accomplish its goals without the increased costs in money and negotiation time related to cooperation.

The third element is perhaps the most subjective: the concept of an independent "reasonable but strenuous exertion." There will be times when an individual agent will indeed be able to accomplish the same goals as a cooperative effort, but the cost in resources would be so great compared to the cooperative effort that it would be beneficial for the nation or company to cooperate.

A simple example of such a situation may be a partnership between two nations to launch a geosynchronous communications satellite, where one of the partners does not have a capable, indigenous rocket. The satellite-building nation may have the technical and economic capacity to build such a rocket to fit their needs, but the infrastructure and development costs in time and effort may simply be too great for the worth of the satellite. They may not even be able to launch heavy rockets due to geographic considerations. For the launching country, additional revenue generated from launch, or access to the satellite's future capability in orbit, may be a beneficial tradeoff even if they could build a similar satellite themselves. Under the Capability Criterion, international cooperation is judged possible and desirable in such a circumstance.

---

[7] Communication from the commission to the council, the European Parliament, the European Economic and Social Committee and the committee of the regions towards a space strategy for the European Union that benefits its citizens, 2011.

What exactly entails a reasonable but strenuous exertion is necessarily left to the decision makers. A detailed quantitative cost/benefit analysis may be required, but a back of the envelope qualitative consideration may work equally well in deciding exactly what reasonable means. What can a nation acting independently reasonably do? At best, this will be an arguable estimate. Regardless, the decision maker must offer a realistic independent alternative to cooperation.

The Capability Criterion is an expression of mutual advantage, based on the premise that no agreement will be made without both parties gaining benefit, with the additional and important point that that cooperation in and of itself does not offer substantial advantage. An actor must benefit by advancing its own space capability (which can be exercised independently), and should not simply use "feel good" internationalism as a legitimate reason for cooperative action.

The best way to explore the Capability Criterion is to apply it to some problems vexing the space community today. Next, we will discuss some problems and apply the criterion to see more clearly whether the criterion helps us gain more understanding of the underlying situation.

Collective Assurance and National Space Policies

In 2011, after the fanfare and applause by many for the new US Space Policy and National Security Space Strategy, the European Union released its long awaited space policy. Despite numerous articles, commentaries, and international discussions about the merits and failings of American space policy statements released in 2006 and 2010, there was very little commentary on the EU's new statement.

The EU space policy is based on years of meetings within the European Commission and its space council regarding the direction for Europe in space. The policy articulates goals and objectives within three main areas: strategic interests, security, and economic prosperity. Throughout the document, strategic language is interwoven with Eurocentric goals and objectives for its industry, economy, civil, and military arenas. The policy indicates that Europeans understand well the political and economic importance of space power as a vital interest, its impact on the everyday life of European citizens, and its affect on Europe's quest for greater security, prestige, and wealth. Interestingly, the order and precedence of the strategic objectives were like a national-focused document with end states reflecting the interests of Europe first, but lacking the global flavor of the 2010 US space policy and follow-on strategy.

The strategic goals of this document are not what many might expect, which would include a US-modeled push for "interdependence," "collective self-defense," and further integration in the "global economy." Rather, the EU produced a highly unilateral document focused on the advancement of European domestic space capabilities. These capabilities aim to enable "economic and political independence" for European citizens, and a greater role for European excellence in space and worldwide. It views space as an area of strategic importance and acknowledges the need for enhanced military capabilities in space, in order to "strengthen its security missions."

The Galileo satellite navigation system is one example among many projects where the European desire is to remain independent. Instead of relying on the United States' Global Positioning System (which even the United States considers a global utility), Europe is duplicating a global navigation system in order to secure their independence from overreliance on US space assets.

Another key topic to note is that "independent access" to space is underscored by the statement that Europe will not rely on any foreign launch or service provider. This is interesting when comparing the EU with current US plans and policy that project reliance on Russian Soyuz for human access to the International Space Station, and American reliance on commercial and foreign partners overall. The US reliance on non-US partners could lead to advantages for foreign commercial entities and hurt the US in terms of the space industry and high tech jobs, an area that suggests a potential strategic contradiction within US policy, one that bears further scrutiny.

Europe's vision for space power advancement includes growth for its domestic space industry and economic capabilities as well. The EU policy states, "a solid technological base [is required] if [Europe] is to have an independent, competitive space industry." To advance the influence of the EU space industrial base globally, innovation must build on innovation, and like US space policy that advocates increased innovation in research and development, the EU policy also advocates innovation, but with a different tone.

To promote "industrial competitiveness" in the marketing of European space technology, it suggests "setting ambitious space objectives" as the key to "stimulating innovation," but contrary to the US approach, it does not call for unlimited funding of STEM (Science, Technology, Engineering, and Math) education initiatives to keep youth excited. The European approach recognizes that beyond mere research and development, their space industrial base will neither innovate nor compete on the world stage without concrete commitment to funded and ambitious objectives in space-related exploration and national security programs. As a result, Europeans desire a strong industry that will provide enhanced prestige and influence necessary for European space efforts to be advanced in multilateral forums.

Our third observation concerns the EU's view of international cooperation. Reading through the document, and reviewing the scant press coverage that was given upon the release of the policy revealed a structure quite dissimilar from US policy. Rather than interweaving international and global themes throughout each sector or mission area, the document focuses on advancing domestic capability and policy for the benefit of Europeans. While Europeans are not anti-international cooperation, they do view themselves as a partner and want to maintain "space dialogues" with their "strategic partners," notably Russia and the United States. The section on international cooperation is rather short, and the overall strategic goal is to use space "as an instrument serving the Union's internal and external policies." The document does, however, acknowledge that space efforts are increasingly not just for individual nations, but in many cases can be achieved through pooling resources.

In contrast, US space policy states that international cooperation in US space programs is a requirement, and it is a directive for all departments to pursue international partnerships in all space mission areas, an entirely different approach than the one proposed by the Capability Criterion framework. The EU, alternatively, appears to see it as something to be considered following the development of domestic capabilities and leadership in critical areas such as positioning, navigation and timing, and space launch, among others. The Europeans intend to maximize their space capability unilaterally in accordance with the Capability Criterion, and want to partner only when it will benefit them.

Among the bolder international efforts is an interest in opening up dialogue with China, and utilizing EU space power to expand European influence in Africa.

In summary, EU space policy is a policy about Europe, its goals and objectives for the Union to gain in space leadership worldwide. Gaining added security, prestige, and wealth in space allows Europe to achieve a "key position" in space power, based on excellence and "increased European capability," a policy directly compatible with the Capability Criterion.

Similarly, many experienced American space professionals, with knowledge of international space cooperation and policy, understand the importance of shaping the strategic space environment to benefit US vital interests, and many wish to get past the perceived international angst that followed the release of the 2006 space policy (see Joan Johnson-Freese's *Heavenly Ambitions: America's Quest to Dominate Space* for a representative example), while maintaining good rapport with our allies.

In our view, a new US national space policy should follow the European lead and apply the Capability Criterion, emphasizing goals and objectives for the development of American leadership through increased capability, ambitious space objectives, innovation, and global competitiveness of the American space industrial base. International cooperation, as the Europeans note, should be best articulated in appropriate bilateral and multilateral agreements, but not in a national space policy. The 2010 US national space policy contains many good points, but it reads more like an international statement of principles than a national strategic document.

Rather than using language such as "collective assurance," "collective self-defense," and "interdependence," and emphasizing a policy of reliance of foreign space capabilities, Europe is pursuing a course of independence and increased European capability to achieve excellence and increased status for the advancement of European space efforts.

As the US current space policy notes, every nation has the right to access and use space. Each nation has the right to develop its own nationally-focused "unilateral" space policies that serve to advance its vital interests in security, prestige, and wealth as the baseline for any international cooperation it chooses to support. Failure to invest in bold, ambitious space efforts with a national tone (in all sectors) in space will not only hurt the US space industry, but will harm its ability to advance its global interests in space, impact its traditional vital interests of independence and achievement, and threaten the very preeminence that America has labored so hard to achieve over the past fifty years.

If the goal of the US is the advancement of global space exploration, then the US needs to observe that other nations and partnerships such as the EU and Russia are taking an alternate path toward increased domestic space capabilities and expanded infrastructure for national interests. The EU is in fact pressing ahead with its goals, and stepped into the leadership vacuum that the US has created as a result of the shutdown of US programs, abandoning US capabilities, and allowing the loss of large numbers of skilled space workers.

In our view, future American space policy and strategy should address international efforts for mutual benefit, but should also focus on advancing American capability to enable a long range strategy for exploration and enhanced military capabilities in space, just as the Europeans are pursuing, and just as the Capability Criterion suggests each nation's space program should.

NATIONAL SPACE EXPORT CONTROLS

For the first time in more than a decade, the US is considering modifying and updating the International Traffic in Arms Regulations (ITAR), a critical set of rules and regulations pertaining to technology exports. The space community, academia, and both the executive and legislative branches of the US federal government are discussing how to improve America's space export control system while maintaining a robust national security and civil space enterprise.

There are three major viewpoints to consider.

First, ITAR itself "poses a potential national security risk," as noted in the latest Quadrennial Defense Review (QDR), due to its complexity, its protection of "everything," and its hindrance to international cooperation and industrial competition.

A second view held by many members of Congress, and dates back to 1999 or earlier, when the current space export control framework was devised and passed in the Strom Thurmond National Defense Authorization Act. It calls for protecting everything that is space and missile related, regardless of its size, commonality, and availability in the global market. All parts and components are to be controlled within the United States Munitions List (USML) as weapons, and not as commodities on the Commerce Control List (CCL).

Third is a thought process advocated by many aerospace and industrial advocacy groups, which promotes moving all commercial communications satellites and their widely available components from the USML to the CCL enabling them to be sold openly. This would give the US space industrial base the added boost it needs to compete in the global space market, and reduce the attractiveness of the "ITAR-free" movement in India, Europe, and China.

Suppose that these three main viewpoints on ITAR are the only relevant viewpoints. Which, then, would be the best way forward, and how can the Capability Criterion help make the determination, as ITAR is nothing if not an issue of international cooperation?

Let's begin with the first. According to the 2010 QDR public report, the current export control system, and not just the part that applies to the space industry, is considered a "national security risk." Previously the concern was that the application of ITAR, with its burdensome restrictions and protections, was causing national security risks because the US space industry was losing revenue and market share, and thus jeopardized its ability to also meet the needs of the US Defense Department over the long term. Now the view suggests that the real problem lies in the system itself, which relies on Cold War-era protection methods, and assumes an economic environment based on Cold War dynamics. This view holds that the current global economic system, which largely shapes relations among states due to the structure of global trade, makes the US current export control system obsolete, and since it is obsolete it results in additional risk to US national security because it diminishes economic security and prosperity.

In addition, it views the risk to security based upon a perceived lack of real engagement in the global commons of space and the utilization of this commons for the benefit of all humanity.

In other words, due to the increasing globalization of the space industry, the United States should maintain its "lead" rather than "preeminence" through increasing participation in the global economy, rather than by protecting its industrial base and ability to produce space capabilities needed to stay ahead of its competitors. In this view, American space power is not just space power; it's space as a foreign policy and strategy tool.

The President addressed this view when speaking on his administration's review of export controls at the 2010 annual conference of the Export-Import Bank and its connection with his National Export Initiative. He stated that changes are needed "for our strategic and high tech industries, which will strengthen our national security," and recently added that his Administration would even consider executing this plan without Congressional approval.[8]

By contrast, the second view that seeks to protect "everything" and requires a rigid licensing and monitoring framework to protect the release of anything space related. This view came about after the Loral satellite incident involving China in the late 1990s, when Loral engineers allegedly

---

[8] http://thehill.com/business-a-lobbying/173425-white-house-export-reform-doesnt-require-congress-to-be-enacted

shared sensitive US rocket and guidance technology to correct deficiencies in the Chinese Long March launch vehicles that Loral was using to launch its satellites in violation of US export regulations. Many politicians expressed concerns that Loral had harmed US national security by sharing technology that could also be used to improve Chinese ballistic missiles, and the uproar in Congress resulted in the so-called Cox Commission.

The Cox Commission and Congress found errors and lapses regarding the control of US technology transfers concerning missiles, and as a result Congress expanded USML, but unfortunately by using vague and broad language:

> ...***all satellites and related items*** *that are on the Commerce Control List of dual use items in the Export Administration Regulations... on the date of the enactment of this Act shall be transferred to the United States Munitions List and controlled under section 38 of the Arms Export Control Act. [Emphasis added]*

It's clear from the language – "satellite and related items" – that the provision now encompasses sensitive space technology as well as commodities like Kapton tape that is easily found on the global space market and widely used in non-military vehicles as well as military spacecraft. The problem, of course, is that while well intentioned, it creates a new problem for American companies, new competitors who step in to fill the void that US suppliers can no longer fill. Whereas before the United States was the leader in the space market for manufacturing commercial spacecraft, now Indian, Chinese, and European firms are profiting. Other nations now rely much less on US spacecraft, launch vehicles, and other components, leading to further shrinkage in the US space industrial base as well as the lower tier supplier base.

Research has shown that it is not so much the ITAR protections themselves that have hurt American firms, but rather the burdensome processes, procedures, and regulatory frameworks (please see the Aerospace Industry Association's Competing for Space, January 2012). Further, money that could be better spent in research, development, and other technology-advancing endeavors is wasted by paying for lawyers and other experts to wade through the bureaucratic red tape to acquire the necessary licenses.

And since the ITAR-free suppliers don't have the increased costs involving protection and mission assurance that American products have, their prices are lower, potentially attracting customers that once would only think of the US space industry to go to Europe to build their satellites, and

China to launch them.

The third idea, which lies at the halfway point between the first two, allows for some items to be moved from the USML to the CCL, which is controlled by the Commerce Department instead of the State Department. Commercial communications satellites are considered by many people to be "safe" to transfer to the CCL, allowing for common components across the space community to be sold around the world. Some critical technologies would still be protected under the USML, while common components would fall under a more permissive system. The goal of this approach is of course to enable US firms to regain lost ground in the space marketplace, and it's currently being explored by a group tasked by Congress in the 2010 National Defense Authorization Act, Section 1248. This congressionally-directed action is tasked to the State Department (which controls the USML) and the Department of Defense, to evaluate the national security risks of removing all space components from the USML.

So what is the best approach for the United States to take to fix the problems caused by ITAR?

The Capability Criterion approach suggests that despite a push for globalization and the trend of national strategy to embrace globalization and the global economic system in trade, the primary responsibility of the US space program and of the export control regime is to protect American sovereignty and provide for the common defense, as stated in the Constitution. In order to do this, America must maintain not just an edge over our peer and near-peer competitors on the world stage but also preeminence and preferably clear leadership in the areas of space.

As one historian noted, John F. Kennedy understood that in order to maintain a leadership position on Earth, a nation must maintain leadership in space, the ultimate "high ground."

To sustain its position as a leading space power, the United States should protect what is proprietary, while also allowing its industry to compete.

Without a strong industrial base that is fully integrated into the planning and strategy-crafting processes of the national security space enterprise, the capacity of US industry to develop high-quality spacecraft and launch vehicles needed to maintain space leadership will decline, and with it, its status as a superpower. While the authors feel the third option is the best overall, the government will necessarily undertake a robust national security risk assessment to determine its effects on US economic leverage and influence.

Do the US need a new export control system? We believe not.

Does ITAR need reform and streamlining? Definitely. ITAR processes need to be reformed to allow US companies, both large and small, to be competitive on the global market without draining their resources in a bureaucratic swamp. And here, the concept of the Capability Criterion can be of use. Adjusting ITAR restrictions to optimize American space capability will mean a balance between overzealous restriction (which can atrophy American space capability) and laissez faire freedom (which can improve others' space capability at the expense of the US while achieving no improvement in American capability). The Capability Criterion can help us make the correct decision.

Leadership through Capability

In a recent article in *The Space Review*,[9] Lou Friedman asserted that, "American leadership is a phrase we hear bandied about a lot in political circles in the United States, as well as in many space policy discussions." He goes on to note that American leadership, "has many different meanings, most derived from cultural or political biases, some of them contradictory." This is true; many nations, organizations, and individuals worldwide, have different preferences, and different views as to what American leadership in space is, and what it should be.

We disagree, however, with Mr. Friedman's assertion that space is "often" overlooked in "foreign relations and geopolitical strategies." Our view is that while space is indeed overlooked in national grand geopolitical strategies by many in national leadership, space is used as a tool for foreign policy and relations more often than not. In fact, the US space program has become less of an effort for the advancement of US space capability and exploration, and is used more as a foreign policy tool to influence the strategic environment toward what President Obama referred to in his National Security Strategy as "The World We Seek."

But should the US shape its future in space using the international collaboration piece as the starting point? We suggest that the goal of the United States should be leadership through space capabilities in all sectors.

America achieved leadership in space because it demonstrated technological skill through the Apollo lunar landings, deep space exploration probes, and exploration of the outer planets. It did not become recognized leaders in astronautics and space technology because it decided to fund billions into research programs in the absence of clearly defined

---

[9] http://www.thespacereview.com/article/1778/1

national objectives.

The US has allowed itself to drift from a traditional strategic definition of leadership in space exploration, as indicated by the decision to shut down the space shuttle program without a viable replacement system, while paying millions to use Russian systems to ferry astronauts and cargo to the International Space Station.

Each nation should have freedom of access to space for the purpose of advancing its "security, prestige and wealth" through exploration and technological achievement.

Maintaining leadership in the space environment is a worthy goal, and space superiority does not require orbital weapons, or preventing other nations from access to space, nor does it preclude international cooperation. Rather, it indicates a desire to achieve goals for national security, prestige, and economic prosperity. The quest for excellence in space is just one part of international space competition that is positive and healthy.

If America wants to retain its leadership in space, it must approach its space programs as the advancement of its national security, prestige and wealth by maintaining its edge in space capabilities and using those talents to influence the international space arena.

Conclusion

The case for using the Capability Criterion to assess proposals for cooperative action in space is hopefully now clear. We believe that the Capability Criterion in action will maximize space development for human civilization because it optimizes space development decisions based on competition and strategy rather than merely on our hopes for the future. Further, if humanity were ever to face a rogue asteroid or another threat from space, the Capability Criterion in action would have yielded the most developed capabilities, enabling all spacefaring nations and indeed all of humanity to have the most advanced space capabilities possible to protect Earth.

Looking at the national and international policy and strategy documents concerning international cooperation in spaceflight, one will see that national sovereignty and interest is not dead, but is alive and well. The Global Exploration Strategy mentioned above speaks of the open nature of agreements in spaceflight, in which nations may come and go as they please. Europeans have expressed interested in participating in deep space exploration with NASA, but also have independent European ambitions;

the Russians and Chinese are the same.

Quality in international partnerships is created through trust and mutual national interests based on the Capability Criterion. While international cooperation is a good thing in space, as the Global Exploration Strategy and Roadmap agrees, national space programs should engage in independent actions when and where they desire advancement of indigenous space capabilities and interests. Space, as mentioned in the 2010 United States National Space Policy is for the use of all nations, but friendly competition among nations and the private sector also fuels the technology enhancements that enable progress, as well as the enhanced prestige and space power capabilities space enthusiasts desire. Utopian notions based on symbols of unity and international cooperation in the absence of strategic goals are a waste of time, and should not be pursued, while spacefaring nations continue the work of real-world space exploration development utilizing the Capability Criterion as the decision criterion.

The Capability Criterion should offer international appeal given the strategic importance of independence and healthy competition. It not only encourages national space capabilities, but also maximizes the overall human space effort by focusing on space development. Perhaps we will reach the goals dreamers have about space, but in order to do that we do not need utopian assumptions, but instead a focus on strategic assessments and cooperative frameworks in real-world documents such as the Global Exploration Strategy and Roadmap. Applying the Capability Criterion to US international cooperation decisions is a first step in ensuring our next steps into space contribute to the best possible outcomes for each nation and for humanity as a whole.

•••

The views expressed in this chapter are solely those of the authors and should in no way be construed as the official policy of the United States Government, the US Department of Defense, or the United States Air Force.

Captain Christopher M. Stone

Christopher Stone is a space operations officer and Flight Commander with the 222nd Command and Control Squadron supporting the National Reconnaissance Operations Center. In his civilian capacity he is the Strategic Integration Analyst for the Secretariat of the Joint Staff, National Guard Bureau.

Captain Stone entered the Air Force in 2004 as a graduate of the University of Missouri. He has served in various assignments, including space policy, strategy, legislative, space operations and international affairs. He has served as a space strategist supporting the Joint Force Air Component Commander (JFACC), Pacific Air Forces as well as serving as Chief of Staff for the Space and Cyberspace Division at Headquarters Air Force-Office of International Affairs. Prior to assuming his current position, Stone was the Space Strategy Planner with the 157th Air Operations Group (PACAF), Missouri Air National Guard supporting the 613th Air and Space Operations Center Strategy Division.

In his civilian career, he has served on the staffs of two United States Senators, the Office of Secretary of Defense (Executive Agent for Space), and NGB-Joint Staff. Additionally, he serves on the Policy Committee of the National Space Society advocating spacepower policy/strategy for the advancement of American space exploration, settlement and security.

Captain Brent D. Ziarnick

Captain Brent D. Ziarnick, US Air Force Reserve, is the Senior Space Duty Officer of the 701st Combat Operations Squadron, the reserve manning unit for the 607th Air Operations Center, Osan Air Base, Republic of Korea. During contingency operations, Captain Ziarnick serves as the Air Force Chief of Combat Operations' senior space command and control representative for the Korean Theater of Operations. Prior to this position, Captain Ziarnick served as a deployed space control officer in the Middle East supporting Operations Enduring Freedom and Iraqi Freedom. On active duty, he was Global Positioning System (GPS) satellite operator, engineer, and tactician.

Captain Ziarnick has been published extensively on space issues in both military and civilian journals, winning multiple awards for his articles.

In civilian life, Captain Ziarnick has served as an engineer and launch planner for Spaceport America in New Mexico, the world's first purpose-built commercial spaceport.

Captain Ziarnick holds a bachelor's degree from the US Air Force Academy in Space Operations, a master's degree in Space Systems Engineering from the University of Colorado - Colorado Springs, and a doctorate in Economic Development from New Mexico State University.

CHAPTER 16

PERSPECTIVES ON IMPROVING UNITED STATES INTERNATIONAL SPACE COOPERATION

JAMES D. RENDLEMAN
USAF, RETIRED

AND

J. WALTER FAULCONER
PRESIDENT, STRATEGIC SPACE SOLUTIONS, LLC

© James D. Rendleman and J. Walter Faulconer, 2010 and 2012. All Rights Reserved. This chapter was previously published in *Space Policy*, Vol 26 Issue 3, August 2010, p 143 under the title, "Improving international space cooperation: Considerations for the USA."

There is a powerful case to be made for the United States to conduct international space cooperation activities. In this chapter, we will discuss how cooperation allows a nation to leverage resources and reduce risk;

improve efficiency; expand international engagement; and enhance diplomatic prestige of engaged states, political sustainability and workforce stability. Unfortunately, although the case for international space cooperation is powerful, the obstacles and impediments to cooperation are substantial, and are manifested through various anti-collaborative behaviors. From a US perspective, cooperation is successfully achieved only after the undertaking and absorbing great expense, and understanding and confronting other obstacles and impediments. To that end, we will examine the challenges posed by technology transfer constraints, international and domestic politics, and exceptionalism perspectives. Finally, depending on the circumstances, four frameworks of cooperation can be employed to overcome these impediments: Coordination, Augmentation, Interdependence, and Integration. This chapter will detail these frameworks and their issues.

The Case for Collaboration

TABLE 1. FACTORS SUPPORTING INTERNATIONAL COOPERATION / COLLABORATION
Cost sharing – not enough money to go alone
Foreign policy goals – space cooperation is part of a larger package
Building political support
Insight into partner capabilities, approaches & plans
Expand scientific/instrument spaceflight opportunities – open the door to more launch opportunities for "ride-sharing"

Interest in international cooperation on space missions is not new. Successful elements of international collaboration have been around since the beginning of the space age. This sentiment to support international cooperation and collaboration is growing. The factors driving a desire for international cooperation are shown in Table 1. Reflecting the desire, the new United States National Space Policy declares international cooperation to be among its key goals, stating in pertinent part:

> "[T]he United States will pursue the following… in its national space programs:… Expand international cooperation on mutually beneficial space activities to: broaden and extend the benefits of space; further the peaceful use of space; and enhance collection and partnership in

sharing of space-derived information."[1]

The new National Space Policy is striking — it expresses strong interest in cooperation throughout the document, and more so than in past policy declarations. As we will discuss, there is considerable justification for this emphasis. The hope and desire in international projects is that one plus one will equal three—that the diverse resources, skills, and technologies of the partners will achieve synergy, adding up to more than the sum of their parts.[2] NASA, commercial, and Europe space activities have already achieved considerable success with their cooperative efforts.

NASA

As an institution, NASA believes in the promise of cooperation. Indeed, its Administrator, Charlie Bolden, has announced that greater international cooperation is coming, affirming "[t]hat's what the president wants to do, and he didn't have to tell me that, because that's what I've been doing all my life." NASA has had a long-standing emphasis on cooperation, as more than half of its forty-two ongoing space and Earth science missions have international participation. Of missions it has under development, nearly two-thirds involve international contributions and participation. Much of the astronomy and astrophysics community is pleased to see NASA leveraging and expanding international investments in its great science enterprise.[3]

COMMERCIAL

A trailblazer in international commercial space cooperation, the International Telecommunications Satellite Organization (INTELSAT) has achieved great success. INTELSAT began in 1964 with 11 participating countries; INTELSAT launched the first commercial international geosynchronous communication satellite the following year. By 1973, there were 80 signatories (member nations of INTELSAT), and the organization was providing service to over 600 Earth stations in more than 149 countries, territories, and dependencies.[4]

---

[1] *National Space Policy of the United States of America*, Fact Sheet, June 28, 2010, p. 4.
[2] See "International Cooperation: When 1+1=3", Toshifumi Mukai, *ASK Magazine*, NASA, Summer 2008, pg. 8.
[3] Remarks by Shana Dale at AAS/AIAA Seminar on the Importance of International Collaboration in Space Exploration, Nov 1, 2006.
[4] Ryan Zelnio, "A model for the international development of the Moon," *The Space Review, December 5, 2005.*

INTELSAT's early substantial financing came from participating national governments. It developed none of its own hardware; instead, INTELSAT purchased system capabilities from global aerospace commercial companies.[5] By the end of the 20th Century, its resource sharing and administrative scheme helped launch and usher in a huge global telecommunications market era. Commercial entities involved in entertainment, news, Internet, government services, and other telecommunications services now spend more on space than all the world's governments combined. With its early quasi-governmental and tax-exempt aspects, complaints arose from competitors, and INTELSAT was pushed to evolve into a purely commercial entity in 2001, Intelsat, Ltd., and it continues to thrive and expand.[6]

EUROPE

The nations of Europe have developed the European Space Agency (ESA) to pursue space activities. It currently has eighteen member states, and integrates countries and their respective space programs to work collaboratively toward common goals.[7] Since 1975, ESA has focused on supporting commercial companies within Europe, doing this by investing in and developing technologies to enable it to compete in the global space market.

---

[5] Ibid.

[6] Recently it consolidated its global dominance in satellite communications by acquiring rival PanAm Sat.
"During the 1990s, there was considerable criticism from new commercial satellite companies focused on the difficulty of competing against an organization with INTELSAT's advantages. The Open-Market Reorganization for the Betterment of International Telecommunications (ORBIT) Act, which was enacted in 2000, provided for sanctions to be imposed upon INTELSAT if it did not privatize in a manner consistent with the terms of the act. On July 18, 2001, INTELSAT transferred substantially all of its assets and liabilities to Intelsat, Ltd.—a holding company incorporated by INTELSAT under the laws of Bermuda—and its wholly owned subsidiaries. By so doing, INTELSAT lost its tax-exempt status, along with other privileges it had been granted as an international organization and under international agreements." "Tax Policy: Historical Tax Treatment of INTELSAT and Current Tax Rules for Satellite Corporations," GAO-04-994, September 2004, p. 1.

[7] ESA is not an agency or part of the European Union (EU) and has non-EU nations as members, though some think that ESA is the *de facto* space agency of the EU. There are ties between the two and the member states, and moves are afoot to better define the status of the ESA with respect to the EU. As security dimensions to European space activities grow, some believe ESA will become more fully integrated into the EU governing structures.

ESA

The European joint venture began in earnest once its founding nations concluded they had no other alternative but to cooperate. Other than France, Germany, Italy and the United Kingdom, individual nations in Europe are generally too small to individually develop a sound and comprehensive space program. The ESA confederation has leveraged resources and proven valuable to its members. Reflecting on this, Jean-Jacques Dordain, Director General of ESA, has stated: "...after more than 30 years we are glad to cooperate because we learned that beyond the difficulties of cooperating, there is the success of cooperating. We know now that it is always easier not to cooperate, but that it is always more difficult to succeed alone."[8]

> *The United States has made clear that it does not seek to "dominate" space and, in fact, has led the way in securing international cooperation in this field.*
>
> Letter from President Lyndon B. Johnson

ESA's complex governance structure has worked; it has rationalized European space efforts, usually allocating expenditures among members based on their relative technical strengths. It has also helped foster standardization and interoperability among Europe's space systems. This success, however, has not always been easy. Member nations have had to reconcile their industrial space priorities with those emphasized by ESA's leadership. The priorities may differ. Mutually destructive squabbles have generated problems. ESA's spending priorities have been criticized; some sniff that expenditures made the framework unfairly favors French, German, and Italian interests in spending; then again, these nations are also among its larger contributors. France's space agency receives a budget which is double the amount it contributes to ESA. However, without the ESA funding framework, France arguably would never have been able to resource the immense Ariane rocket system and fund development of the French spacecraft buses used by Europe's EADS Astrium and Alcatel Alenia Space. Cooperation has been a tremendous boon to France's national space aspirations, as well as for the rest of Europe. ESA has created an environment for space activities in which a wide variety of European consortiums and industrial concerns are able to participate.

The NASA, commercial and European space successes in

---

[8] Jean-Jacques Dordain, "International Cooperation in Space," remarks at 40th Anniversary of the Universities Space Research Association (USRA), March 26, 2009.

international cooperation have a number of bases. International cooperation allows states to leverage resources and reduce risk; improve efficiency; expand international diplomatic and other engagement; and enhance prestige of engaged states, political sustainability and workforce stability.

RESOURCE AND RISK SHARING

Cost motivations dominate the calculus on whether a state or commercial entity should engage in international space efforts. Why? Most space endeavors are terribly expensive and capital intensive, and as a result, are highly debated, especially the returns on investment, except in the most authoritarian states. International cooperation offers the potential to reduce the burdens to gain access to space by even the poorest of nations.

With cooperation, a spacefaring state can draw in outside resources. Given the large costs involved in accessing the space domain—satellite system research, concept development and system design, manufacture, launch and operation—cooperation is needed by all but the largest spacefaring nations. Cooperation spreads the resource investments and expenditures among nations and entities. Cooperation also reduces exposure by spreading the risk of failure. Per-partner utility of international cooperation invariably increases as per-partner costs decrease.[9] There is therefore a strong incentive to engage in cooperative activities when they provide this savings and leverage. This is especially compelling for nations whose resources are insufficient to achieve any substantial space operational and technical goals. As an example of this reality, even ESA has engaged the United States and Japan to join them in what were previously traditional European-only science missions as a way to rescue the European mission portfolio from increased cost growth.[10]

EFFICIENCY

International cooperation offers the opportunity to improve the efficacy of expenditures, which is a significant cost consideration. With cooperation, resources can be rationalized, standardized, and made interoperable to bring about the best and most efficient use of research,

---

[9] D. A. Broniatowski, G. Ryan Faith, and Vincent G. Sabathier, "The Case for Managed International Cooperation in Space Exploration," *Center for Strategic and International Studies*, Washington DC, 2006.

[10] "Fighting Inflation, ESA Science Candidates Pushing the Cost Curve Could be Saved by U.S. and Japanese Roles," *Aviation Week and Space Technology*, December 7, 2009, Pg. 46.

development, procurement, support, and production resources. Cooperation can also foster more effective operations. Thus, if a hypothetical space partnership involves two nations, one with sophisticated remote sensing engineering capabilities, and the other, spacelift, a rational approach would allocate program activities in accord with these strengths. As an example of this allocation, the two primary instruments to help locate water and other resources, onboard Chandraayan, India's first satellite to the Moon, were contributed by the United States. Incredibly, the US payloads cost more than what India spent building and integrating the launch vehicle and the balance of the spacecraft. Still, the United States benefited because it saved on its launch costs and was able to join in India's scientific mission to the Moon.

International cooperation can rationalize resources and provide for much needed programmatic redundancy. For example, following the 2002 loss of the Space Shuttle Columbia, the International Space Station (ISS) program was able to continue on track by leveraging the transportation capabilities of the Russian Soyuz spacecraft system. Without this, the ISS program could have failed in the midst of the second extended US space shuttle program's mission stand-down.[11] Programmatic redundancy can reduce per-partner cost by creating a higher net reliability than that would otherwise result by depending on a single resource provider.[12]

Standardization of hardware, software, procedures, and the like also helps achieve close practical cooperation among partners, and improves programmatic efficiency. This enables an efficient use of resources and reduction of operational, logistic, technical, and procedural obstacles. International partnerships usually begin their efforts by standardizing administrative, logistic, and operational procedures, and originators of standardizing systems and procedures often become the de facto leaders of collaborative efforts. It is for these reasons that the United States, and sometimes the Europeans, exercise leadership among cooperating major spacefaring nations by defining standard interfaces for space systems.[13]

Finally and closely related to standardization, interoperability is essential. "Designing for programmatic redundancy provides a strong argument for interoperability between nations' space exploration assets, as this would allow nations to substitute each other's critical capabilities with

---

[11] The other stand-down began in 1986 after the explosion of the Space Shuttle Challenger.
[12] D. A. Broniatowski et al., "The Case for Managed International Cooperation...," *supra*.
[13] *Ibid*.

relative ease."[14] Nations whose space systems are interoperable can operate together more effectively. Designing for interoperability enables spacefaring partners to substitute each other's critical capabilities with relative ease.[15] This also provides much needed redundancy in event one nation cannot supply a key service or component for any number of reasons.

In the end, rationalization, standardization and interoperability provide important capabilities. Space programs can use them to: efficiently integrate and synchronize operations; enable data and information exchanges; share consumables and resources; enhance effectiveness by optimizing individual and combined capabilities of equipment; increase efficiency through common or compatible support and systems; and assure technical compatibility by developing standards for equipment design, employment, maintenance, and updating them. With rationalization, standardization, and interoperability, nations that are likely to join a partnership can better prepare to perform their respective responsibilities.

Figure 1. US President Barack Obama escorts Indian Prime Minister Manmohan Singh to a White House meeting.

INTERNATIONAL ENGAGEMENT

International cooperation on space activities presents a unique opportunity to develop dependencies among nations that may obviate conflict. Such cooperation may foster the understanding, and, indeed, friendship, that can reduce the perceived need to prepare for doomsday scenarios where one imagines or projects the technologies that an adversary could develop, regardless of the technical merit or reality of the paranoia.

International cooperation now extends to a whole range of scientific endeavors. This sharing and cooperation among space programs harkens back to the best spirit and intentions of the Outer Space Treaty, in which

---

[14] *Ibid.*
[15] *Ibid.*

the preamble calls for space to be used for "peaceful purposes."[16] This objective has been the hope expressed by many nations since the beginnings of the space era.

The full realization of cooperation's promise began to be more fully realized with the end of the Cold War. Space and Earth science research and space exploration activities were no longer bound and subjugated by an overarching competition between two superpowers. Capitalizing on the opportunities and leveraging the expertise of other nations, the global scientific community rushed into the new post-Cold War, multi-polar world creating numerous international space alliances and partnerships.[17]

> *"The space program plays a positive role in enhancing American influence and prestige, especially with our Pacific and European allies... It also serves to demonstrate America's continuing commitment to technological, economic and political leadership. The space program is an excellent vehicle for cooperation with longstanding allies, such as Western Europe and Japan, and for the development of new ties to Eastern Europe and the Soviet Union. Space cooperation with the Soviet Union...can play a positive foreign policy role by contributing to better East-West relations...."*
>
> "State Department Helps US Space Program Meet Future Challenges," State Dept. Dispatch, Reginald Bartholomew, Under Secretary for International Security Affairs, December 24, 1990.

The United States is continuing this trend by reaching out to large global space (and nuclear) powers like India and China, both growing economic and engineering powerhouses, in the hope such engagement shapes their future space and engineering activities in positive directions. Reflecting on the growing spirit of collaboration, President Barack Obama and India Prime Minister Manmohan Singh agreed November 24, 2009 to expand cooperation to civil space, less than a week after Obama returned from Beijing, where he and Chinese President Hu Jintao pledged to expand dialogue between US and Chinese space agencies.

---

[16] See generally, *Treaty on Principles Governing the Activities of States in the Exploration and Use of Outer Space, including the Moon and Other Celestial Bodies*, other known as the *Outer Space Treaty (1967)*. Article III of the Treaty declares that states parties must conduct their space activities "in the interest of maintaining international peace and security." The treaty's preamble also recognizes "the common interest of all mankind in the progress of exploration and use of outer space for peaceful purposes."

[17] *Approaches to Future Space Cooperation and Competition in a Globalizing World*, National Research Council, 2009, pg. 1.

DIPLOMATIC PRESTIGE

Cooperation provides opportunities for a nation to demonstrate its organizational, educational and technical prowess. For example, India has used its recent launches to host payloads from a number of international partners, and has been recognized for leadership.

South Korea is leveraging Russian launch technology in attempt to join the small and select group of nations that have successfully launched and orbited satellites. This cooperation is an important part of fulfilling a national dream to become a "top ten" space fairing nation. Russia and China work with much of the global spacefaring community to launch satellites.

Ultimately, support for cooperation and collaboration increases when the perceived utility and diplomatic prestige derived from cooperation increases. A demonstration of the utility and diplomatic prestige gained from space cooperative endeavors can be seen in the 1975 Apollo-Soyuz space link-up, the 1995 Space Shuttle - Mir docking, and ISS programs. According to James Oberg, these programs' true and complex diplomatic utility was not made apparent for many years:

Only with the Soviet program at a standstill did Moscow agree to fly a joint orbital mission. Its fallback position was that if it couldn't be Number One in space, it could at least pose as the equal partner of the new Number One, the United States. It was better than letting on how far behind its space program had fallen.

The cooperation resumed only as the USSR was collapsing in the early 1990s. Shuttle-Mir and the critical role of Russia in the International Space Station were enabled by the rise of a freer, more democratic Russian society, not by inertia from decades-old space handshakes.[18]

POLITICAL SUSTAINABILITY

International cooperation provides a wonderful capacity to increase a nation's political will to sustain and fund space programs and associated budgets. As noted, cooperation provides a spacefaring state the basis to draw on additional resources when its own are not adequate to achieve desired space goals and visions. Cooperation also enables a space programs to hunker down and increase chances to survive attempts to be reined in even when faced with contentious and devastating cost-growth or budget realities (something nearly all space programs invariably face). Thus, within the United States, a civil space program can usually win a bit of

---

[18] James Oberg, "The real lessons of international cooperation in space," *The Space Review*, July 18, 2005.

sanctuary from cancellation threats or significant budget reductions to the extent that Congress and the administration feel compelled to not break, stretch, or withdraw from international agreements the program is associated with.

Significant political good will is often generated by funding these programs. To find an example of the power of this good will, one only need look to the politics surrounding NASA's manned programs. Money has continued to be allocated to the program even when the perceived justification for a substantial or expansive manned program has collapsed. Similarly, some argue the political and diplomatic integration of Russia into the ISS program may well have saved the ISS and Space Shuttle programs from cancellation.[19] The pressure to continue international cooperative efforts is often tremendous.

Once cooperation has commenced, canceling a program becomes inconsistent with political sustainability as long as the utility cost associated with the loss of diplomatic benefits and the negative effects on reputation of terminating an international agreement is larger in magnitude than the utility cost that must be paid to maintain the system... The corollary to this is that there is a high cost to be paid by any nation that chooses to unilaterally withdraw from an existing cooperative endeavor. This cost comes in the form of damage to the departing nation's reputation or credibility. In general, any unilateral action sends a signal that the actor is an unpredictable and therefore an unreliable and possibly disrespectful partner. This tends to sabotage the possibility of future cooperation. As such, there is a long-term benefit to maintaining cooperation, even when the immediate cost may seem to call for terminating it.[20]

Of course, if significant cooperation has never occurred, commencement of a new or improved relationship is thought to be a defining moment, delivering specific political rewards and diplomatic utility. This is why the pronouncements on space cooperation made by President Obama and Chinese officials during his November 2009 visits have been watched with special interest. The same attention is being paid to the overtures made with the Indian government and its space community.

---

[19] In the year before Russia was introduced as a partner, the ISS was saved by a single vote in Congress. See D. A. Broniatowski et al, "The Case for Managed International Cooperation...," *supra*.

[20] D. A. Broniatowski et al, "The Case for Managed International Cooperation...," *supra*.

Obstacles: If international collaboration makes sense, why is it so difficult to achieve?

TABLE 2. Barriers/Factors Detracting from International Cooperation
Expense
National pride, prestige
National security
Economic development
Complexity - different languages/cultures make complex projects even more difficult
Government vs. commercial interest

International civil space cooperation is admired as a noble and worthy goal, but space program managers confront a far different reality from that facing the diplomats and national policymakers as they seek to shape such efforts. Hopes for cooperation can be overwhelmed by competing interests and priorities, and also reduced or constrained by budgets. These anti-collaborative behaviors and factors shown in Table 2 are demonstrated by two cases.

Case 1

"Fly me to the moon": Over the last several years five different spacecraft have been sent to the moon by the United States, European Space Agency, China, Japan and India; each mission essentially has performed the same basic science missions.[21] Expenditures for these repetitive efforts totaled between $2 and 3 billion dollars. More astounding, scientific data from several of the missions hasn't been shared. Were these just expensive stunts—or better characterized as lost opportunities? Perhaps better science

> **Myth**:
> - *International cooperation saves money*
>
> **Reality**:
> - *It costs time, manpower, and other resources to deal with partners*
> - *There are significant obstacles and impediments to international partnerships*

---

[21] See for example, India – Chandrayaan; Europe – Smart-1; China – Cheng'e; United Kingdom – MoonLite; Russia; Japan – Selene.

and exploration could have been achieved if these activities had been consolidated into a single mission, with the excess funds spent on other scientific objectives. Now we hear the South Koreans, Brazilians, Iranians, and others want to launch their own moon missions; the rationale for these moon missions are draped in the words of great tribal patriots, and described in hushed reverent tones, emblematic of the best expressions of national pride.

CASE 2

"Up, up, and away": The international launch market is well over capacity for launching the current and foreseeable demand for communications, remote sensing and navigation satellites. Yet eight different countries continue to subsidize their own launch capability and other nations are developing their own launch capability. The United States prohibits US civil and commercial spacecraft from being launched on Chinese launchers. The European Space Agency demands that European satellites be launched on the Ariane launch vehicle. These directions are driven by important national or regional interests. However, there may be no easy way to foster improved international cooperation if such "protectionist" behaviors stand in the way.

Of course, regardless of the potential for success, the cynic knows that a nation's decision to engage in space cooperation is very much a political decision. Nations pick and choose if, when, where, and how they expend their national treasure. They choose the manner and extent of their foreign investments for reasons both known and unknown to other nations. Most states make their choices based on perceptions of their own national interests. The only constant is that a decision to "join in" cooperation is, in every case, a calculated political decision by each potential member of a commercial partnership or alliance, or inter- or quasi-governmental structure. Indeed, large private commercial investments are nearly always controlled at a national level, usually by the force of domestic (municipal) law, regulation, or licensing.[22]

National decision-making influences and shapes commercial and government entity governing structures. Accordingly, some space capabilities will be funded, developed, and offered, if and only if, they are strictly operated and controlled under specific national direction and within

---

[22] The *Outer Space Treaty* and associated *Convention on International Liability for Damage Caused by Space Objects(1972)* also known as the *Liability Convention* provide an underlying basis for this regulation, to address liability issues of launching states arising out of their activities.

strategic national guidelines. Reflecting these interests, military space cooperation tends to occur only when overarching national security military and intelligence community interests are satisfied. In contrast, international civil cooperation generally wins internal national political support for a different set of reasons: that is, if the cooperation generates national diplomatic prestige, provides for political sustainability, or enables workforce stability.[23]

The obstacles and impediments to cooperation are substantial. They include cost, technology transfer constraints, international and domestic politics, and exceptionalism perspectives.

International cooperation can be too expensive. Despite desires to save money, international cooperation successes are often secured only at a tremendous expense. For example, one can point to the ISS as a stunning blend of international politics, technology, and cooperation. While the ISS' research capabilities and benefits have been much ballyhooed and trumpeted, the system has turned out to be a very expensive offering on the altar of international cooperation. Billions of dollars have been squandered on it. The ISS success has been forged only at the detriment of other much more scientifically productive projects such as robotic spacecraft missions, space exploration, and aeronautics science and technology programs. Unfortunately, very little scientific research on the Station has been planned and executed.[24] The technical deficiencies in the ISS design limit its utility. The station has a need for high levels of dangerous and risky, crew-based maintenance, accomplished via recurring extra-vehicular activities (spacewalks). The high inclination of the station's orbit has also led to a higher cost for US-based Space Shuttle launches to the station.[25]

The need to support the ISS has gobbled up moneys needed by other programs, and at the same time helped justify continuing other NASA

---

[23] D. A. Broniatowski et al., "The Case for Managed International Cooperation...," *supra*.

[24] "The world's most expensive scientific laboratory installed additional solar panels yesterday, capable of producing 100 kilowatts or so of additional power for experiments. The panels cost $372 million to build, and about three times that much to send up to the ISS. Stand by for important new results. The only unique feature of a space environment is micro-gravity. One of the things you could study in micro-gravity is cavitation in spherical drops of water. A paper just published in Phys. Rev. Lett. reports important new insights from such studies except the experiments weren't done in space. They were done on a European Space Agency aircraft flying in parabolic arcs." Robert Park, "Space: International Space Station Unfurls New Solar Panels," *What's New*, September 15, 2006.

[25] The U.S. space shuttle launches were used to support heavy lift requirements, bringing up key elements to the station during its construction.

programs that provided only marginal value for the investment. For example, during the 1990s, to continue its success in obtaining funding for the Shuttle program and manned spaceflight, NASA switched from funding rationales that argued the reusable Shuttle spacecraft provided flexibility and cost savings to new ones that emphasized that the system was vitally needed to service and supply the ISS. This funding strategy had the unfortunate effect of siphoning off staggering amounts of moneys that could have been used to fund cutting-edge astronautics and aeronautics science and technology programs. Senior technologists within NASA saw the damage that was being done to its science and technology portfolio but could do little to fight the machinations of the manned spaceflight cabal.

Had NASA abandoned the Shuttle program, declined to help form the ISS as it was conceived and is being executed, and instead flown traditional government and commercially available expendable boosters, significant and draining spending might have been avoided, or, more realistically, better used. This would have freed the then unused funds for other initiatives and perhaps spawned a more balanced, scientifically-based civil space program.[26] Similarly, by using expendable rocket options, the US domestic commercial booster industry could have been stimulated, with more resources directed to lowering the cost of space access.[27]

First Launch	Country
1957	Russia
1958	United States
1970	Japan
1970	China
1979	European Space Agency
1980	India
1988	Israel
2009	Iran
soon	Brazil

---

[26] See generally, "Criticism of the Space Shuttle program," *Wikipedia, the free encyclopedia*, http://en.wikipedia.org/wiki/Criticism_of_the_Space_Shuttle_program, citing Roger Handberg, *Reinventing NASA: Human Spaceflight, Bureaucracy, and Politics*, Greenwood Publishing Group (2003).

[27] *Ibid.* See also Launius and Howard E. McCurdy, *Spaceflight and the Myth of Presidential Leadership: and the myth of presidential leadership*, University of Illinois Press. (1997), pp. 146–155.

As it turned out, Space Shuttle features that have been argued and described as vital to the ISS support have proved superfluous. The Russians have demonstrated that its expendable launch vehicles and unmanned supply systems have sufficient flexibility and robustness to sustain much of the station's needs.

In response to some of these arguments, true-believers for manned space activities argue that criticism of the ISS is plainly short-sighted. Some of these proponents are satisfied proffering a minimalist argument that just achieving human spaceflight is a singular wonderful end in itself. More pragmatic advocates countenance a more balanced view that manned space research and exploration, and the international cooperation efforts, have produced billions of dollars' worth of tangible benefits for all of mankind. Indeed, NASA's Innovations Partnership Program distributes a wonderful and glossy annual report, Spinoff, that trumpets and celebrates space technologies that, "through productive partnerships with industry, entrepreneurs, universities, and research institutions have resulted in products and services that elevate health and public safety; augment industrial productivity, computer technology, and transportation; and enhance daily work and leisure."[28] NASA argues the many benefits and indirect economic return from spin-offs of human space activities has been many times the initial public investment. These claims are not without their detractors; the authors only remember drinking Tang™ as an orange juice substitute during their childhood.[29]

Other US space programs with significant international content, such as the James Webb Space Telescope (JWST), suffer resource and expenditure problems. ESA will provide the Ariane V launcher to lift the JWST to orbit, instead of a domestically produced Evolved Expendable Launch Vehicle (EELV) nominally procured through the United Launch Alliance.[30] Launching on the Ariane V was originally intended and

---

[28] *Spinoff,* National Aeronautics and Space Administration, 2009, p. 7.
[29] Of course, let's be honest. The authors would probably feel differently about Tang if they were selected for service with the astronaut corps. Yum!
[30] The JWST is a partnership between ESA, NASA and the Canadian Space Agency. Formerly known as the Next Generation Space Telescope (NGST), the JWST is due to be launched in August 2013, and it is considered the successor of the NASA/ESA Hubble Space Telescope. The ESA financial contribution to JWST will be about 300 million Euros, including the launcher. Other European institutions will contribute additional Euros, all in return for flying the Mid Infrared Camera Spectrograph (MIRI) payload. According to a formal agreement, ESA will manage and co-ordinate the whole development of the European part of MIRI and act as a sole interface with NASA, which is leading the JWST project. MIRI will be built in cooperation between Europe and the United States (NASA),

described as a way to help NASA avoid costs. Unfortunately, the expected savings will never materialize. They have been lost because the JWST prime contractor did not contemplated use of the Ariane system. As a result, costs to integrate the JWST on that launch system have skyrocketed.[31]

Though cited approvingly in this paper, the admirable success enjoyed by ESA has not come cheap. The Ariane V spacelift system, while proficient, is very expensive to build, sustain, and push through a launch campaign. Of course, getting the 18 member governments of ESA to set common objectives, pool resources and make their industry work together has been a never ending task. Wise space professionals know the reality behind the motivation for Europe's cooperation— much of ESA's successes are really about ensuring full employment within Europe, especially among France's aerospace workforce.

Finally, the costs associated with terminating cooperation can be huge. Such moves can risk alienating key allies.[32] This all serves as a logical consequence to the rule that cooperation improves a program's political sustainability and the space community's workforce stability. There is no easy way to back out of cooperative relationships once they have been initiated.[33] The end result of this is that one may choose to endure the high price and continue even failed cooperative efforts.

---

both equally contributing to its funding. MIRI's optics, core of the instrument, will be provided by a consortium of European institutes. In addition to MIRI, Europe through ESA is also contributing the NIRSPEC (Near-Infrared multi-object Spectrograph) instrument.

[31] The program's funding crisis surfaced when Northrop-Grumman advised NASA that it would need an additional $270 million to make changes to the program as requested by the agency. Those changes included adjustments to the telescope's instrument module and ground test equipment, Mohan said. NASA had also advised Northrop-Grumman to plan to launch Webb on a European Ariane 5 rocket rather than a U.S. Evolved Expendable Launch Vehicle. In the case of Webb, "changing one thing tends to ripple more than it does on other programs," Mohan said. Ben Iannotta, "Webb Telescope Cost-Control Effort Focuses on Schedule, Requirements," *Space.com*, August 22, 2005, https://www.space.com/spacenews/archive05/Webb_082205.html.

[32] Such happened when the United States nearly a decade back suspended an exception to the International Trafficking in Arms regulations (ITAR) that Canada enjoyed for many years. The exemption was only regained after considerable effort by the Canadian Government.

[33] "If it were necessary to cease cooperation, a mutual choice to do so would likely mitigate many of the negative reputation effects, because there would be no unilateral actor to whom one could assign blame." D. A. Broniatowski et al, "The Case for Managed International Cooperation…," *supra*.

TECHNOLOGY TRANSFER CONSTRAINTS

Designing, manufacturing, and operating increasingly interoperable platforms, performing cooperative planning, and executing satellite operations are complicated by US law and policy that imposes controls on the release of sensitive technologies and operations. Indeed, important technologies and information relating to some space operations and technologies are often determined by the US government to be non-releasable, even to allies and close partners.[34] This is not just a US phenomenon; other nations have their own laws and policies that clamp down on technology transfers and specific relations with other nations.

Important portions of US technology transfer "releaseability" law and policy arise out of the Arms Export Control Act (AECA).[35] The AECA governs the sale and export of defense articles and services and related technical data, and serves as part of a statutory scheme to ensure compliance with technology control regimes. The regimes seek to slow the proliferation of missile and other technologies used to deliver weapons of mass destruction. Designated controlled articles, technologies and services are identified in the US Munitions List (USML), which is contained within the fearsome International Traffic in Arms Regulations (ITAR). Under the AECA, spacecraft, space related articles and services are specifically designated to be subject to export control. Exports of space articles, services and related technical data must therefore meet US national security interests. Proposed recipients must offer assurances they will protect them before transfers are made. Approvals usually require substantial paperwork and training.

Unfortunately for the United States, the export and technology control rules are driving small suppliers out of the export marketplace as they lack the economies of scale to properly respond to legal requirements.

> "Politically and technologically, the United States could gain from leading an international cooperative program to advance in space exploration. But for such a space program we will have to learn how to pursue 'shared' goals, which would give the United States less latitude in setting the program objectives..."
>
> Exploring the Moon and Mars: Choices for the Nation, Office of Technology Assessment, July 1991

---

[34] Other nations also secure their technologies for comparable diplomatic, military and economic reasons.

[35] See *Arms Export Control Act of 1976*, Section 38, as amended, P.L. 94-329. The AECA's 22 U.S.C. § 2778 provides the authority to control the export of defense articles and services.

This is damaging US economic security interests since small companies are usually the engine of innovation within the US economy and especially its space community. International partners are also wary of the legal rules and procedures. Nearly all members of the space community, foreign and domestic, find that the AECA rules are quite burdensome and onerous. The requirement for the assurances, and the threat for US criminal liability and prosecution arising out of them, is generally agreed to have cost US industry billions of dollars in sales in the international space marketplace. The US communications satellite industry is losing significant market share to international competitors who claim their systems, products and services are "ITAR-free."

Of course, other US laws, regulations, and policies apply to exports of space data, hardware, and services.[36] The new administration and the Congress are reviewing them along with the AECA releasability rules. Some believe the President will push approvals for some transfers back to the Commerce Department, where approvals were issued for communications satellite technology transfers until the Chinese scandals of the late 1990s.[37]

VOLATILITY IN INTERNATIONAL AND DOMESTIC POLITICS

National political processes can bring uncertainty to international agreements. For example, in 2004, President George W. Bush unveiled his Vision for Space Exploration, which put a near-term emphasis on returning

---

[36] See generally, National Security Decision Memorandum (NSDM) 119, "Disclosure of Classified United States Military Information to Foreign Governments and International Organizations" and Executive Order (EO) 12958, "Classified National Security Information," April 17, 1995, as amended by EO 13292, "Further Amendment to EO 12958, as Amended, Classified National Security Information," March 25, 2003, and by other executive orders. See also, the *Export Administration Act of 1979*, P.L. 96-72. The *Export Administration Act of 1979* (EAA) governs the export of most dual-use unclassified articles and services (having both civilian and military uses) not covered by the AECA. The EAA controls exports on the basis of their impact on national security, foreign policy, or supply availability. With the expiration of EAA in 1994, the President declared a national emergency and exercised authority under the International Emergency Economic Powers Act (P.L. 95-223; 50 U.S.C. 1701 et seq.) to continue the EAA export control regulations then in effect by issuing EO 12924 on August 19, 1994.

[37] See *Final Report of the Select Committee on U.S. National Security and Military/Commercial Concerns with the Peoples' Republic of China*, May 25, 1999 (the *Cox Report*). Technology transfer abuses identified within the *Cox Report* caused the Congress to tighten space systems export rules in with the *Strom Thurmond National Defense Authorization Act, Fiscal Year 1999*, P.L. 105-261, October 17, 1998.

humans to the moon. International partners, especially in Europe did not immediately adopt this policy because they were more interested in performing Mars missions. However, after four years of international workshops, bilateral meetings, then intense hectoring and haggling, a collective "global vision" was forged with the US prospective partners, especially ESA. ESA then cajoled its members to support the Vision. Then, just as ESA was announcing that its membership had synched its planning and programming roadmap to match the Bush Vision, the United States led by a newly elected internationalist President, announced interest in a radically different space vision, that is, the one recently identified and described by the Augustine committee. The United States has now abandoned the Vision's "lunar base" concept and moved to a "flexible path" to manned space exploration. The change devastated the ESA partners.

Similarly, after the fall of the Soviet Union, US companies were encouraged to work with Russia to help them on their course to capitalism. The US Government was motivated to ensure the Russian scientists and engineers were working on non-threatening, yet productive, activities. When the RD-180 Russian rocket engine was selected by Lockheed Martin to serve as the first stage engine for their new Atlas V rocket as part of US Air Force's EELV program, US Congress hailed the choice as a great achievement in international cooperation. Several years later, however, when the Atlas V was being looked at as a launch vehicle to support human missions to the ISS, NASA was criticized for considering a vehicle with Russian engines. The new-found opponents to the RD-180 decried the possibility that Russia could stop selling the engines to the United States; they argued that US designed and manufactured engines should be selected.

The volatility problem cuts both ways in terms of US partnerships — not all partners work predictably with the United States. For example, NASA Administrator Sean O'Keefe cancelled the Crew Return Vehicle (CRV) initiative in 2001 after explaining it was cheaper in the long run to buy Russian Soyuz capsules as escape pods for the ISS, instead of investing $1.5-3.0 billion to develop and build a US rescue capability.[38] Once the Space Shuttle program came to an end in 2011, the United States was obliged to rely on the Russians to transport astronauts to the ISS for at least

---

[38] Hearing Addresses NASA Budget Request, Shuttle Investigation, *American Institute of Physics Bulletin of Science Policy News*, March 25, 2003. Bruce Moomaw, "The Science of Spending Billons," *Space Daily*, Sept 21, 2002. Congressman Criticizes O'Keefe, *Space Today.net*, April 18, 2002.

seven years (according to the Augustine Committee's Report).[39] Now a number in Congress are unhappy to find out Russia has nearly doubled the cost of an escape pod capsule to almost $65 million each.[40] Soviets rebranded as Russian capitalists have learned the lessons of capitalism all too well.

EXCEPTIONALISM

Exceptionalism is the perspective that a country, or society, holds that it is unusual or extraordinary in some way.[41] Many nations throughout history have made claims of or exhibited the hubris of exceptionality: the United States, China, India, Britain, Japan, Iran (Persia), Korea (both South and North), Israel, the USSR, France and Germany.[42] The term "exceptionalism" can also be used to describe a nation's desire to remain separate from others.[43] There is oftentimes a strong and intense political and cultural pressure to go it alone, to demonstrate a nation's prowess and strength—to show a nation has joined the leaders of the world. That may explain the following news report:

> SEOUL (Reuters) - South Korea plans to launch a lunar probe in 2020 and make a moon landing by 2025 under a new space project that will develop indigenous rockets to put satellites into orbit, the Science Ministry said on Tuesday. The lunar probe program will be based on a rocket South Korea is developing at a cost of 3.6 trillion won ($3.9 billion) in the next decade. South Korea is behind regional powers Japan and China in the space race. China became only the third country to launch a man into space on its own rocket in 2003 and put its first lunar probe into orbit in early November. Japan's first lunar probe began orbiting the moon in October, four years

---

[39] *Seeking a Human Spaceflight Program Worthy of a Great Nation, Review of U.S. Human Spaceflight Plans Committee, Final Report*, October, 2009, p. 10.
[40] The cost of Soyuz was $50M per launch (1999 dollars) - www.astronautix.com/lvs/soyuz.htm. Russian is now charging NASA $51M per seat per astronaut. Tudor Vieru, "Russia Wants $51 Million for Seat on Soyuz - NASA will have to pay to get to the ISS," *Softpedia.com*, May 14, 2009.
[41] "Exceptionalism," Wikipedia, http://en.wikipedia.org/wiki/Exceptionalism, accessed December 5, 2009. See also, Michael Kammen, "The Problem of American Exceptionalism: A Reconsideration," *American Quarterly*, Vol. 45, No. 1 (Mar., 1993), pp. 1-43.
[42] *Ibid.*
[43] *Ibid.*

behind schedule due to technical glitches.[44]

The desire for exceptionalism accounts for this above report's stated national objective for the Republic of Korea. The South Koreans are working hard to achieve success with their space programs; indeed, their space community leaders have told the authors that they take great pride in the successes they have already gained, and hope to gain. We take them at their word. Note, however, how the Reuters article puts the Korean space effort in juxtaposition with its traditional rivals and enemies, China and Japan.

> *Cooperation vs. leadership:* During one panel discussion at the International Space University's 13th Annual Symposium, members argued that the United States should provide more leadership with regards to space "traffic control" to prevent satellite collisions. Only moments later, however, during the next panel, its members complained that the United States "took charge" much too often instead of working as a collaborative member (e.g., comparable to a member of the European Space Agency).

Exceptionalism also explains a desire by China for a national manned spaceflight, space station, and moon programs; the desire by a wide variety of nations to develop spacelift and on-orbit capabilities; and, of course, the desire by other overachieving states and individuals to launch their own missions to the moon.

Unfortunately, exceptionalism pressures can cause inefficiency, with tremendous duplication and overlap occuring in global space science and other missions. The reported problems can also generate considerable mistrust. For example, there has been much discussion about inviting China to participate on the International Space Station. Unfortunately, in its single-minded zeal to forge a unique world-class military and space program China has generated considerable angst and distrust among the international community. This is underscored by China's program's secrecy and the recent, alarming ASAT test that contaminated low earth orbit with thousands of pieces of space debris that will pose a threat to space systems for well over a hundred years.

---

[44] Jack Kim, "South Korea eyes moon orbiter in 2020, landing 2025," *Reuters*, November 20, 2007, http://www.reuters.com/article/scienceNews/idUSSEO24596320071120, accessed November 29, 2009.

Improving International Space Cooperation

When attempting to achieve success with cooperative efforts participants must find utility arising out of their efforts. To maximize this utility it is important to consider the types of cooperation frameworks that exist. According to Ryan Zelnio, there are four framework types: coordination, augmentation, interdependence, and integration.[45]

Coordination

Each country operates a separate program independent of others but coordinates on technical and scientific matters. According to Zelnio, "This model of cooperation is inviting in that it is easy for people to agree to, as it allows each country to maintain its total independence and manage its own contributions. The disadvantage of this is that often countries push programs that greatly overlap the efforts pursued by other countries, causing much duplication of efforts."[46]

Coordinating groups exist, and they have achieved success in improving international dialogue on scientific efforts. For example, both the Committee of Earth Observing Satellites (CEOS) and Global Earth Observation (GEO) promote sharing of earth observation data and coordination of such missions. In important moves, some coordination activities are now occurring and expanding on US, European, Russian, China, and other proposed global precision navigation and timing satellite programs (GPS, Galileo, GLONASS, etc.)

The recent duplicative, overlapping and expensive missions to the moon serve as imperfect examples of this framework. Worthwhile cooperation has failed to occur on several of the missions and some participants have yet to share the data gleaned from their observations, even though the data appears to involve no real national or economic security matters. The problem in failing to share occurs many times as an outgrowth of the exceptionalism, xenophobia, and paranoia some countries exhibit about their technology programs.

Augmentation

Cooperating countries provide important elements of the project of the prime country but are not on the prime's program critical path.[47] The United States often employs the augmentation framework. The new US

---

[45] Ryan Zelnio, "A model for the international development...," *supra*.
[46] *Ibid.*
[47] *Ibid.*

National Space Policy states that a fundamental principle of US space activity that it is "committed to encouraging and facilitating the growth of a US commercial space sector that supports US needs, is globally competitive, and advances US leadership in the generation of new markets and innovation-driven entrepreneurship."[48] Consistent with this policy and its predecessors, the United States has led the world in space exploration for five decades. The augmentation framework is consistent with this leadership imperative and, more importantly, reflects the country's tremendous resource investments in space activities.

The disadvantage of the augmentation framework is that the bulk of the costs fall upon the prime country. The United States usually accepted these burdens because it nearly always assumes the prime risks of each mission and wants to control them. Using the augmentation framework allows the United States to exercise centralized control over a mission's critical resource, schedule, technology development, and operational paths. Given the allocation of risk, marginal or minimal contributors to space efforts are not usually given veto power over the mission decisions. US space exploration efforts like the JWST and Cassini follow this model, with MIRI and Huygens payloads only being contributed by ESA, respectively. They do not control the essential mission activities and decisions. The framework has also been followed with Space Shuttle program, however, with only small payloads and invited international astronauts being flown on-orbit.

Some question whether the augmentation framework really provides for true cooperation. For example, D. A. Broniatowski, G. Ryan Faith, and Vincent G. Sabathier contend:

> ...[T]here are diplomatic drawbacks to insisting on sole control of the critical path. By restricting international partners to noncritical-path items, a nation is sending a signal indicating a lack of trust and confidence in the partner's capabilities and unwillingness to rely on that partner. Rather than committing to work through problems, the nation is hedging bets in case the partner "fails." This sort of partnering is, in effect, not truly cooperative, because the requirement that one nation possess all of the critical-path capabilities is an implicit statement that such a nation can complete the system under its own power and therefore does not need its partners. As such, there is no true programmatic incentive for the cooperation to happen. From a practical standpoint, this structure endows the nation that

---

[48] *National Space Policy of the United States of America*, supra., p. 3.

maintains the critical path with all of the decision-making power, thereby making the partner nations utterly dependent and essentially irrelevant... [T]he argument that international cooperation reduces cost must also be seen within the context of the critical path. A partner who provides a component that is off the critical path is not genuinely reducing the cost for the integrator nation. On the other hand, such cooperation does not negatively affect the employment associated with the space exploration system. Instead, this nation is providing a capability that is, by definition, unnecessary to the minimal operation of the system. It is an extraneous capability... [I]t is receiving a capability that it would not have had otherwise. This form of cooperation therefore creates a natural hierarchy of partner nations among those who have the most control of the critical path; the most de facto decision-making power; and those who provide the extraneous capabilities but have little in the way of programmatic utility and contribute little in the form of decision-making.[49]

INTERDEPENDENCE

Cooperation occurs on the critical path as well as on functional systems with each participant still controlling their component part of the project.[50] The United States and Russia employ the interdependence framework to cooperate on the ISS program. Each serves as a prime resource contributor, and this has satisfied each nation's desire to exercise leadership over the enterprise and protects equities. Unfortunately, as noted earlier, the ISS framework has proven to be extremely costly. With the ISS, neither principal is able to keep the other from slipping their contributions and causing significant delays and cost increases. For example, the ISS partnership suffered a lengthy stand-down in Space Shuttle flights after the disintegration of the Columbia Shuttle.

The European Union's Galileo precision navigation and timing satellite constellation program also suffers from using an interdependence framework. Its partner nations have, from time-to-time, unilaterally withheld contributions causing the overall program to slip. This brings to mind an observation by Broniatowski et al.: "A nation should not be "held hostage" by the policy, schedule, or budgetary difficulties of its partners...

---

[49] D. A. Broniatowski et al., "The Case for Managed International Cooperation...," *supra*.
[50] Ryan Zelnio, "A model for the international development...," *supra*.

Too many cooks spoil the broth."[51]

INTEGRATION

Full cooperation with a pooling of resources on shared and joint research and development. This framework spreads the financial costs, and can utilize the industries of multiple nations while maintaining a single entity to control the critical path."[52] Both ESA and Intelsat successfully employ the integration framework, with the latter doing the better job of it. While ESA has had some failings, its successes have attracted partners in and outside of Europe, including NASA. According to Jean-Jacques Dordain:

> *"Inside Europe we have developed a strong cooperation with the European Union (EU) of 27 countries. Together with Norway and Switzerland, which are not members of the EU, but members of ESA, there are today 29 European countries cooperating in space. These countries approved a European Space Policy in 2007 and two flagship programs: Galileo and GMES (Global Monitoring for Environment and Security). Connecting ESA with the EU means connecting space with the European citizens, because the EU is in charge for European policies for European citizens. The two "children" of the ESA/EU relationship, Galileo and GMES, are connecting space with European Transport Policy (Galileo) and with European Environment and Security Policy (GMES)."*

> *"Outside Europe, ESA is practically cooperating with every space faring nation in the world. Obviously with the USA, which formed the "technical culture" of our space activities, but also with Russia (in science, the ISS program and launchers), with Japan, China (in science), India or Canada. ESA also cooperates more and more with non-space powers in order to share some data with those that have pressing needs and no space capabilities. ESA and the French Space Agency CNES are co-founders of the Charter on Natural Disasters. The members of this Charter make their space systems available to civil protection organizations across the world in case of natural disasters. This European initiative has now become a global initiative since Canada, Japan, India, the USA, Argentina, China have joined*

---

[51] D. A. Broniatowski et al., "The Case for Managed International Cooperation...," *supra*.

[52] Ryan Zelnio, "A model for the international development...," *supra*.

the Charter."[53]

The primary negative with the integration framework is that it requires acceptance of maximum levels technology transfers. This is often difficult and complex to achieve as national policy makers may disagree with sharing of technologies that offer a unique security or economic edge. The European Union has been able to internally mitigate some of these technology transfer concerns by applying its own unique free market laws and regulations.

Concluding Thoughts

The case for cooperation is strong and powerful. Each nation engages in international cooperative activities because it is in its best national interest to do so. Cooperation enables states to leverage resources and reduce risk; achieve efficiencies; improve global diplomatic and other engagement; and enhance diplomatic prestige, political sustainability and workforce stability. Given these benefits, space leaders must organize their programs to allow for cooperation following some of the recommendations in Table 4.

TABLE 4: Recommendations for Productive International Cooperation In Space
Address technology transfer and releasability concerns early in the process – develop architectures for international environment
Diplomacy, knowledge of partners, and rapport is vital
Employ respect, reciprocity and transparency
Apply patience - build in time for the team to "build trust", "earn trust", "maintain trust" and deal with communications issues

Although powerful, cooperation's success is often achieved only at great expense. The other obstacles and impediments to cooperation are substantial. Releasabiliity constraints, international and domestic political volatilities; and exceptionalism must be confronted early by program

---

[53] Jean-Jacques Dordain, "International Cooperation in Space," remarks at 40th Anniversary of the Universities Space Research Association (USRA), March 26, 2009.

manageres to reduce their risk and expense. Significant harm can occur if these obstacles and impediments are ignored, so a space system architect contemplating cooperative activities helps by developing a program that attacks them up front.

Differences among partners must be understood and respected. Leaders help by attempting to foster conditions that allow for reciprocity and transparency when they can occur, although these conditions will vary. Rapport, knowledge of partners, and patience among the assembled team members will also be essential. Successful international partnerships tend to have unifying attributes—fundamental trust with aligned goals. Without trust and common goals, relationships become inefficient and often fail.

In the end, there isn't single recipe to achieve success. Depending on the relationships, mission needs, and relative strengths of the partners, coordination, augmentation, interdependence, and integration frameworks can each be employed to overcome problems and achieve success. Each of these frameworks requires an early definition of common objectives—this enables participants to successfully develop mission architectures and reduce coordination costs. Partners must define these objectives together and stress the benefits of the relationship that is being built. By keeping the benefits of cooperation in mind, system architects will each achieve solid success.

•••

COLONEL JAMES D. RENDLEMAN, USAF, RETIRED

Col James D. Rendleman, USAF, retired (BS, Chemistry, University of North Carolina - Chapel Hill; MBA and MPA, Golden Gate University; JD, Whittier College School of Law; LLM, University of San Diego School of Law). Mr. Rendleman was commissioned a second lieutenant through the Air Force ROTC at the University of North Carolina - Chapel Hill. He served in a wide variety of science and technology, engineering, management, and policy positions within the Air Force laboratories and space acquisition community, Headquarters, Air Force Space Command, the Air Staff, and the National Reconnaissance Office. He is a level-3 space professional and trained director, space forces. He served as study director for The National Academies study of the US Aerospace Infrastructure and Aerospace Engineering Disciplines. An attorney and member of the State Bar of California, Mr. Rendleman engaged in law practice as a partner, solo practitioner, and associate with firms in Los Angeles, San Francisco, and Napa, California. He is a member of the American Institute of Aeronautics and Astronautics' Legal Aspects Aero and Astro Technical Committee and International Activities Committee, and is an elected member of the International Institute of Space Law. He taught management theory for Cerro Coso Community College and Golden Gate University. He also taught space law, policy, command and control, international cooperation, and missile defense for the National Security Space Institute.

WALTER J. FAULCONER

Mr. J. Walter Faulconer (BS, Space Sciences, Florida Institute of Technology; MS, Systems Management, University of Southern California). Mr. Faulconer is president, Strategic Space Solutions, LLC. He has spent over three decades in the aerospace industry providing executive leadership, creative program management, systems engineering, and business development to civilian, commercial, and national security space customers. Before assuming his present duties, Mr. Faulconer was business area executive for civilian space at John Hopkins University's Applied Physics Laboratory (APL), leading 500 scientists and engineers on cutting edge space programs such as the MESSENGER Mercury orbiter mission; the TIMED study of Earth's upper atmosphere, ionosphere, and thermosphere; the twin spacecraft STEREO study of the sun and its coronal mass ejections; and NEW HORIZONS, the first mission to Pluto. Before APL, he worked at Lockheed Martin Space Company and Martin Marietta Astronautics Company as a director, program manager, flight test manager, systems engineer, and mission operations lead on a wide variety of classified and unclassified programs dealing with space launch, space control, missile defense, Space Shuttle, and other missions. Mr. Faulconer is an Associate Fellow of the American Institute of Aeronautics and Astronautics, a member of the board of directors for the American Astronautical Society, and a member of the International Astronautical Federation and National Space Society. He has taught graduate level business marketing and strategy at John Hopkins University, Whiting School of Engineering.

CHAPTER 17

MODELING THE NATION STATE IN SPACE COMMERCE AND THE PRINCIPLES OF SOVEREIGN SURVIVAL

THOMAS E. DIEGELMAN
NASA

AND

THOMAS C. DUNCAVAGE
NASA

© Thomas E. Diegelman and Thomas C. Duncavage, 2012. All Rights Reserved.

INTRODUCTION: TERRESTRIAL GOES CELESTIAL

The argument in this chapter states that celestial expansion will follow the same principles as those which describe the survival and success of nations. We explore three key principles to which all nations must adhere if there are to endure. The argument does not overlook the role of umbrella organizations such as the United Nations, but it does imply and indeed relies upon the view that nation states are the legitimate political authorities, both terrestrially and celestially.

NATION STATES: THE THEORY OF THE THREE PRINCIPLES

The circumstances under which nation states were formed have differed throughout history, but every nation was and is sustained on the same three immutable pillars, and every successful nation state is ultimately successful or fails because of these same three:

(1) national sovereignty;
(2) national security and defense; and
(3) national economy.

These pillars are more than just characteristics of success, they are operational identities that emerge and continue while playing in the competition for survival against external rivals and internal dissent. When the aggregate strength of the pillars of various nations is compared, the result shows the international balance of power. A weak or deficient pillar will cause the other two to weaken, thereby affecting the international power balance, and ultimately may cause end stage atrophy. Complete removal of any of the pillars and the nation collapses. Such has been the case throughout the history of civilization, and there are no exceptions. The three pillars model may seem to be an oversimplification of varying historical circumstances, evolving cultures, and specific historical circumtances, but it holds up to scrutiny nonetheless.

ROOTS OF THE THEORY: GAMING AND BEHAVIOR

Conceptually, the theory of games results in defined behaviors that, in aggregate, reflect nation states and their governance models, which is well stated in this classic game theory paper:[1]

> "What economists call game theory psychologists call *the theory of social situations*, which is an accurate description of what game theory is about. Although game theory is relevant to *parlor games*

---

[1] "Economic and Game Theory: What is Game Theory?" by David K. Levine, Department of Economics, UCLA, http://www.dklevine.com/general/whatis.htm

such as poker or bridge, most research in game theory focuses on *how groups of people interact.* There are two main branches of game theory: cooperative and non-cooperative game theory. Non-cooperative game theory deals largely with *how collectively and individually intelligent individuals interact with one another in an effort to achieve their own collective goals."*

A clever Cold War Era board game called "Summit" iIllustrates this. The objective of the game was national survival, which required the acquisition of the principle building blocks that all nations throughout history require: the pillars, here identified as military bases (national security and defense), factories (national sovereignty), and steel mills (national economy). The board and pieces are shown in Figure 1.

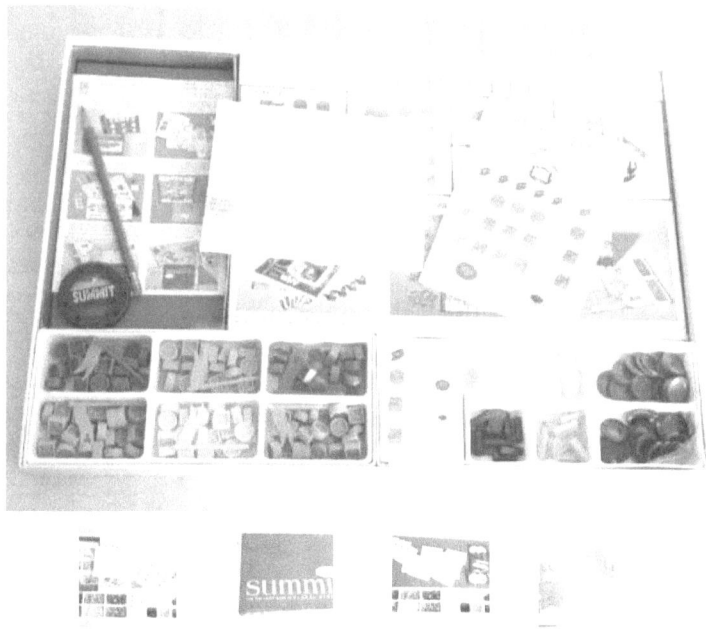

Figure 1: Vintage 1961 Milton Bradley Summit Global Strategy Game

The objective of the game was of course to maximize the security or longevity of the country. Factories were protected by military bases, while military bases were built by steel mills, and fueled by the wealth created in factories. To grow the economy, war was generally avoided because it required the combatants to "pitch pennies" against one another until one "lost" or became penniless.

This game was quite successful in its time not only because it reflected the Cold War dynamics of the 1960s, but also because of its intrinsic complexity and adherence to the reality of international affairs.

FAST FORWARD TO INTERNATIONAL COMMERCE: THE RULES FOR OUTER SPACE REMAIN UNCHANGED

The three pillars model suggests that national economic growth and power, scientific and technical leadership, and national security remain the three components essential to any national space program, regardless of the country. Since the world-captivating success of the Apollo missions, followed by the creation of a winged, reusable space vehicle, the United States has enjoyed a singular position of leadership in human space flight.

But this is no longer the case from a technological perspective, and in recent years the American human space flight capability has become the most expensive and the most controversial aspect of the NASA enterprise. Nevertheless, as a percentage of the national GDP based federal budget, NASA's share has decreased by more than a factor of ten since 1972; today it is considerably less than a single percent.

Books, treatises, dissertations and speeches have been written to explain why and how this happened, and the essential truth is that human space flight missions, particularly those related to beyond low Earth orbit exploration, have become less important in the United States' national sovereignty, national security, and national economic growth.

While there is broad agreement that the work done by NASA is worthwhile, it's also commonly held that it's not important enough to be funded to the degree necessary for human space flight. In the days of Apollo, human space flight to a great extent defined the national identity; today this is certainly not the case.

Our three pillars model suggests that NASA has become progressively disconnected from the "triad" over the three decades since Apollo, and the previously held view that "the US must be the leader in space" no longer resonates in the public arena largely because NASA seemingly cannot answer the question, "So what if we don't?"

Meanwhile, China has successfully rendezvoused two spacecraft in orbit, and is progressing in its capability toward a moon landing. Yet to date this has not stirred the kind of fear that the space successes of the USSR evoked in American mindset during the 1950s and '60s.

NASA: A BELLWETHER FOR THE UNITED STATES FUTURE – INCLUDING INTERNATIONAL CELESTIAL COMMERCE

The current NASA budget is short by about $3 – 5 billion per year to continue to modestly build on its capability to voyage beyond low Earth orbit (LEO). This figure is based on the programmatic outlays that have been trimmed to meet Congressionally-mandated funding limits. These monies would fund research, development, facilities maintenance, infrastructure, and information technology funds that have been cut to fit under the congressional budget cap, equivalent to $1/6^{th}$ of NASA's current budget, or 10% of the Department of Energy's funding

The $3 - 5 billion is a minimum. A figure of about $7 – 10 billion per year would probably be necessary to restore NASA's capabilities. Anything less than this funding level will sacrifice American leadership in human space flight in less than a decade, perhaps permanently.

To gain this additional support, the essential message that NASA must articulate is a simple and understandable human space flight vision that relates the significance of the three pillars to its own short and long term plans and activities. These plans must be not only motivating and compelling, but factual and substantive, and they must be articulated so as to resonate with political leaders and *the American public.*

The connection of national interests to local economic and technological development for the creation and growth of wealth in the private sector must be a major focus. A model such as is being proposed here, one simple enough to demonstrate that international commerce has made the United States wealthy, would illustrate that the practice of wealth creation through trade is now being extended to the next commercial frontier, space, and that to remain viable as a nation, the US must fully participate in the growth and development of commerce in space.

THE TRIAD MODEL IN BALANCE

A high level approximation of the decision process is intended to illustrate how this logic applies to policy discussions.

1. Framing the central importance of the model illustrates how it applies to the core responsibility for the ultimate "common good" – that of survival.
2. Further illustrated with historical examples, the model could be utilized in policy decision making, by employing historical

scenarios to illustrate the underlying principles builds confidence, and provides a basis from which to extrapolate from the present to the future.
3. Illustrations through examples of current decisions confirm the model as one of utility.

Use of the model requires only two assumptions.

First, there must be a defined set of national goals that are expected to remain valid for the foreseeable future, defining a context in which the survival of the state as an entity is accomplished by the execution of policy and actions that assure sovereignty, security, and economic competitiveness. How these are attained and at what cost may indeed change with changing conditions, but essentially there must be a national will to flourish for future generations which frames decisions and scenarios over a span of 50 years or more.

The second assumption is that there is an accurate assessment, apolitical and unaligned, of the proposed decision paths, which may necessitate that reassessments be performed on the measured data as conditions change.[2]

The decision pathway chart shown as Figure 1 illustrates in a very simplified way the character and caliber of the thinking that is necessary to use the triad model as a decision making support tool that helps to identify the best course of action, certainly in support of sound policy decisions.

Application of the model suggests that when policy errors are made, a vital and economically strong nation will auto-correct.

The starting and end point for the model is the revenue or treasury of a nation. If the coffers are empty at the state level, personal wealth is also likely to be long since gone, and the prospects for the future are not bright. If there is surplus capital, then society must decide where to invest it. Should it be directed toward social programs, or military endeavors, or other investment in the future – and in what proportional amounts? Given that there are sufficient funds for high priority tasks, the question then becomes: What is the priority? (Note that the model deals only with the amounts, not the associated justifications.)

---

[2] "The Sexual Counterrevolution" , by Mary Eberstadt, *National Review*, April 2, 2012, p. 34-6

This is illustrated in a very simplified form in Figure 2, a generic flow chart showing the probably decision pathways that may lead to a sound policy decision.

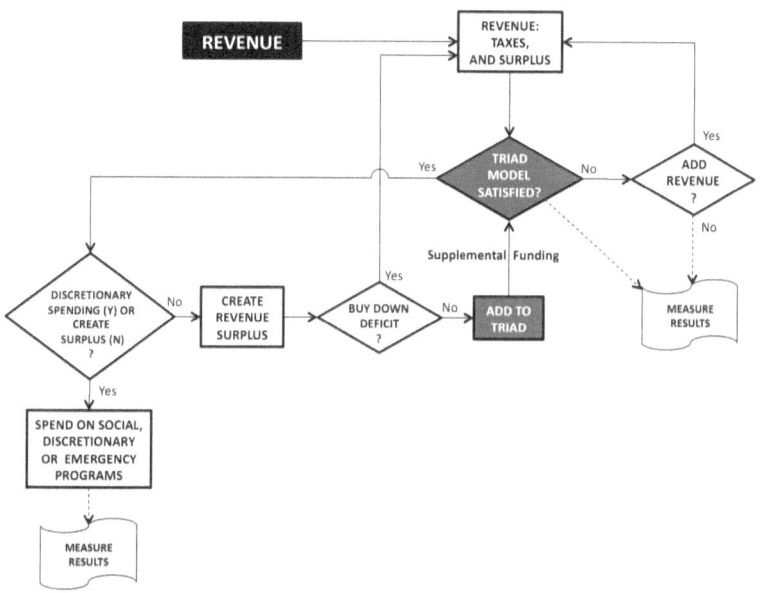

Figure 2: The Basic Decision Tree Model

As a matter of discussion or debate in the decision making process, we recognize that there are varying levels of inquiry that become pertinent, including Value, Fact, Definition, and Policy. For example, we seek to identify where the "value" lies, and to sort the "facts" from the myths, to "define the pertinent concepts clearly, and finally to identify the best course of action for our policy to pursue. Sound decision making requires clear thinking at all four levels, ultimately in service to the search for the best policy. This is the goal of the model.

An historical example is the isolationism of the United States in the 1930s, which did not require massive outlays of defense dollars, but a few years later Pearl Harbor ushered in unforeseen and essentially unlimited funding of military capabilities.

Another excellent historical example is the conversion of the British Navy and shortly thereafter its merchant marine fleet to the use of oil, and away from coal, in the early 1900s. This was a high stakes gamble that, due to the geopolitical situation Britain found itself in, could not be avoided. The only certainty was that if this gamble was not taken, its preeminence as a world power would have be lost. We explore this in detail as an illustration of the model functionality.

Example 1: Geopolitical Background: Great Britain's Slow, Imperceptible Slide from Preeminence

The following is paraphrased from *Oil and the Origins of World War I* by F. William Engdahl:[3]

> That the sun had begun to set on the British Empire was almost unnoticed in 1850, but by 1873 it was evident after the onset of a Great Depression that year. By the end of the 19th century, though the City of London remained the undisputed financier of the world, British industrial excellence was in decline. The decline paralleled the equally dramatic rise of a new industrial Great Power on the European stage, the German Reich. Germany soon passed England in output of steel, and in quality of machine tools, chemicals, and electrical goods. With Africa and Asia territories long since claimed by the other Great Powers, and especially by Great Britain, German policy set out to develop a special economic sphere in the imperial provinces of the debt-ridden Ottoman Empire. The policy was termed "penetration pacifique," economic dependency which was sealed with German military advisors and equipment. Initially the policy was not greeted with joy in Paris, St. Petersburg, or London, but it was tolerated. What Berlin and the Deutsche Bank did not disclose was that they had secured mineral rights along the path of the Baghdad railway, and that their geologists had discovered petroleum in Mosul, Kirkuk, and Basra.
>
> Once Germany launched a significant effort in shipbuilding, however, Great Britain faced a significant strategic challenge to its global empire. To sustain its competitiveness, Great Britain was obliged not only to develop new ship's hulls and engines, but in keeping with our model, a total commitment to the concept of global

---

[3] http://oilgeopolitics.net/History/Oil_and_the_Origins_of_World_W/oil_and_the_origins_of_world_w.HTM, By F. William Engdahl, 22 June, 2007, Oil and World War I: Oil and the origins of the 'War to make the world safe for Democracy.'

power and economic dominance to be enforced by the ships of the Royal Navy, thereby fusing national sovereignty, defense, and economic development.

The conversion of the British Navy under Churchill to oil from coal was a high risk strategy, as England had abundant coal but no oil reserves of its own. Increasing German assertiveness, including the decision in 1900 to build a modern navy that could rival England's. This set the stage for the outbreak of a war in August 1914, the real significance of which was a colossal and tragic struggle to determine which nation would succeed the ebbing power of the British Empire.

The resolution of that epic struggle required another quarter century and a second world war before the victor was undeniably established. All this is said fully recognizing a historical fact: Approaching the end of the 1890's, Britain was in all respects the pre-eminent political, military and economic power in the world.

Engdahl's synopsis states that Britain increased its dependence on imported goods following the introduction of free trade. From 1883 to 1913 the Sterling value of her imports rose by 84%. The real effect of the shift to import dependence was obscured by the phenomenal success of earnings from invisibles. In 1860 Britain led the world in coal production, the raw material feeding her industry and fueling her navy, with almost 60% of the total. By 1912 that fell to 24%. Similarly, in 1870 England enjoyed an impressive 49% share of total world iron forging output, but by 1912 it was 12%. Copper consumption in the UK, an essential component of the emerging electrification transformation, went from 32% of world consumption in 1889 to 13% by 1913.[4] Clearly the final quarter of the nineteenth century marked the beginning of the end of Britain's hegemonic position as the world's dominant economic power. The revolution in technology marking the shift from coal to oil power was the final blow to the preeminence of the British Empire. After the 1890s, though little publicized, and even today not well understood, the search for secure energy in the form of petroleum would become of paramount importance to Her Majesty's Navy and Her Majesty's government. A global war for control of oil was shaping up.

---

[4] Cited in Sonderabdruck aus der Frankfurter Zeitung, Gegen die englische Finanzvormacht, (7 November, 1915), Frankfurt am Main, Druck & Verlag der Frankfurter Societsdruckerei GmbH.

A Revolution in Naval Power

In 1882, petroleum was of little commercial interest, as the development of the internal combustion engine had not yet revolutionized world industry. But one man did understand the military and strategic implications of petroleum for future control of the world seas. In a public address in September 1882, Britain's Admiral Lord Fisher argued that Britain must convert its naval fleet from bulky coal-fired propulsion to the new oil fuel, insisting that oil-power would allow Britain to maintain decisive strategic advantage in future control of the seas, control that was vital to the economic health of Britain.

The argument about fuel was not a matter of national pride nor technological prowess, but simply raw performance: oil significantly outperformed coal. A battleship powered by diesel engines burning petroleum issued no tell-tale smoke, while a coal ship's emission was visible up to 10 kilometers away, meaning the oil provided a significant stealth advantage. Further, it required 4 to 9 hours for a coal-fired ship's engine to reach full power, while an oil-burning engine required a mere 30 minutes, and could reach peak power within 5 minutes thereafter, an 18:1 time-to-readiness ratio, and roughly 5:1 ratio on battle readiness. To fuel a battleship using oil required the work of 12 men for 12 hours, while a coal ship required the work of 500 men and 5 days, a 40:1 manpower ratio and 6:1 time-to-load ratio.

For equal horsepower propulsion, the oil in an oil-fired ship required 1/3 the engine weight, and almost one-quarter the daily tonnage in fuel, a critical weight factor for a fleet whether commercial or military, a 3:1 advantage in size and therefore usable volume afloat, and a 4:1 advantage in energy density. The radius of action of an oil-powered fleet was up to four times greater than that of the comparable coal ship,(8) a 4:1 payload to mass ratio.

The thought of abandoning the security of domestic British coal fuel in favor of reliance on foreign oil was a strategy embedded in risk. But Lord Fisher pushed the risky oil program through with one argument: "In war speed is everything." Winston Churchill had by then replaced Fisher as First Lord of the Admiralty, and was a strong advocate of Fisher's oil conversion strategy.

In 1913 less than 2% of world oil production occurred within the British Empire,5 so conversion of the British fleet to oil dictated that securing large oil reserves outside Britain was to become a national security priority and an essential factor in Britain's grand strategy and its

---

5 Anton Mohr, *The Oil War*, Harcourt, Brace & Co., 1926, p. 118-120.

geopolitics. Ironically, on the eve of the assassination of the Austro-Hungarian Archduke in Sarajevo, agreements had finally been reached between the Germans, the British, and the Turkish parties over oil rights in Mesopotamia, but too late to avoid the debacle of the Great War.

Given its diplomatic, military and security focus, would England risk a world war? The answer is yes; that is historically seen.

The next question is why. Why would England risk a world war in order to stop the development of Germany's industrial economy in 1914?

Eric J. Dahl offers his answer:[6]

> The ultimate reason England declared war in August, 1914 lay fundamentally in "the old tradition of British policy, through which England grew to great power status, and through which she sough to remain a great power," stated Deutsche Bank's Karl Helfferich. "England's policy was always constructed against the politically and economically strongest Continental power. Ever since Germany became the politically and economically strongest Continental power, England did feel threatened." Germany more than any other country challenged England's global economic position and its naval supremacy. Once that happened, English-German differences were unbridgeable, and susceptible to no agreement in any one single question. Helfferich sadly noted the accuracy of Bismarck's declaration of 1897, "The only condition which could lead to improvement of German-English relations would be if we bridled our economic development, and this is not possible."[7]

England chose deficit spending and curtailed domestic programs, reflecting her focus on all three pillars of our model. The result was that Britain continued as a world power for 50 years beyond its prime, riding largely on the strength of the decision to convert its fleet to oil.

In summary, then, the Coal-to-Oil Scenario was:

1. Inherently risk intensive, and very high stakes.
2. It appeared counter-intuitive; that is, going from a plentiful fuel to a rare one seems counter to reality from a resource

---

[6] Dahl, Eric J., "Naval innovation: from coal to oil," *Joint Force Quarterly*, Winter, 2000.
[7] Helfferich, Karl. *Der Weltkrieg: Vorgeschichte des Weltkrieges*, Ullstein & Co., Berlin, 1919, p. 165-6.

perspective, but it is justified from a performance perspective.
3. The switch cost jobs, which made it highly unpopular domestically.
4. Return-on-Investment (ROI) did not exist unless the strategic question was framed as the very existence of the British Empire and Great Britain itself.
5. The veracity of the strategy could not be proven without involvement in a conflagration the magnitude of the First World War (which it was).

The decision demonstrated the focus on strategic fit against tactical domestic politics.[8]

EXAMPLE 2: THE CONSTELLATION PROGRAM AND ITS CANCELLATION – TESTING THE TRIAD WHEN THERE IS NO BALANCE

A more recent example of the strategic calculus is the recent cancellation of NASA's Constellation Program. This program was considered to be the resumption of the 1960's initiative to explore the universe on behalf of all mankind, and the stated goals of the program were to gain significant experience in operating away from Earth's environment, develop technologies needed for opening the space frontier, and conduct fundamental science.[9]

What is conspicuously missing from this list is any connection or relevance to the American economic engine, which would ostensibly provide the funding for the Constellation Program. While this is the most obvious flaw in the program as constituted, it could be argued that there were also technical flaws.[10]

Our proposition is that the Constellation program was fatally flawed because it was "triad" deficient; it neither grew the economy, nor defended the nation, nor promoted sovereignty.

(1) It did not address access to space for other than scientific research, on the assumption that science for the sake of science

---

[8] http://en.wikipedia.org/wiki/Business_process
[9] http://en.wikipedia.org/wiki/Constellation_program
[10] http://chapters.nss.org/ny/nyc/Shuttle-Derived%20Vehicles%20Modified.pdf; Greg Zsidisin 25 October 2003, A presentation to the New York City chapter of the National Space Society.

was sufficient for an investment of that magnitude. The public did not agree.
(2) The economic benefits to the nation, to the world, and to the science community were not only unarticulated, but also undefined.
(3) It purposely excluded the involvement of commercial interests.
(4) It was an "Apollo on steroids" approach, which 35 years after Apollo, held little interest for any of the key stakeholders, including:
   a. Technology companies
   b. The investment sector that could benefit from derived technology
   c. The fickle American public, which expressed a "been there, done that" response.
   d. The major aerospace companies, for which the core technologies were Apollo or Shuttle derived, with little or no new intellectual property anticipated.
   e. And there was no international sense of excitement for collaboration or competition.
(5) It contradicted the long held premise of rocketry development heading toward "Celestial-Terrestrial" horizontal take-off and horizontal landing vehicles with orbital payloads, vehicles that would provide access for everyone to space and the stellar oceans. It instead proposed a continuation of vertical takeoff and landing, the most expensive and dangerous approach.

Consequently, there should not have been any surprise when the Constellation program was cancelled, because there wasn't a plan that assisted the developers of this program to undertand the necessity of widespread national, congressional and industrial support.

Nevertheless, the program did have its merits, aspects that did (although weakly) support the triad model:

(1) It transitioned the workforce from the shuttle program to the Constellation program to make maximum use of the existing trained and highly valuable expertise.
(2) It leveraged the existing shuttle propulsion contractors to build the Constellation rockets, as well as preserving the legacy and knowledge of ballistic capsule technology.
(3) It utilized many existing technologies, that while not future oriented, provided low-cost interim solutions until funding could

be identified for superceding technology.
(4) It matched the proposed NASA funding levels without a need for increases or movement of funding levels between years.
(5) It kept in place the existing support base in Congress by maintaining NASA much the same in terms of core competencies and staffing levels.

Support equals funding, but as Constellation had far too little support to weather the congressional environment, it proved (yet again) an axiom that has been true in space endeavours for more than 50 years:

"No bucks means no Buck Rogers"

It appears that NASA needs to propose ideas that will most surely not see fruition in order to promote the necessary introspection within policy circles to begin this invogoration process that could have saved Constellation. One such idea for example, would be to capitalize more than superficially on the merits as given above by proposing heavy shuttle *derived* legacy.

Summary: Integrating NASA into the Model
How NASA Can Succeed Long Term

How might NASA realign itself to accommodate the realities of this political and fiscal dilemma?

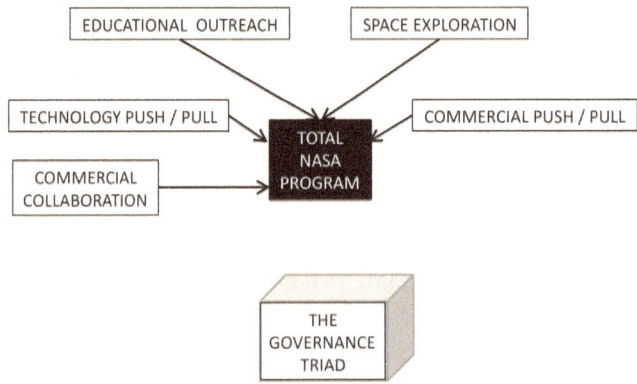

Figure 3: NASA: Currently Non-Aligned with Triad Governance

Could NASA make a compelling and politically supportable and sustainable argument for a policy that established a priority sufficient to compete against other needs and requirements? NASA was established as a civilian agency in 1958 by the National Aeronautics and Space Act of 1958 (as Amended), with the intent of maintaining civilian control of space access *for non-military uses.*

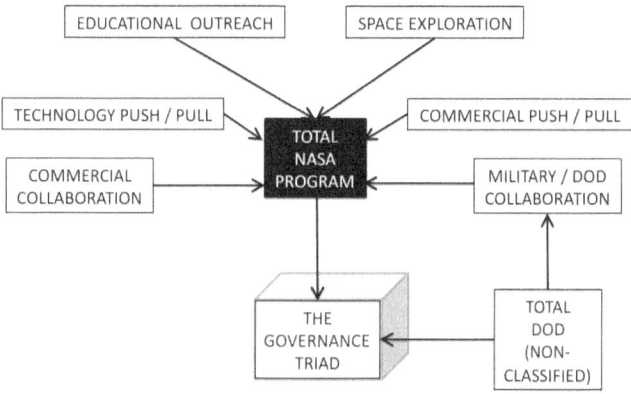

Figure 4: Proposed Alignment of NASA and Triad Goals

Copied directly from the leglislative language, the excerpt below clearly shows that *international collaboration is expected, that military collaboration is acceptable* given control lines are clear, and that environmental considerations are well within scope as well of any potential NASA program.

Sec. 102.

(a) The Congress hereby declares that it is the policy of the United States that *activities in space should be devoted to peaceful purposes for the benefit of all mankind.*

(b) The Congress declares that the *general welfare and security of the United States require that adequate provision be made for aeronautical and space activities.* The Congress further declares that such activities shall be the responsibility of, and shall be directed by, a civilian agency exercising control over aeronautical and space activities sponsored by the United States, except that activities peculiar to or primarily associated with the development of weapons systems, military operations, or the defense of the United States (including the research and development necessary to make effective provision for the defense of the United States) shall be the responsibility of, and shall be directed by, the Department of Defense; and that determination as to which such agency has responsibility for and direction of any such activity shall be made by the President in conformity with section 201(e).

(c) The Congress declares that the general welfare of the United States requires that the National Aeronautics and Space Administration (as established by title II of this Act) seek and *encourage to the maximum extent possible the fullest commercial use of space.*

(d) The aeronautical and space activities of the United States shall be conducted so as to contribute materially to one or more of the following objectives:

(1) The *expansion of human knowledge* of the Earth and of phenomena in the atmosphere and space;

(2) The improvement of the usefulness, performance, speed, safety, and efficiency of *aeronautical and space vehicles;*

(3) The *development and operation of vehicles* capable of carrying instruments, equipment, supplies, and living organisms through space;

(4) The *establishment of long-range studies* of the potential benefits to be gained from, the opportunities for, and the problems involved in the utilization of aeronautical and space activities for peaceful and scientific purposes;

(5) *The preservation of the role of the United States as a leader in aeronautical and space science and technology* and in the application thereof to the conduct of peaceful activities within and outside the atmosphere;

(6) *The making available to agencies directly concerned with national defense of discoveries that have military value or significance, and the furnishing by such agencies, to the civilian agency established to direct and control nonmilitary aeronautical and space activities, of information as to discoveries* which have value or significance to that agency;

(7) Cooperation by the United States with other nations and groups of nations in work done pursuant to this Act and in *the peaceful application of the results thereof;* and

(8) The most effective utilization of the scientific and engineering resources of the United States, *with close cooperation among all interested agencies of the United States in order to avoid unnecessary duplication of effort, facilities, and equipment.*

(e) The Congress declares that *the general welfare of the United States requires that the unique competence in scientific and engineering systems of the National Aeronautics and Space Administration also be directed toward ground propulsion systems research and development.* Such development shall be conducted so as to contribute to *the objectives of developing energy- and petroleum-conserving ground propulsion systems and of minimizing the environmental degradation caused by such systems.*

Certainly NASA already does, without controversy or issue, exercise its enabling legislation to collaborate with other nations and with private entities to enhance national sovereignty, national security and defense, and the nation's economy. But just as clearly, NASA does not seem to be choosing its major mission objectives in keeping with these three principles, nor is it effectively advocating for its role in this regard.

Both deficiencies must be overcome for NASA to regain the support of the American people such that it could aspire to missions of the magnitude of the cancelled Constellation program. What is required? In summary, the triad decision model should be applied, the plan must be constructed, and then it all must be eloquently but succinctly explained to the public and the legislature.

- A connection to the mission of NASA and the military in a way that the American people recognize the "dual use" technology that goes to the "black" (classified military) world and to the consumer / citizen world at *a cost lower than if not integrated.* In addition, if it were purely military in funding and in application, the benefit to the general populace would be zero - it would not be in the marketplace and thus not available to consumers. This is how national sovereignty aids and abets economic growth through research - and subsequent application.

- By spawning technologies that can be either or both in application, the national defense is achieved at a lower cost and with an increased pool of technical talent, much like the "X-Program" boom years of aerospace in the 1950s and 1960s.

NASA and the successful lunar missions would not have been possible without these one-off research programs that trained STEM (Science, Technology, Engineering and Mathematics) into seasoned designers.

- And perhaps most importantly, for federal dollars expended, product is being produced, and benefit derived. As has been discussed in the media, social programs have significant constituency. A similar constituency must be grown here for its technology benefits, because the free market works best when it is controlled by the consumer. Hence pumping federal dollars into social engineering programs, while popular, does not address true growth in the national economy via the passing of technology from the high risk takers - NASA and the military - to the high financial risk takers - commercial companies.

NASA should be leading a national discussion on the role of space as it relates to issues including the national deficit, economic stability and growth, and national defense. Neither NASA nor the nation can ill afford a "lead from behind" posture, suggesting that a leadership policy based on all three elements of the triad model is essential.

From such a national discussion a solid policy platform could be developed from which the nation could meaningfully answer the questions, "What of we do?" and "What if we don't?" It is then, and only then, that it becomes possible to engage in dialog and in cooperation with global partners while maintaining American preeminence in the pursuit of new goals and ambitions in space.

...

THOMAS E. DIEGELMAN

Tom Diegelman has been in the aerospace community for over 35 years, involved in the research, development and operation of training simulators, ground based flight control installations and facility operations. Tom started his career with Cornell Aeronautical Laboratory as a research engineer, working on early versions of shuttle handling quality study simulations and shuttle shock tunnel testing.

Tom moved to Houston in the late 70's to join Singer / Link Flight Simulation and worked in the Shuttle Mission Training Facility (SMTF) as a model developer, and later a manager of simulation projects. In 1988, Tom joined NASA to lead the $170M redesign of the SMTF. Assignments at NASA / JSC include projects in advanced mission control technology, technology development, and facility operations control. He served as Facility Manager for Mission Control for 3 years before accepting an account manager position in the Technology Transfer Office, developing partnerships and Space Act Agreements.

The design of the training facility for the Constellation Program culminated his nearly 30 years of experience at JSC in the Jake Garn astronaut training facility. Tom most recently became the ISS Vehicle Safety Engineer for Communications and Tracking Subsystem.

Tom was elected to Seabrook City Council in 2006, and served two terms, during which he worked closely with the Port of Houston Authority on the Seabrook / Bayport Terminal Facility issues. Tom is a member of Baytran, a non-profit organization promoting inter-modal transportation solutions in the Houston / Bay Area, and continues to be involved in local, state and federal government on behalf of the space technology community.

A dedicated writer, he is coauthor of two chapters in the previous volume in this series, *Space Commerce*, and also author of Chapter 6 in the present volume.

THOMAS C. DUNCAVAGE

Tom Duncavage holds a Bachelor's degree in Physics from Providence College. He is a graduate of the United States Navy Test Pilot's School, Patuxent River, Maryland and is a National Security Fellow of Syracuse University's Maxwell School. He has over 3000 tactical and flight test hours in over 20 US and foreign military and civil aircraft. Mr. Duncavage has served in various positions of technical, managerial and tactical leadership throughout his 31 year professional career both as a Marine Corps officer and as a federal civil servant. He is the founder of NASA's Concept Exploration Laboratory, a recipient of the NASA Exceptional Service Medal, an associate member of the William P. Hobby Center for Public Policy, and currently has served as the

NASA Visiting Executive to the Bay Area Houston Economic Partnership. Tom is currently manager for the Johnson Space Center Engineering Concept Exploration Lab (CEL).

CHAPTER 18

OVERVIEW OF SPACE PROGRAMS OF
BRAZIL, INDIA, AND ISRAEL

Although we were not able to obtain chapters written specifically by experts on the space programs of these nations, in the interest of providing a sense of their efforts, the following pages present concise overviews of current activities based on media sources.

BRAZIL

The Brazilian Space Agency, in Portuguese: *Agência Espacial Brasileira* (AEB), is the civilian authority in Brazil responsible for the country's growing space program. It operates a spaceport at Alcântara and a rocket launch site at Barreira do Inferno. The agency has given Brazil a leading role in space in Latin America and has made Brazil a partner for cooperation in the International Space Station.

Brazil's space program was controlled by the country's military until 1994 when it was transferred into civilian control.

It suffered a major setback in 2003, when a rocket explosion killed 21 technicians. Brazil successfully launched its first rocket, a VSB-30 on a sub-orbital mission on October 23, 2004 from the Alcântara Launch Center. Several other successful launches have followed.

On March 30, 2006, AEB astronaut Marcos Pontes became the first Brazilian and the first native Portuguese-speaking person to go into space, where he stayed on the International Space Station for a week.

The Brazilian Space Agency has pursued a policy of joint technological development with more advanced space programs. Initially it relied heavily on the United States, but after meeting difficulties with technology transfers, Brazil has begun working with other nations, including China, India, Russia and Ukraine.[1]

ALCÂNTARA LAUNCH CENTER

According to Douglas Messier, writing for *The Space Review*, the Brazilian spaceport Alcântara Launch Center (Centro de Lançamento de Alcântara, or CLA, in Portuguese) could become one of the busiest launch sites in the world, and one of the most lucrative. Alcântara is sequestered from the surrounding cities and only accessible by air or boat. Launching a rocket near the equator decreases the amount of fuel needed as the rotation of the Earth at the equator gives launch vehicles a boost relative to sites further away from the equator, resulting in as much as a 500 km/h increase in velocity. This increase in velocity decreases the fuel cost necessary to place an object in orbit, and as the Alcântara site lies a little less than 3 degrees away from the equator, it is the perfect place for launching geosynchronous satellites.

INTERNATIONAL PARTNERSHIPS

Brazil has launched sounding rockets, but its does not have rockets that can send a payload into orbit. It is now undertaking cooperative programs with Ukraine and Russia to build six rockets that could launch everything from small satellites into low Earth orbit (LEO) to heavy geosynchronous communications satellites.

Ukrainian Space Agency head Yuriy Alekseyev said recently that more funding is required to complete the project. "Today around $280 million has been spent and around $260 million more will be required. Unfortunately, Brazil has invested $50 million more in the project compared to Ukraine," he said. Ukrainian Prime Minister Mykola Azarov was recently quoted as saying that the partners are looking to complete the launch complex at Alcântara by 2014.[2]

---

[1] http://en.wikipedia.org/wiki/Brazilian_Space_Agency
[2] Messier, Douglas. "Will a new space power rise along the Atlantic?" *The Space Review*, August 15, 2011.

Brazil is also interested in partnering with its South American neighbors. Argentina recently proposed the creation of a South American Space Agency, an initiative which was supported by host Brazil during a Defense seminar in Sao Paulo.[3]

Israel is also in talks with Brazil, with a focus launching satellites from Alcântara, as Israel's geographic position makes domestic launches difficult. A 2000 agreement between Brazil and the United States also allows for private US corporations to launch from Alcântara if they desire.

In 2009, Brazil and Russia agreed to create and launch five rockets as part of the Southern Cross Project. The largest, Epsilon, will carry a payload weight of four tons. The Southern Cross project will create five additional launch pads and oversee up to a dozen launches a year. The project is on pace to make its first launch in 2022.

Creating the workforce

Brazil's Science without Borders programs aims to spend $2 billion to educate 75,000 students who would pursue advanced degrees in engineering and physical sciences at home and abroad, with hope that some of those students will return to Brazil and provide the necessary workforce.[4]

India

The Indian Space Research Organisation (ISRO) is the primary space agency of the Indian government, and is amongst the six largest government space agencies in the world, along with NASA, RKA, ESA, CNSA and JAXA.

Established in 1969 and headquartered in Bangalore, ISRO superseded the erstwhile Indian National Committee for Space Research (INCOSPAR). ISRO is under the administrative control of the Department of Space, Government of India.

India's first satellite, Aryabhata, was built by ISRO and launched by the Soviet Union in 1975. Rohini, the first satellite to be placed in orbit by an Indian-made launch vehicle, SLV-3, was launched in 1980. ISRO subsequently developed two other rockets, the Polar Satellite Launch

---

[3] "Argentina, with Brazilian support proposes a South American Space agency" *Mercopress*. September 1, 2011.

[4] Vernese, Keith. "Could Brazil be the next space superpower?" March 9, 2012. http://io9.com/5891721/could-brazil-be-the-next-space-superpower

Vehicle (PSLV) for putting satellites into polar orbits, and the Geosynchronous Satellite Launch Vehicle (GSLV) for placing satellites into geostationary orbits. As of early 2012, the Polar vehicle has a strong of 19 consecutive successful launches.[5] These rockets have launched numerous communications satellites, Earth observation satellites, and, in 2008, Chandrayaan-1, India's first mission to the Moon, and the first spacecraft to detect water there. ISRO's satellite launch capability is mostly provided by indigenous launch vehicles and launch sites.

Over the years, ISRO has conducted a variety of operations for both Indian and foreign clients. Future plans include indigenous development of GSLV, manned space missions, further lunar exploration, and interplanetary probes. ISRO has several field installations as assets, and cooperates with the international community as a part of several bilateral and multilateral agreements.[6]

New support for Indian science

A recent feature article in *Science* magazine described past and current developments in Indian science in detail.

> "After the nation's first atomic bomb test in 1974, the United States and other countries slapped sanctions on India that squeezed its supply of high-tech equipment and materials. Over the next 3 decades, India grew an indigenous civilian nuclear power industry and a space program on par with those of leading nations. In 2008, a landmark civilian nuclear pact between India and the United States beckoned Indian scientists in strategic sectors to come in from the cold; access to imported precision instruments is allowing India to make up ground in areas such as nanotechnology and supercomputing.
>
> "Now the government intends to lift all disciplines on a rising tide. At the Indian Science Congress in Bhubaneswar last month, Prime Minister Manmohan Singh pledged to hike R&D expenditures during the 5-year plan that begins this spring, from around $3 billion last year to $8 billion in 2017.
>
> "The windfall is meant to turbocharge initiatives under way to create elite research institutions, bring expatriate Indian scientists home, enrich science education, and equip smart new laboratories.

---

[5] Bagla, Pallava. "*Ad Astra*, With a Uniquely Indian Flavor." *Science* Magazine, Vol 335, February 24, 2012, p 906.
[6] http://en.wikipedia.org/wiki/Indian_Space_Research_Organisation

Over the next 5 years, an estimated $1.2 billion in public funds will be funneled to a new National Science and Engineering Research Board modeled after the US National Science Foundation.

"Researchers will have to clear some daunting hurdles, though. India's legendary bureaucracy can snarl grant proposals and expenditures in red tape for months. The anticipated R&D budget boost 'will be useless if structural reform is not undertaken,' warns vaccine specialist Maharaj Kishan Bhan, secretary of the Department of Biotechnology, the central government's main conduit for supporting applied biology in India. Another woe is that scores of universities are deteriorating or riddled with corruption. Says Raghavendra Gadagkar, a sociobiologist at IISc Bangalore. 'Our system creates followers, not leaders. That's our biggest problem.'"[7]

AN INTERVIEW WITH THE PRIME MINISTER

Science Magazine editor-in-chief Bruce Alberts, editor Richard Stone, and correspondent Palleva Bagla's interview with Indian Prime Minister Manmohan Singh was published in *Science* on February 24, 2012. These are some pertinent excerpts from their discussion.[8]

Q: You mentioned your feeling that China has overtaken India in science. Are you competing with China?

M.S.: *Well, we are competing, yes and no. India and China are engaged in a stage of development where we have both to compete and cooperate. We are the two largest developing countries and the two fastest growing countries. China is our great neighbor. Now, we've had in the past problems way back in the 1960s, but we are finding pathways to promote cooperation.*

Q: India has invested very large amounts of money in space.
M.S.: *And it has paid off.*

Q: The country wants to put astronauts in space. Indian astronauts from Indian soil using Indian rockets. Is that something you support?

---

[7] Stone, Richard. "India Rising." *Science* Magazine, Vol 335, February 24, 2012, p 904.
[8] Bagla, Pallava and Richard Stone. "India's Scholar-Prime Minister Aims for Inclusive Development." *Science* Magazine, Vol 335, February 24, 2012, p 907.

M.S.: *We supported the Chandrayaan lunar missions. And satellite technologies, rocket technologies—those are, I think, highly favorable outcomes of the Indian space program, and we need to do more.*

Q: But what about the astronaut program? The Indian Space Research Organisation is asking for $2.5 billion. You talk of inclusive growth. In that inclusive growth, how does human space flight fit in?

M.S.: *Ultimately science and technology must be viewed as an instrument of raising the standard of living of our people. Now, if information technology can be seen to promote the development of our country, particularly in the inclusive style of development, I think people will see space technology also as a new way of dealing with the ancient scourges of poverty, ignorance, and disease. Science and technology are the ultimate salvation for finding meaningful new pathways of developing our economy.*

ISRAEL

Israel's efforts in space became evident in 1988 with the launch of Ofeq 1 by the Shavit launcher, enabling the country to become the eighth country to launch a self developed satellite with its own rocket. Geographical constraints and safety considerations have led the Israeli space program to focus on very small satellites with payloads of high sophistication.

Currently Israel is developing its third generation of satellites. In addition to panchromatic satellites for Earth observation and picture downloading, a new radar satellite is currently being developed that will have a SAR payload capable of taking images at all weather conditions.

The vision of the Israeli Space Agency (ISA) stems from the understanding that Israel should promote innovative scientific projects based on international collaboration. Amongst the projects is the renewal of the TAUVEX (Tel Aviv University Ultra Violet Experiment), a UV telescope for astronomical observations which is intended to be hosted on the Indian Geo-Synchronous satellite G Sat-4, and would be jointly operated and utilized by Indian and Israeli scientists.[9]

---

[9] Adapted from
http://www.most.gov.il/English/Units/Israel+Space+Agency/About+ISA.htm

THE ISRAELI GOVERNMENT PERSPECTIVE

The following excerpt from a *Scientific American* blog authored by John Matson describes his conversation with Daniel Hershkowitz, Israel's Minister of Science and Technology.[10]

> "Hershkowitz notes that the Israeli space program is a minuscule operation compared to NASA or the European Space Agency—not surprising for a nation with about the land area and population of New Jersey. 'The Israeli Space Agency does not have its own industries,' Hershkowitz says. 'It's just a very small body that coordinates in the activities of the other industries, and also coordinates between the civilian and the military applications.' To do that, he says, the agency has an annual budget of about $50 million.
>
> "Israel's presence in space is defined primarily by a network of Earth observation, communication and reconnaissance satellites. But Hershkowitz notes that his nation takes the overall enterprise of scientific research quite seriously. Israel leads the world in terms of percentage of GDP spent on research and development, and he notes that by some criteria its space program is fairly advanced.
>
> "As in most countries of the world that have space programs, things started from the military, mainly observation satellites. The focus on surveillance continues today. Israel is certainly not alone in accessing the ultimate high ground for observation purposes, nor in entangling its defense needs with its peaceful aims. NASA's Space Shuttles ferried 10 or so secret Department of Defenses payloads to orbit during the 1980s and 1990s, even as the shuttles carried out other unclassified missions for scientific aims. Hershkowitz estimates that 'close to half of what we invest nowadays in space has to do with scientific applications and civilian applications,' such as monitoring water pollution and soil conditions for agriculture. But he acknowledges that Middle East turmoil will ensure that reconnaissance remains a top priority.
>
> "Israel fielded its first astronaut, Israeli Air Force colonel Ilan Ramon, in 1997, but the mission ended in disaster when space shuttle Columbia broke up over Texas in 2003, killing all seven crewmembers. According to Hershkowitz, Israel has no immediate plans to recruit a second astronaut.

---

[10] Matson, John. "Israel's Science Minister on Space Technology–for Peaceful and Militaristic Aims." *Scientific American*, May 10, 2012.
http://blogs.scientificamerican.com/observations/2012/05/10/israels-science-minister-on-space-exploration-for-peaceful-and-militaristic-aims/

"Usually the public is very fascinated by human missions, and by astronauts. But you know, one of the reasons that the United States has decided to abandon its human programs and does not use the Shuttles anymore, and as I said, when they have to send astronauts to the International Space Station they use Russian shuttles, the reason is that I would say scientifically and even technologically, manned missions have ceased to be interesting. Of course, you know, it's nice to have an astronaut, but it really doesn't help the strategic needs of the state of Israel. It is possible that in a certain stage we will have another Israeli astronaut ... but we don't have right now much interest in that. We have other priorities."

A CIVILIAN SPACE PROGRAM

Nevertheless, in 2010 Israel announced plans to invest $77.5 million over five years to accelerate its civilian space program. Israeli President Shimon Peres, a strong supporter of Israeli aerospace initiatives for years, said he expects the industry to develop into a major source of business, Haaretz reported Friday.[11]

Peres and Prime Minister Binyamin Netanyahu tasked government officials to develop a national space program to help the 25 Israeli firms in the civilian space sector expand their market. The plan was drafted by a team of scientists and economists that includes Prof. Haim Eshed, head of the Defense Ministry's Space Division, and the director-general of the Science and Technology Ministry, Menahem Greenblum.

The international space industry is undergoing major changes, including privatization, Israeli officials noted, adding the civilian space market is worth an estimated $250 billion a year. Sources within the Defense Ministry told Haaretz Israel could capture up to 5 percent of the market. Despite Israel's advanced technology, sales of its space platforms over the last 20 years have totaled less than $2.5 billion.

An article in the Jerusalem Post notes that the multi-year plan calls for the government to annually increase support for space research and development. This investment would focus on new platforms – primarily Israel's niche market in "mini satellites" – intended to yield billions in sales.[12]

"We have the assets, but we are not marketing them," said Eshed.

---

[11] http://www.space-travel.com/reports/Israel_to_launch_civlian_space_program_999.html

[12] Katzlast, Yaakov. "PM set to okay space R&D program." *The Jerusalem Post*, August 13, 2010.

The plan envisions stepped-up sales beginning by 2015. However, Israel will not sell its top-of-the-line payloads and platforms; these it will retain for the Israel Defense Forces.

MINIATURIZATION OF SATELLITES

Much of the investment will focus on the miniaturization of satellites and their payloads. Israel is also developing nano satellites, and plans to launch the "Incline," its first nanosatellite, which will weigh a mere 12 kg. This prototype will serve as a relay for data transfers, but could also carry miniature cameras in the future.

Recently ISA and CNES of France reached an agreement to jointly work on a new project to design and construct an innovative micro-satellite and a ground station for scientific purposes. The micro-satellite would incorporate a multi-spectral camera for Earth observations and a ground station for imaging processing, and demonstration of an electrical propulsion thrust system examining its maneuverability potential.[13]

•••

---

[13] Adapted from
http://www.most.gov.il/English/Units/Israel+Space+Agency/About+ISA.htm

Part III

... and Beyond

CHAPTER 19

THE 100 YEAR STARSHIP ENDEAVOR

JEFFREY NOSANOV
NASA, JET PROPULSION LABORATORY

AND

MICHAEL POTTER
DIRECTOR, PARADIGM VENTURES

© Jeffrey Nosanov & Michael Potter, 2012 All Rights Reserved

1. INTRODUCTION

The 100 Year Starship Study (100YSS) is an unprecedented, multi-disciplinary initiative that challenges humanity to identify commercially sustainable ways to design, develop, and build a manned vehicle for interstellar space travel within 100 years, by the year 2111.

Aside from the obvious technical challenge, there are numerous organizational challenges to mounting a multi-generational endeavor to launch a human Starship mission. First and foremost is the need to develop a clear vision and mission statement for such an ambitious undertaking; the scope of such a mission is truly a global undertaking.

Nevertheless, current and past multinational space projects demonstrate that there are additional obstacles inherent in such large scale international projects. Particularly challenging is the difficulty of building international consensus on the very need to embark upon a multi-generational mission, and one whose core technology does not yet exist. In the current globally hard-pressed financial environment, it will be difficult to obtain political support for a single massive project towards this goal.

This suggests that it will be critical to move forward step by step, perhaps with shorter-term robotic analog Starship missions and commercial demonstrations. Spreading knowledge of the threat of near-Earth objects, as exemplified in December 2011 with the near-collision with Asteroid 2005 YU55, shows that planetary defense should be an increasingly important driver of space investment, technology development, and exploration over the coming century.

In their recent book *The New Universe and the Human Future: How a Shared Cosmology Could Transform the World*, authors Nancy Ellen Abrams and Joel R. Primack argue that, "...we are living in a cosmically pivotal moment today, and we have a higher level of responsibility than any generation that came before us... We and our children may be the most significant generations of humans that have yet lived."

If it happens, such a venture into interstellar space may well prove to be an event of exactly such significance.

2. VISION

Some Native Americans believe that in making important decisions, one has the responsibility to think seven generations ahead. While this is an admirable and visionary core principle, it may not be sufficient to motivate the interstellar stage of humanity's movement into space.

Yet just such a multi-generational challenge was issued in 2011 by NASA and DARPA, the United States Department of Defense Advanced Projects Research Agency, in the form of the "100 Year Starship Study" (100YSS). DARPA is the source of funds and NASA will administer the award. This request for proposal was organized as a business plan competition for an organization that can, without government funding, lead a technology research program for a century towards the goal of launching of a manned Starship by the year 2111.

The solicitation introduces the challenge as follows:

"The 100 Year Starship™ (100YSS™) is a project seeded by the Defense Advanced Research Projects Agency (DARPA), with NASA Ames Research Center as executing agent, to develop a viable and sustainable non-governmental organization for persistent, long-term, private-sector investment into the myriad of disciplines needed to make long-distance space travel viable. The goal is to develop an investment vehicle—with the patronage and guidance of entrepreneurs, business leaders, and technology visionaries—which provides the stability for sustained investment over a century-long time horizon, concomitant with the agility to respond to the accelerating pace of technological, social, and other change."

The unique nature of the 100YSS Study has led to speculation regarding DARPA's actual intentions, and a summary of DARPA's history may offer some visibility into this unusual context.

DARPA's mission is to develop technology for the military, but this definition omits important historical context. The real key to developing an understanding of DARPA's activities is to consider the concept of "Technological Surprise," which refers to activities of foreign nations that demonstrate a surprising technological capability, and one that domestic scholars or authorities could not have predicted.

DARPA's mission is described on its website as, "To maintain the technological superiority of the US military and prevent technological surprise from harming our national security by sponsoring revolutionary, high-payoff research bridging the gap between fundamental discoveries and their military use."

In fact, the agency that became the modern DARPA was created in response to the technological surprise of the launch of Sputnik in 1958. The initial stated goal of DARPA at the time was to maintain technological superiority for the United States, and to avoid any such technological surprises in the future. This function evolved over time to include creating such technological surprises for the United States.

To fulfill its goals, DARPA is charged to look beyond contemporary needs and capabilities as defined by the military itself. In this regard, military historian John Chambers has noted that, "none of the most important weapons transforming warfare in the 20th century – the airplane, tank, radar, jet engine, helicopter, electronic computer, not even the atomic bomb – owed its initial development to a doctrinal requirement or request of the military." And to this list, DARPA would add unmanned systems,

Global Positioning Systems (GPS), and Internet technologies.[1]

DARPA is perhaps most famous for its support of the early Internet. Indeed, DARPA (then called ARPA) was instrumental in supporting the early computer networking research of the 1960s, which initially networked together two computers at two California universities. Other projects have ranged from autonomous cars with no human driver (DARPA Grand Challenge) to the 15,000 mph unmanned bomber aircraft (Hypersonic Research Program) with additional projects including natural language interpretation algorithms (eventually made available to the public in Apple's Siri virtual personal assistant software), and robots that refuel themselves by foraging natural resources (EATR).

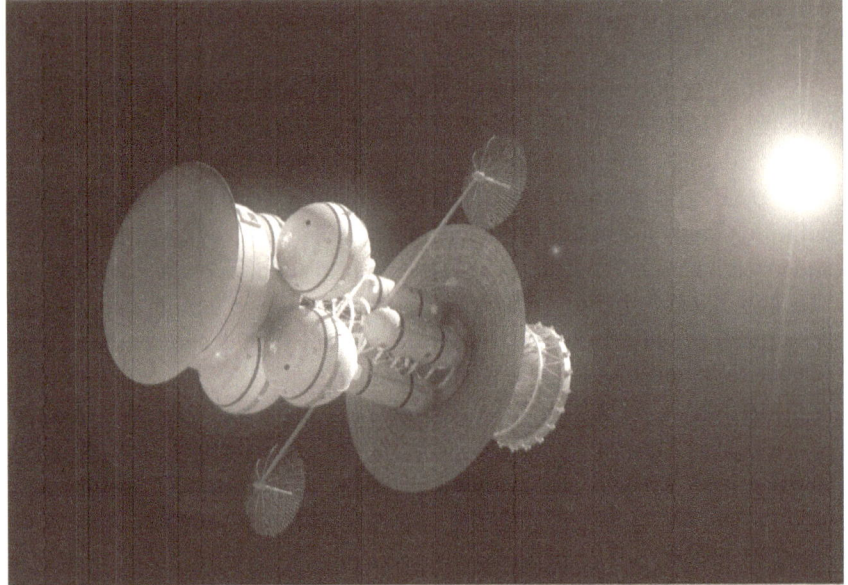

Figure 1: Icarus Pathfinder
This illustration by Adrian Mann shows a concept for an interstellar spacecraft. Reprinted by permission.

In keeping with its history, the purpose of the DARPA 100 Year Starship Study is to challenge academia, industry, and the public to identify methods to improve humanity's technological, economic, and sociological capability to a level such that we are capable of building and launching a manned, interstellar Starship in a commercially sustainable way.

---

[1] Chambers, John, ed., *The Oxford Companion to American Military History*. Oxford University Press, 1999. p 791.

After the 100 Year Starship proposal due date, DARPA subsequently announced two additional challenges: software to enable reassembly of documents that had been shredded to be put back into legible form, and methods to reuse otherwise defunct satellites or their components in orbit.

What is unique about the 100YSS is that DARPA primarily undertakes technical challenges. If the challenge itself does not identify technical goals, then such goals are generally expected to be included as part of the response. Further, DARPA typically focuses on relatively short term technology development efforts, and its flat organizational structure is intended to enable the rapid deployment of technical approaches to problems. DARPA embraces high-risk, high-reward positions.

The 100 Year Starship stands out in that it deviates from the normal DARPA model by asking for an organizational plan instead of a technical plan. Instead of the "standard" DARPA approach of a technology development challenge with a 2 to 4 year goal, this is a business plan competition for an organization that will endure for a minimum one-hundred-year lifetime, an organization that is intended to enable an unprecedented technical achievement on a colossal scale. It further aims to do so with a relatively small award that was eclipsed by the cost of the study itself. Hence, DARPA spent more than the actual award amount on running the program that would identify the recipient. For example, DARPA had a 3 day conference with over 1000 attendees at a major hotel, and also had to run their program internally for several years. This undoubtedly cost more than the US$ 500k they eventually plan to award.

3. Challenges

In a recent article about the 100YSS project, Lou Friedman, former Executive Director of the Planetary Society wrote, "Interstellar flight is a vision for some and science fiction for others. I recall a discussion I had with Freeman Dyson a few years ago about whether we were further from interstellar flight than was Leonardo Da Vinci from the airplane."[2]

This illustrates the colossal scale of the technical challenges involved in interstellar flight. Da Vinci's time was three centuries prior to the initial attempts at powered flight: Are we that far from mastering nature to the point at which she will permit interstellar flight?

Among technical experts as well as well-informed generalists there is

---

[2] Friedman, Lou. "Mind expansion," *The Space Review*, Monday, November 21, 2011.

a tremendous uncertainty as to the possibility, or even the practicability of placing almost any time scale on such a massive speculative event. In a world where technology is changing exponentially, the better argument might be about when the technology will be available, not so much about if the technology will ever exist.

Recent discoveries of Earth-like planets orbiting other stars may provide a powerful near-term motivation to begin the technical work necessary for interstellar flight. The discovery of "other Earths" with even the possibility of human-compatible atmospheres suggests that such a planetary diaspora may become a realistic option at some point in the not-so-distant future.

Scientific motives or the need for self-preservation may also contribute to public and commercial support for this goal. Indeed, many scientists, notably Stephen Hawking, have observed the need for humanity to eventually become a multi-planet species in order to avoid extinction.

4. Structures & Longevity

The authors of this chapter are part of the "Global Starship Alliance," one of the teams that submitted a proposal to DARPA for the 100YSS project.

[Editor's note: A definitive announcement has not been made as of the publication date of this volume, but the authors' team appears to be the competition runner-up.]

In preparing our ideas we examined some very long-lived organizations, seeking to identify factors that may contribute to establishing a long-term visionary organization that can endure a 100 years plus or preparation for a mission that may then last hundreds more.[3]

Hudson's Bay Company was incorporated in 1670 by charter, and remains the oldest commercial corporation in North America, now at 342 years of operations. A fur trading business for much of its existence, today it owns and operates retail stores throughout Canada.

The Dutch East India Company was chartered in 1602 when the States-General of the Netherlands granted it a 21-year monopoly to carry out colonial activities in Asia. It was also the second multinational corporation established in the world, and the first company to issue shares

---

[3] Most of the facts related to Hudson Bay Company, the Dutch East Indies Company, and the National Geographic society were established via Wikipedia search.

of stock. It remained an important trading concern and paid an 18% annual dividend for almost 200 years. Weighed down by corruption in the late 18th century, the Company went bankrupt and was formally dissolved at age 198 in 1800.

Founded in 1888 (and now 123 years old) the National Geographic Society, headquartered in Washington, D.C., is one of the largest non-profit scientific and educational institutions in the world. In 2010, the Foundation's endowments totaled more than $130 million, and the income from these endowments is used to support the creation and dissemination of educational resources, professional development for teachers, public awareness, policy reform and other programs that enhance geographic education.[4]

These organizations share the fact that they are concerned with the exploration and development of new frontiers, which by definition also necessitates a risk-tolerant posture, and they also share the fact that they demonstrate a close exploration and development relationship between private entities and governments. These factors will also be characteristics of the 100YSS.

And so will international cooperation. During most of the short 50-year history of human space flight, international space cooperation was primarily dictated by Cold War and US-Soviet superpower politics. But as the Cold War ended, the International Space Station became the largest international science and technology project ever undertaken, a global effort involving many nations and decades of effort and investment.

It is clear that that similar motivations will be central if and when an interstellar mission is undertaken. However, 100YSS will also have to be "sold" to the public based on shared global imperatives and justifications, such as establishing the enabling technologies for planetary defense, or as the central vision and imperative for humanity becoming a multi-planetary species.

The actual technical implementation of a 100YSS is undoubtedly an international undertaking. Indeed, most "mere" space science missions in our time have international participation. An interesting question raised by the 100YSS process is the extent to which government structures and limitations can hinder attempts at truly humanity-wide projects. For

---

[4] Interesting to note that science fiction writers Jerry Pournelle, Larry Niven and John Barnes also used the National Geographic Society as the legal and financial vehicle for their fictional Herot starship mission, writing in great detail about how to handle and pay off the finances of such a mission while in flight and also to measure the benefits of such a ventures.

example the United States has a body of law called the International Traffic in Arms Regulations that governs the sharing of technical information with foreign nationals. The United States State Department maintains a list of technologies that fall under this protection. This includes most spacecraft related technologies regardless of their intended application. This means that it is very difficult for the international technical community to even talk about the technical issues with American experts. There are several current proposals working their way through the American political system to ease these restrictions on benign technology items. However it remains to be seen whether the political will can be wielded to make such a change. Any international project along these lines will need to navigate these waters.

The international cooperation paradox suggests that there is a logical construct for all international parties to work together to achieve a resource and talent intensive endeavor that is for the greater good of all of humanity. However, this seems to often be at odds with the logical construct, that seems, inherent in the human condition. And that is a condition where politics, bureaucracy and petty self-interest sometimes prevails over the great good of the community and humanity.

The April 24th, 2011 audacious announcement of a new company, called "Planetary Resources," focused on the mining of asteroids, by a consortium of international billionaires, may symbolically have defined a new era, where international commercial space, may one day rival the past cooperation of governments and nation states.

5. Conclusion

There are a number of critical and fundamental challenges to mounting a multi-generational endeavor to launch a human Starship mission. Foremost is the need to develop a clear vision and mission. There are both national and international challenges related to developing a consensus about the need, utility, and the desirability of embarking on such a starship mission.

It may be particularly difficult to build an international consensus on the need to embark upon a multi-generational mission when its core technologies do not yet exist. In the current financial environment, allocation of significant resources for a huge commercial, technological, and scientific project is not likely to resonate nor find sustainable support without also providing significant and on-going near-term benefit. Consequently, when DARPA and NASA announced the 100YSS initiative, it was expected by many in the space community that there would be only a modest response.

In reality, however, this bold announcement inspired hundreds of visionaries in the space community, and they formed dozens of different teams that are responding to the DARPA request for proposal. As noted, the authors of this chapter are part of one of those teams.

After long consideration, we proposed and continue to advocate the following recommendations:

The creation of an organization that we provisionally refer to as the "Open Source Starship Alliance," which has brought together all of those scientists, technologists, and visionaries from the many teams that bid on the DARPA project to participate on an ongoing basis, in a community dedicated to the efforts and the vision of 100 Year Star Ship. Essentially, the DARPA 100YSS process has created a focal point and a catalyst for bringing many like-minded people together who share an interest in this ambitious project. It may turn out that this emerging community may be one of the greatest value drivers of the 100YSS initiative.

A robotic analog 100YSS mission to serve as a "precursor mission" for the ultimate goal of a human starship mission, a robotic craft on an interstellar journey as a first test of technologies and to identify unknown challenges.

The creation of numerous, smaller, commercial-focused missions that would create the seeds of a long-term commercially sustainable space ecosystem that will eventually be necessary for sustainable robotic and human starship missions.

Critical imperatives driving the 100YSS include the need for planetary defense, and the imperative for humanity to become a multi-planetary species.

In the recent New York Times bestselling book *Abundance*, world renowned aerospace engineer Burt Rutan is quoted as saying, "Revolutionary ideas come from nonsense. If any idea is truly a breakthrough, then the day before it was discovered, it must have been considered crazy or nonsense or both – otherwise it wouldn't be a breakthrough."[5] This paradox of innovation and perception of nonsense before breakthrough has, throughout history, bedeviled policy makers and legislators who are in the business of allocating public resources for science and technology projects.

We have discussed the challenges related to the longevity of

---

[5] Diamandis, Peter and Steven Kotler. *Abundance: Why the Future will be Much Better Than You Think*, Free Press, 2012, p. 229.

commercial organizations. But do civilizations themselves have the endurance that would be required to send a starship into the heavens? Civilizations themselves cannot become great without great ideas, and certainly starship technologies, among the only ideas that can save our species from eventual extinction, are some of the greatest ideas of all human history.

A presenter at the 100YSS Symposium held in Orlando, Florida, September 29 – October 2, 2011, described this underlying paradox in economic terms:

> *"The future value in starship technology is nearly infinite due to its capacity to save the human species from extinction and perpetuate the species beyond the limited capacity of Planet Earth. However, the present value of such technology is very low due to its extreme cost and limited near term application."*

At a recent gathering of space visionaries, glasses were raised and a toast was proposed to the 100YSS initiative: "To the stars, or under the table!" an expression of ultimate commitment to this goal.

As of publication date of this book, the current status is that the Jemison Foundation seems to have won, as news was leaked on Dec. 31, 2011 that a decision had been made, and leaks over the next week confirmed the Jemison Foundation had been selected. However, as of April 19, 2012, there has been no official confirmation from DARPA, nor any transfer of funds.

The 100YSS challenge offers humanity the opportunity to choose a clear and compelling vision and a path to the stars, and just as importantly, to leave behind the current political muddling and chaos that holds back our planet's shared potential.

•••

Jeffrey Nosanov

Jeffrey Nosanov currently works at NASA's Jet Propulsion Laboratory (JPL) for the Radioisotope Power Systems (RPS) program. In 2011 he represented the RPS program at the United Nations Committee on Peaceful Uses of Outer Space meeting in Vienna, Austria. He works on proposals and business strategy for JPL. He was the first person in the United States to earn an LL.M. degree in Space and Telecommunications law, following his J.D. at New York Law School at which he was a public interest fellow. For the LL.M., he wrote a thesis on the International Traffic in Arms Regulations (ITAR) and their impact on the competitiveness of the United States Space Industry, which was published in the journal *Astropolitics*.

Michael Potter

Michael Potter serves as Director of Paradigm Ventures a family investment firm focused on high technology ventures. Previously Potter was Vice Chairman, founder and President of Esprit Telecom plc., a pan-European competitive telecommunications services provider. During his eight years at Esprit, the team grew the company to 1,000 employees in more than nine European countries and a market capitalization of a billion US dollars. He was formerly an international telecommunications analyst at the Center for Strategic & International Studies (CSIS) in Washington, D.C. Potter was also Vice Chairman of the founding Board of the European Competitive Telecommunications Association (ECTA).

As a first-time documentary filmmaker he created the award winning film, *Orphans of Apollo*. Potter previously worked on the 13-part WGBH Series, *War & Peace in the Nuclear Age*. He is a member of the Board of the Trustees of ISU, and was one of the founders of the ISU Scholarship fund which has raised scholarships fund for more than 15 scholars. He is on the Board of Directors of the Manna Energy Foundation, a non-profit foundation that is installing clean water solutions in high schools in Rwanda, on the board of advisors of Odyssey Moon, the first entrant into the Google Lunar Xprize competition.

Michael is a Senior Fellow at the International Institute of Space Commerce and a director Global Connect. His articles on high technology business and policy have been widely published.

He received his MS degree from the London School of Economics, his BA from California State University at Sacramento, and a certificate in Space Studies from the International Space University.

CHAPTER 20

AN INCOMPLETE SPECIES:
UNFOLDING OF SPACE LAW TO SUPPORT THE SURVIVAL OF HUMANKIND AND ITS UNIQUE ENVOYS MIGRATING OFF-EARTH; THE CONTINUING EVOLUTION OF HUMANITY, SOCIETY, TECHNOLOGY, AND LAW

GEORGE S. ROBINSON

© George S. Robinson, 2012. All Rights Reserved.

INTRODUCTION

All living species are transitory. Depending upon the definition of "technology" in specific contexts, they either evolve biologically or biotechnologically to meet ever-changing external and internal conditions, or they ultimately become extinct. Even the humankind sentient "essence," i.e., the whole seemingly being greater than the sum of its parts, is constantly in evolutionary transition. If not, it too becomes extinct, or perhaps may be transferred and incorporated in a similar fashion by a subspecies or totally different species; perhaps even by a biotechnological or totally technological post human ... beyond cyberpersona.[1] No species has "lasted forever"... yet!

Empirically-based human laws that have been formulated to enhance transitory and specific interests in space exploration, migration, and settlement, lack the essential and underlying philosophic construct, i.e., survival of the species, its descendants, and the evolutionary odyssey of their "essence."[2] This survival odyssey is the real empirical foundation of space law, specifically the survival of humankind and of its biotechnological descendants.

Moore's Law and the Singularity Principle[3] remind us that time and

---

[1] "Cyberpersona" is a component of "cyberspace" and includes a person's identification when on the network encompassed by e-mail address, computer IP address, cell phone number, and the like. Cyberpersona exist in cyberspace, which is a "global domain within the information environment consisting of the interdependent network of information technology infrastructures, including the Internet, telecommunications networks, computer systems, and embedded processors and controllers." Cyberspace has been defined in many different ways by both the private and public sectors, but on May 12, 2008, the U.S. Undersecretary of Defense officially signed a document defining cyberspace as in the preceding quote, and it is considered the same as land, sea, air, and space in the context of national defense strategy interests. In 2009, the Chairman of the U.S. Joint Chiefs of Staff "approved the definition of Cyberspace Operations as including computer network operations and activities to operate and defend the Global Information Grid."
[http://www.signal.army.mil/ArmyCommunicator/2011/Vol36/No2/2011Vol36No2Sub06.pdf].
The terms *cyberspace, cyberpersona, and cyberlaw* are discussed in greater detail in the ensuing sections of this article.

[2] For a working legal definition of "essence" in a jurisprudential context, *see* H.C. Black, *Black's Law Dictionary*, 7th ed. (St. Paul, Minn: West Group Publishers, 1999), s.v., "essence."

[3] For a working definition of Moore's Law, *see* fnt. 22, *infra*. The Singularity Principle is subject to multiple definitions and interpretations. In many ways, it represents the classic scientifically theoretical "work in progress." Nevertheless, despite being defined in many different ways for many different purposes, "Singularity" for the present discussion is described as having been invented or conceived initially by Vernor Vinge as "the fundamental discontinuity in history created by the technological invention of smarter-than-human intelligence. Other definitions have included a time of exponentially faster technological progress (even faster than now, that is), or the positive-feedback effect created by enhanced intelligences working out improved methods of intelligence enhancement. The core idea remains the same: There is a massive discontinuity approaching, a Singularity, within human history. This has to do with the rise of smarter-than-human intelligence, the ability of technology to alter human nature, the final conquest of material reality through nanotech, or some other fundamental change in the rules." More recent definitions may render that of Verner Vinge somewhat

technological evolution are compressing the time available for humankind's intentional evolution and space migration for purposes of promoting the survival of the human essence. The question posited is, "What is the relationship between individuals, society, civilization, a species, the solar system, the galaxy, and the constantly evolving empirical nature of the 'known' universe and potential parallel universes?" This is the question upon which the humankind evolutionary odyssey pivots, and the principal objective of this chapter is to examine jurisprudential issues that are applicable to a technologically altered species that is living and functioning off-Earth.

The humankind species is now shaping its ongoing survival by changing the life-support capabilities of Earth to the point where off-Earth migration may not be optional, but rather necessary for the continued evolution of the species, and the continuation of its odyssey to survive and inch toward understanding its own "essence."

I. Natural Law Theory, Jurisprudence, and some Fundamental Questions

The driving factors that underlie the evolution of all living systems also underlie what traditionally has been referred to as Natural Law, or Natural Law Theory. From Natural Law humans have derived jurisprudence, concepts aimed at individual and collective survival. Legal philosophies and the concepts of jurisprudence are not the products of a spiritually amorphous understanding of acceptable individual and collective behavior patterns formulated out of some ethereal or self-serving sense of "morality" or "ethic."

To the extent that the relevant empirical underpinnings are not yet known or quantifiable, laws remain "best guess" theories upon which individual and societal behavior is directed; it is a transitional framework.

The formulation and application of space law principles, domestic, international, and trans-global, are no exception. These laws should be, indeed, must be designed to enhance the survival of humankind and its descendants off-Earth; in space. They must reflect the underlying philosophic construct of survival of the humankind species and its evolving essence, however that latter term is defined and applied at any given time

---

obsolete, *See*, therefore, E.S. Yudkowsky, "The Singulartarian Principles" (2001, rev. 5/14/ 2001) at [http://yudkowsky.net/obsolete/principles.html].

and in any given context. But do current and presently anticipated space laws really reflect this imperative?

The path from the simplest form of carbon-based life on planet Earth to the extraordinarily complex, spacefaring, modern human, Homo sapiens sapiens, has been indirect and seemingly meandering.[4] The phenomenon appears astonishing and bewildering, and was explained only by the development of technology and scientific methodology sufficient to describe the continuously unfolding complexity of the human organism.

To envisage and design a reasonable future for the human species, both Jurisprudence and Natural Law must be aligned with the currently prevailing direction of evolution. We must not lose sight of a basic dictate, the absolute requirement to adjust and adapt to change, and thus to survive ... or become extinct. Our current dilemma is that human technology, as both the problem and the answer, is compressing the time available to the human species to adjust its survival strategies.[5]

Some of the questions that have to be addressed to design properly the enduring survival-strategy requirements or jurisprudence include:

---

[4] For an excellent discussion of the various factors involved in human evolution and what yet remains to be deciphered, *see*, generally, by R. Potts and C. Sloan, *What Does it Mean to be Human*, (National Geographic publication as the official companion book to the Smithsonian Institution, National Museum of Natural History's David H. Koch Hall of Human Origins, Washington, D.C. (2010).

[5] In this general context, Dr. Lyn Margulis of the University of Massachusetts, Amherst, may be considered by some traditional biological evolutionists as somewhat limited in her views, perhaps even eccentric. Nevertheless, her views regarding the origins of all life forms, including *Homo sapiens*, as deriving from a single bacterium and its subsequent colonizing proclivity, are based upon what she terms an evolutionary symbiotic partnership. Every life form, including modern humans, is considered by Dr. Margulis literally to be a "community of symbiotic bacteria." She believes a great deal of empirical evidence exists that modern humans are "mammalian weeds" and "[l]ike many mammals, we overgrow our habitats and that leads to poverty, misery, and wars." *See*, therefore, "The Discover Interview: Lynn Margulis – Q+A," by Dick Teresi in *Discover* magazine (April 2011, pp.66-71). The importance of this perspective relates to human population dynamics run amok, as reflected in the exponential human population growth shortly reaching seven billion individuals in a relatively abbreviated time. Humankind carrying capacity of Earth, all involved factors considered, makes humankind off-Earth migration critical to survival of *Homo sapiens sapiens*, its biotechnological descendants, and any hope for the continuing evolution and odyssey of the species' "essence." *See*, generally, by L. Margulis, *Symbiotic Planet: A New look at Evolution* (Barnes & Noble, 2003). For an interesting perspective on symbiotioc evolution, *see* by R. Morrison, *The Spirit in the Gene: humanity's proud illusion and the laws of nature* (Comstock Book – 1999).

1. What is the evolutionary direction of humankind?[6]
2. Is it really understood, and does the current body of space jurisprudence embrace, catalyze, promote, and enhance it, or does it compromise the evolving pursuit of that evolutionary direction?
3. If technology is an integral biological component of humankind evolution (as it clearly was of its protohominid ancestors), is a next step in human evolution and survival the biotechnological adjustment necessary for space migration and permanent off-Earth habitation?
4. If and when that is achieved, will humankind habitation on Earth cease?
5. And if so, what time frame would be realistic for this?
6. What is the current scope and projected direction of Natural Law Theory[7] leading to a suitable jurisprudence, all in turn yielding to international, global, and transglobal law?
7. What, ultimately, may the evolving global and transglobal jurisprudence look like as it relates to space exploration, migration, and settlement?
8. Do, or will, these questions and issues lead global or transglobal jurisprudence support or secure survival of the humankind, transhuman, and post human species?[8]

---

[6] Collectively, *Homo sapiens* is referred to generally as humankind. However, as occasionally used herein, the term is considered a characterization of the biotechnologically integrated transitioning of the species into what is referred to as "transhumans," hence the emphasis of *kind* in human*kind* in certain specific contexts. *See, therefore,* by G. Robinson and R. Lauria, "Legal Rights and Accountability of Cyberpresence: A Void in Space Law/Astrolaw Jurisprudence," in Annals of Air and Space Law Vol. XXVIII (2003) at pp. 311-326.

[7] In general terms, the theory of "natural law," or *jus naturale*, derived from philosophical musings in the Antonine Age, when Roman jurists attempted to formulate a system of core principles underlying rules to guide human behavior in a societal setting. These core principles were considered to be integral components of human *nature*, or human *biology*, shared by all humans. *See* Black's Law Dictionary, Fourth Edition, p. 1177 (1957). Interestingly, there is/will be a distinct point of departure between humans, transhumans, and post humans when the "nature," both biological and/or ephemeral and those that are generally thought to be non-empirically based at this point in time (such as morality, ethic, and the like), of *Homo sapiens sapiens* transitions into a totally separate and distinct set of core principles directing the survival of a biotechnologically integrated or totally separate and distinct entity with "judgmental capabilities" and non-biological metabolic and self-replicating characteristics, i.e., post humans.

[8] *Op cit., supra,* Note 6, Robinson and Lauria, for a general discussion of the foundations of current human evolution leading to biotechnological descendants of *Homo sapiens sapiens*.

9. What is the current status of "space law" as it pertains to the biotechnological evolutionary potential of the "essence" or "nature" of Homo sapiens sapiens? Is it an obstacle to evolutionary potential and survival?
10. What is the role of current and projected cyberspace, cyberpersona, and cyberlaw?
11. Does cyberspace require a totally unique cyberlaw for a projected sovereign cybergovernment and cyberpersona?
12. Does presently developing space law, in an international context, suggest a reasonable likelihood of the survival of the humankind species and its evolving essence, as opposed to a short-term competitiveness?
13. Is there room for a unified global effort to ensure space migration and permanent survival off-Earth at the same time as both governmental and private commercial competitiveness is encouraged through space treaties and international governmental and private sector collaboration and agreements?
14. Do both approaches, governmental and commercial, support humankind evolution, adjustment, and permanent survival off-Earth?
15. Does the body of unfolding space law and its implementation in the form of treaties, conventions, bilateral/multilateral operating agreements, customary international space law, and relevant domestic laws, constructively and timely influence the ongoing survivability of humankind and its biotechnological descendants, i.e., transhumans and post humans?

This rather massive collection of questions unfortunately is only the tip of the iceberg, but for purposes of this chapter it is sufficient to provide us with plenty of working material.

II. Biotechnology and the Emerging Transhuman

Technology in a very significant part has allowed the evolution of Homo sapiens to reach its current species survival status. Technology will allow its further continuing evolution, or alternatively, lead to its extinction.

Based on our present understanding, it is reasonable to assume that from the very emergence of the first humans, technology has been the primary factor in hominid adjustment, survival, and the ongoing evolution for Homo sapiens and its immediate predecessors on the evolutionary bush. It has been a rapid journey, as only about five millennia have passed between the time when metal was used in the first human tools, and the launch of the first human-made rockets into outer space, followed in an evolutionary instant by the space shuttle and then the International Space Station.

For this discussion, "technology" is defined as an "applied science, and a scientific method of achieving a practical purpose."[9]

"Biology" is defined as "a branch of knowledge that deals with living organisms and their vital processes."

When addressing the phenomenon of human evolution enhanced or permitted by the integration of biology with technology, i.e., "biotechnology," the result might be defined as an organism (or various of its organ and organelle components) integrated with technology to create a benefit to an individual or its interactive society.

But just what is the core benefit, the underlying objective which that integration serves? Answer: Survival. No matter how temporary or prolonged, if a biotechnologically enhanced organism performs more favorably than an alternative genetic coding, it is by definition survival enhancing, and hence evolutionary.

The modern era of biotechnology is generally considered to have begun with the discovery of deoxyribonucleic acid structure (DNA) by James Watson and Francis Crick.[10] The discovery of restriction enzymes

---

[9] *See*, therefore, "technology" defined in Webster's Ninth New Collegiate Dictionary at p. 1211 (1991). Note that lower primates also possess technological capabilities, as do most life forms in specific contexts, but not of the sophisticated capability of *Homo sapiens sapiens* to enhance a given technology exponentially, and even create a technology with the capacity to self-evolve. *See* section IV B, infra pp.: "Defining 'Technology' and its Potential Role in Ongoing Humankind Evolution." In this context, it also should be noted that humankind technology is increasingly enhancing the role and characteristics of autocatalysm in humankind evolution. It is important to know the role of technology in that evolution, and then establish values and laws that direct the use of the technology in humankind's own biotechnological evolution. *See*, therefore, by S. Kauffman, *At Home in the Universe: The Search for the Laws of Self-Organization and Complexity* (Oxford University Press – 1991).

[10] In February 1953, James D. Watson and Francis H.C. Crick announced that they had determined the double helix structure of deoxyribonucleic acid (DNA), the molecule that contains human genes. "Though DNA was discovered in 1869, its crucial role in determining genetic inheritance wasn't demonstrated until 1943. In the early 1950s, Watson and Crick were only two of many scientists working on figuring out the structure of DNA. California chemist Linus Pauling suggested an

in the 1960s by Werner Arber opened the pathway to a number of additional biotechnology pursuits, in particular recombinant DNA technology.[11]

The principle issue addressed in this discussion is the expanding role of biotechnology with respect to human DNA, i.e., on human evolution, particularly in an off-Earth space setting, and how the manipulation of the human genome[12] and specific genetic structures impact the process of humankind evolution and survival. Given the current state of technology and the direction of its development, we must assume that new subspecies of humans will soon emerge. For example Homo alterios spatialis, i.e., humans altered specifically for permanent survival off-Earth, will be created to carry, and be capable of carrying, the essence or nature of "humanity" into space. We also should anticipate that new entities will indeed be created through biotechnological integration to form a completely independent post-human species.[13]

---

incorrect model at the beginning of 1953, prompting Watson and Crick to try and beat Pauling at his own game. On the morning of February 28, they determined that the structure of DNA was a double-helix polymer, or a spiral of two DNA strands, each containing a long chain of monomer nucleotides, wound around each other. According to their findings, DNA replicated itself by separating into individual strands, each of which became the template for a new double helix. In his best-selling book, *The Double Helix* (1968), Watson later claimed that Crick announced the discovery by walking into the nearby Eagle Pub and blurting out that 'we had found the secret of life.' The truth wasn't that far off, as Watson and Crick had solved a fundamental mystery of science--how it was possible for genetic instructions to be held inside organisms and passed from generation to generation." *See,* therefore, February 28, 1953, This Day in History, presented by Travelers, online at http://www.history.com/this-day-in-history/watson-and-crick-discover-chemical-structure-of-dna.

[11] Werner Arber, a challenger of Charles Darwin, a Swiss microbiologist, and Nobel Laureat (along with fellow laureate physiologists Daniel Nathans and Hamilton O. Smith), discovered what is termed "restriction enzymes" and their applications to molecular genetics. Restriction enzymes cut DNA at specific places called restriction sites, allowing researchers to work with small sections of genes and to carry out recombinant DNA work, a process that launched the modern genetic revolution.." *See,* by J. Bergman, "Werner Arber: Nobel Laureate, Darwin Skeptic" at [http://www.icr.org/article/werner-arber-nobel-laureate-darwin-skeptic/]. The application of this technique to transhumans and post humans, particularly in an off-Earth synthetic life support environment, is particularly significant.

[12] Simply defined, the "genome" is the sum of all the genetic information contained in the chromosomes of any organism, including its genes and DNA sequences. Physically, a chromosome is defined as a threadlike linear strand of DNA and associated proteins that carry the genes and their functions in single or multi-celled organisms, the nucleus of which is surrounded by a membrane.

[13] *Op cit, supra* note 6, Robinson and Lauria. "Legal Rights and Accountability of Cyberpresence: A Void in Space Law/Astrolaw Jurisprudence," in *Annals of Air*

The incipient capability of genetic re-engineering notwithstanding, the basic principle of biological evolution is natural selection. This has been defined as those species best suited to survive in a given environment, and thereby most likely to reproduce successfully and pass along to the next generation certain characteristics that promote survival in those specific environmental dictates.

III. POPULATION DYNAMICS, EVOLUTION, AND SPACE LAW

For those left on the Earth, what will be the motivation to underwrite human space exploration with the objective of permanent off-Earth settlements?

Leaving aside for the moment the role of technology and biotechnological integration in human and humankind evolution and survival, there are four basic elements that seem to drive biological evolution.

The first is "mutation," which is described as "changes in the sequencing of DNA in a cell's genome." Changes in sequencing may have no manifest effect, and become "trash" genes with no harmful influences, but which may nevertheless play a role in mutation and survival purposes at some unknown point in the future. Change in sequencing may also result in immediate benefit to the survival of a cell, perhaps the cell's host organ, and even the organism.

But the change may also have a detrimental effect, long or short term, such as death of the cell or its host, and perhaps over a period of time even the decline and death of the particular society in which the organism exists, or even the extinction of the entire species, given sufficient time and perhaps negative environmental changes and influences. Nevertheless, mutations, even though often resulting from internal or external environmental changes and influences, are generally random and occur independent of any prevailing survival needs of the individual cell, organ, or organism. More succinctly, mutations do not occur as a result of their usefulness or potential harmfulness to the host.

From an evolutionary and/or survival perspective, mutations, however they occur (sexually, asexually, or epigenetically) that are passed

---

and Space Law, Vol. XXVIII, pp 311-326 (2003); and by G. Robinson, "Space Law: Addressing the Legal Status of Evolving 'Envoys of Mankind,'" in *Annals of Air and Space Law*, Vol. XXXVI, pp 447-512 (2011).

on to succeeding generations have the potential to impact survivability of the individual and its succeeding generations.

Perhaps equally as important as mutations in species survival is "migration," which often is referred to as "gene flow." Here, individuals of a species drift to and mate with individuals of other populations. If the genetic structures are the same, or essentially the same, the ensuing offspring remain primarily the same. Individual specimens of each subpopulation or societal grouping mate "randomly" for the most part, but geographic distances and ecological barriers may make it difficult if not impossible to mate with specimens of other subpopulations. Passing on mutations that may prove beneficial for survival (or may prove fatal) may require movement to a new geographic and survival-beneficial ecotone.[14] Movement to and survival in such a new ecotone frequently does not result from, or require, a mutation.

Merely moving to a more survival-beneficial environment, such as one that is less physically competitive, or more biochemically favorable, or possessing greater nutrition resources and opportunities, etc., may also promote the survival of the genome of an individual or its group.

All of these forms of movement are referred to as "migration."

We must likewise assume that humankind migration off-Earth to promote the ultimate survival of the species and its biotechnologically evolving descendants is also inevitable. (Note that its descendants may include completely independent technological surrogates embracing and perhaps enhancing the principles of "human nature" or its "essence," i.e., robots).

Another primary component of biological and biotechnologically-integrated evolution is a phenomenon referred to as "genetic drift." This process results primarily in smaller subgroupings in which individual and collective genetic frequency[15] in the populations are random occurrences, and do not necessarily result from mutation, migration, or natural selection.

The fourth component of evolutionary change is an occurrence with which all students of biology are familiar, the differential reproduction of

---

[14] An "ecotone" is a transitional area between two or more communities of different species, but with each specimen containing the characteristics of its participating species, i.e., where two different species survive successfully together over a given period of time.

[15] Gene frequency is defined as the frequency of occurrence of an allele in relation to that of other alleles of the same gene in a population. An allele is an alternative form of a gene (one member of a pair) that is located at a specific position on a specific chromosome.

genetically-varied individuals in a population, resulting in some individuals with greater fitness for survival in a specific geographic and specimen-interactive environment. This favors some offspring to perpetuate their genetic structures.

While these four principles of biological evolution are extraordinarily complicated in cause and effect manifestations, far beyond the brief characterizations presented here, their functioning in the context of human evolution beyond Earth leads us to posit the evolution of post-humans, whose survival potential off-Earth is so significantly enhanced technologically as to define a separate species or sub-species. Leaving aside for the present the potentials offered by, and the significance of, telepresence, teleportation, cybernetics and cyberpersona in space migration technology. We must also posit that post-humans will be accountable not to our present notions of jurisprudence, but to non-human formulated jurisprudential concepts and laws, i.e., the laws created by and for a new species taxonomically classified as simply Homo alterios spatialis, or perhaps cyber spatialis... or some similar innovative taxonomic model.

This presents an issue worth exploring, namely that of the transition of one legal regime to the next, a process that history has shown is often the source of sometimes severe conflict.

IV. THE HUMAN ESSENCE

Leaving aside for the moment the evolving potential DNA evidence of past breeding between modern humans and Neanderthals, each human living today is a biological specimen of currently "modern" humankind, i.e., Homo sapiens sapiens.

In attempting to define the "essence" of being human, or what constitutes "human nature," the almost universal tendency is to raise humankind significantly, and even in a misleading fashion, above its biological origins and subtending biochemical/biophysical dictates. This certainly can be seen in all domestic and international jurisprudential concepts and implementing positive laws with roots in Natural Law Theory.[16]

---

[16] *Op cit, supra* note 7, wherein Natural Law Theory is defined and discussed. Note, also, that transhumans (and perhaps even post humans) deriving from hominid reproductive organs and technology also function in a fashion consistent with humankind definitions of Natural Law Theory.

Principles, values, specific words such as "ethical," "moral," "fair," etc., and concepts such as "human rights" and "private ownership" versus "common heritage," subtend most pevailing domestic legislation and international agreements, conventions, and treaties. But even with specific definitions in context, these words and phrases are designed to protect the parochial economic and other physically-oriented survival interests of the societies represented in these negotiated or dictated documents. Only recognizing these underlying biochemical and biophysical motivating factors in negotiating any type of agreement between two or more individuals (including societies, civilizations, and even the incipient stages of interspecies communication) will the basic principle(s) underlying Natural Law Theory be recognized. And this recognition is critical to any type of functional order in the current (and future) level of globalism being shaped by very sophisticated and complex electronic communications.

It is imperative that jurisprudents of any persuasion not rely on wishful thinking, such as the formulation of laws and relationships based on ephemeral definitions inherent in "moral" and "ethical" behavior that do not address the requirements of ultimate "physical" and "essence" survival. The meaning of the odyssey of Homo sapiens sapiens and its technologically altered offspring rests on the physical and sentient[17]

---

[17] For purposes of the present discussion, "sentient" can be defined as "responsive to or conscious of sense impressions." [*Webster's Ninth New Collegiate Dictionary*, p. 1073 (1991).] Perhaps more accurately in the present context, emphasis should be placed upon a receptive sensitivity to abstract perception and reasoning. A *non-empirical* definition of "abstract," beyond simply "the whole being greater than the sum of its parts," is a "thought...apart from any particular...material object[s]." *Webster's New World Dictionary of American English* (1988) at p. 5. Nevertheless, "thought" is still only energy in the form of organized information. Beyond rather ephemeral lay definitions, the empirical foundations of what constitutes "sentience," "conscious," and "abstract" are being pursued in a secular fashion by neuroscientists, psychoneurophysiologists, and the like. David Engleman, a neuroscientist at the Baylor College of Medicine in Texas, asserts that "consciousness" is only a tiny fraction of the brain's function. For Engleman, individual perceptions, thoughts and beliefs are the final results or consequences of the seemingly endless interactions of billions of brain cells, interstitial fluids and hormones, whole body sensory systems, and the like. Actions taken by individuals are totally dependent upon the functionality of these assets working together. But resultant decisions are undertaken without awareness of these interrelated cerebral and whole-body organism activities. There is a significant gap between physiological awareness and actions taken on the basis of resultant specific or flexible knowledge. Consciously induced knowledge makes consequent actions almost endlessly adaptable...without tapping into the billions of neurophysiological underpinnings. This allows for *Homo sapiens sapiens* to have a flexibility capability to react to any existing or rationally perceived environment with which they are confronted...individually and collectively. Query: Can this

survival of the species and its biotechnological descendants, not on the ephemeral and trasitory social, cultural, or religious concepts.

Admittedly, this concept in itself presents a significant challenge to a great many individuals, but this is what science teaches us.

Humans are now, as in the past, using and fine-tuning their technological capacity to adapt to changes in Earth's natural phenomena. These changes include the planet's shifting orbital characteristics in relation to the Sun, as well as its other solar system and galactic physical relationships. Those technologies and their evolutionary fine-tuning consistent with Moore's Law[18] enable identification both of hostile and also survival-conducive environments, leading to the capacity for migration and adaptation to new environments, both on Earth and off.

Homo sapiens sapiens is faced with a pressing and relatively imminent migratory quest for survival. This time the direction is off-Earth. Although periodic natural disasters, now including those made by humankind as a result of its evolving technologies, may inhibit humankind population growth on Earth, such phenomena may also force relatively smaller humankind populations to focus on new technologies that enhance human and humankind survival capabilities required for space migration.

Also necessary for the success of this migration will be the formulation of the relevant space-related jurisprudence and its implementing laws, including domestic, international, global, and

---

form of conscious awareness be reproduced artificially...whether through biotechnological integration or simply technologically? *See*, generally, by D. Eagleman, *Incognito* (Pantheon Books – 2011). Also, note should be made of the growing empirical data regarding sentient and cognitive awareness in taxonomically designated lower orders of animals, such as cetaceans, i.e., water adapted mammals such as whales, dolphins, and porpoises. In the context of ongoing research into the potential of dolphins, possessing, for example, a complex and sophisticated intelligence comparable to that of humans, and which allows an equally as complex interspecies communications with humans, *see*, for example, the work of John Fraser, D. Reiss, P. Boyle, K. Lemke, J. Sickler, E. Elliott, B. Newman, and S. Gruber, "Dolphins in Popular Literature and Media," *Society & Animals* 14:4 (2006).

[18] Moore's Law refers to the observation made in 1965 by Gordon Moore, co-founder of Intel, that the number of transistors per square inch on integrated circuits had doubled every year since the integrated circuit was invented. Moore predicted that this trend would continue for the foreseeable future. In subsequent years, the pace slowed down a bit, but data density has doubled approximately every 18 months, and this is the current definition of Moore's Law, which Moore himself has accepted. Most experts, including Moore, expect Moore's Law to hold for at least another two decades. *See* online at http://www.webopedia.com/TERM/M/Moores_Law.html.

ultimately transglobal legal regimes. As noted in the introduction, the humankind species is now shaping its ongoing survival requirements by changing the humankind life-support capabilities of Earth to the point where off-Earth migration may not be optional, but rather necessary for the continued evolution of the species, and the continuation of its odyssey to survive and inch toward understanding and defining its own "essence." This is the odyssey of all life, beginning with single cell life forms, the "shoulders" upon which Homo sapiens sapiens and its evolving transhumans and post humans stand.

Survival has always been the objective of problem-solving hominids (and, indeed, that of all life forms regardless of their problem solving capabilities). Environmental changes frequently create the need for adaptations in life-style surroundings. Frequently, as noted above, adaptation is accomplished through migratory practices, moving to a more favorable environment or ecotone. In addition, perhaps a change in diet and/or new expressions of existing genetic codings and sequencing, as well as relevant corrective health technology may be available at the critical time.

For humankind, in addition to adaptations by anatomical and other biosystem characteristics, survival migration is, and has been in the past, enhanced by its sentient characteristics, its problem-solving capability, and its technology. In many ways, evolution of specific cultural characteristics can be said to result from weather and climate changes that forced migrations, or conversely that enabled long-duration settlement that led to agricultural practices based on seasonal predictability. On Earth, humankind currently lives in an interglacial climate that, leaving aside negative human induced alterations, has contributed significantly to the human population explosion and on-Earth migrations, and that may also eventually force humankind migration off-Earth

V. THE ROLE OF TECHNOLOGY IN HOMINID MIGRATION AND EVOLUTION, AND IN ITS SECULAR SEARCH FOR THE SPECIES' INDIVIDUAL AND SHARED "ESSENCES"

In the greater context of the universe and possible parallel and succeeding universes, Carl Sagan put the question concerning the significance of humankind's odyssey quite simply. "Who are we? We find that we live on an insignificant planet of a humdrum star lost in a galaxy tucked away in some forgotten corner of a universe in which there are far more galaxies than people."

Many believe our species has stopped evolving, and that there is no need for biologically or biotechnologically evolved changes. In this view, the conditions forcing genetic and epigenetic changes necessary for evolution, adjustment, and survival no longer prevail, and that what constitutes the "essence" of being human, or "human nature," has reached its ultimate pinnacle or end point as necessary for continuation of its existence ad infinitum.

Natural Law suggests that this is an incorrect formulation. Perhaps humans may not evolve, or need to evolve, within the extant ecosystem, but all environmental characteristics will change, and sooner or later create fatal conditions for human survival, whether on Earth or off. Put simply, modern humans have not yet reached an evolutionary end point.[19]

Humankind is indeed experiencing a rude awakening regarding the impermanence of atmospheric and geologic conditions. We are experiencing increasing reliance on humankind technology to help us adapt to conditions on Earth, and Natural Law indicates that to survive and perpetuate itself ad infinitum will require that we venture off planet Earth.

At the same time, humans are currently disrupting traditional understandings of the essence of humanity itself, of what it means to be human, both physically and in terms of sentient or abstract perception properties. These inroads are direct results of biotechnological integration

---

[19] A fundamental precept of those advocating long term and permanent off-Earth human*kind* exploration, migration, and settlement is that it must be in the form of a new species with unique cognitive, and/or sentient abstract perception characteristics. Nothing is forever, of course. Not even *Homo sapiens sapiens*. The species must evolve to survive off-Earth in a completely synthetic and alien life support environment, amidst extraordinarily destructive effects of gravity free environments, solar and deep space radiation, psychology of long-term confinement, logistics required for organic life forms, and the like. This likely will result in a requisite biotechnological and artificial intelligence re-engineering of these space envoys of humankind that will result in a significantly altered "essence" of being a human. It also may require a totally new and different energy form, such as telepresence or some other form of teleportation, which in turn will raise the issue of the "legal" relationships between and among the designer/manufacturer and programmer/operator. As noted in the ensuing discussions relating to cyberlaw and its evolution, the conflicting interfaces may not be between entities of differing cultures dictated by their specific ecosystems, but rather between and among totally different forms of independent, unique entities with abstract perception and societally interactive requirements for survival.

of humans in cyber persona in cyberspace.[20] This leads us to posit an expansion of Natural Law to encompass not only a physical journey beyond the Earth, but also a cyber, informational, and perhaps even metaphysical one as well. So our evolutionary path is perhaps branching in multiple directions, into the physical including the biological and biotechnological, and into the physical and non-physical as well. Fascinatingly, all of this is occurring at the same time, i.e., now.

Some explorers are developing the empirical underpinnings describing what it means to transition into a separate and distinct post-human species. At the same time, the phenomena of telepresence and teleportation[21] are assisting in identifying precisely what constitutes the distinct individuality of each human, and in understanding that the individual is but one component in a larger and more complex organism with a greater sense of awareness and sentient capabilities. Therein rests a significant confrontation with the evolving tenets of humanism[22] and of religions, and the seeds of one an important future social/cultural challenge.

---

[20] "Cybernetics" is a term coined by American mathematician Norbert Weiner in 1948. While the word has many definition variations depending upon context, it usually refers to the general analysis of control systems and communication systems in living organisms and machines. Analogies in cybernetics are drawn between the functioning of the brain and nervous systems and the computer and other electronic systems. The science overlaps the fields of neurophysiology, information theory, computing machinery, and automation. *See* definitions online at http://encyclopedia2.thefreedictionary.com/Cybernetics. While "cyberperson" has many definitions in the context of cybernetics, it might be defined generally as someone who "continually searches for theories, models, paradigms, metaphors, images, [and] icons that help chart and define the realities that we inhabit" [http://amiquote.tumblr.com/post/2658717182/timothy-leary-on-cybernetics-and-a-new-global]. The word "cyber" means *pilot*, and "cyberspace" also seems to have many definitions, depending upon context and user intent. However, the coining of the word "cyberspace" is credited to William Gibson, who used it in his science fiction book, *Neuromancer*, written in 1984. Gibson defines cyberspace as "a consensual hallucination experienced daily by billions of legitimate operators, in every nation, by children being taught mathematical concepts... A graphical representation of data abstracted from the banks of every computer in the human system. Unthinkable complexity. Lines of light ranged in the non-space of the mind, clusters and constellations of data" (New York: Berkley Publishing Group, 1989), p. 128.

[21] For definitions and descriptions of "telepresence" and "teleportation," *see* fnts. 38 and 39, *infra*.

[22] "Humanism" has many definitions, again, in differing contexts. For the present discussion, "humanism" may be said to be a characteristic typical of *Homo sapiens;* a reflection or component of human "nature," however that term may be defined empirically.

VI. "Technology" and its Role in Ongoing Humankind Evolution

The history of modern human evolution is a history of ever receding horizons. We stand on the shoulders of our single cell predecessors. Our evolutionary successors may never look back to our own shoulders.

As noted in section II, above, Webster defines "technology" as "an applied science, or a scientific method of achieving a practical purpose."[23]

Technology itself is a dual driver in the evolutionary process. It is, first of all, our improving understanding of technology that has enabled humanity to develop the tools to understand Natural Law itself. We may cite as examples gene sequencing mentioned above, or the tools of physics such as particle accelerators that reveal the previously hidden nature of fundamental physical particles, or the means and methods of computing technology, which result in the computational tools that are essential to both gene sequencing and particle physics. Technology is also likewise the application of technology that is precisely developing and defining what a trans-human or post-human organism may be, such as neuro-and bio-integrated prosthetics, not to mention cyber-worlds, and of course robots.

VII. Space Law Versus Cyberlaw and Post Human Law

Will astronaut/cosmonaut "Envoys of Humanity" continue to be subject to prevailing tenets of Earth-defined space law, or will they transition into a new subspecies of Homo sapiens, or perhaps into a truly independent species? In the context of this question, it is necessary to assess the present state of evolving space-related robotics. Although "Robonauts" designed and programmed in large part to conduct mechanically-oriented tasks in space[24] can do so relatively endlessly, the real problem may rest in the step-by-step ongoing fine-tuning of what characterizes "human essence," and then incorporating it into such robonauts.

Given exponentially increasing sophistication of self-replicating and

---

[23] *Webster's Ninth New Collegiate Dictionary* (1991), at p. 1211. *See*, also, *op cit., supra*, note 9.

[24] At this point in space exploration and migration, robonauts have been designed primarily to work with and assist humans working and/or living in space. The succeeding generations of robonauts may well approach ever more closely to independent thinking plus a rather unique "essence" peculiar only to them.

even more sophisticated artificial intelligence in the form of post humans migrating to, and settling permanently in, space, it is essential to determine when and under what circumstances such entities are no longer humankind's "envoys" in space, as reflected in the spirit and intent, as well as wording, of the Outer Space Treaty of 1967.[25]

Reference to "envoys" in the Outer Space Treaty is in the context of representing the interests of all humans while those envoys, those messengers or agents "of all mankind," are in space, i.e., the "common heritage of all mankind" as specifically referenced in the treaty. The concept of space and its components being the "common heritage of all mankind" was introduced into this treaty primarily to assert among the signatories that individual sovereign nations and individual persons or public/private organizations would not be allowed to claim ownership rights over any areas and components of off-Earth space. With the planting of the American flag on the lunar surface by Neil Armstrong during the Apollo 11 mission, significant effort was made to assure the international community that the act did not represent the traditional planting of the flag on Earth as a symbolic gesture asserting sovereign control or ownership of that portion of Earth's land mass. The flag was planted as symbolic achievement of a lunar landing on "behalf of all mankind."

It will also be necessary to determine when the traditional dictates of space law are no longer responsive or applicable to post-humans functioning off-Earth and/or in cyberspace. Put differently, what are the determining factors regarding when such entities become subject to the law as it pertains to humankind and transhumans, and when they are solely subject to their own jurisprudential regimes of "cyberlaw" or "post-human law"?

Individuals engaged in telepresence[26] and teleportation[27] will

---

[25] Treaty on Principles Governing the Activities of States in the Exploration and Use of Outer Space, including the Moon and other Celestial Bodies, Jan. 27, 1967, 18 U.S.T. 2410, 610 U.N.T.S. 205 commonly referred to as the Outer Space Treaty of 1967.

[26] "Telepresence" also lends itself to varying descriptions of the same concept. For present purposes, it can be defined as "the projection of a user's sensory, cognitive, and motor capabilities to a distant environment. Alternatively, the distant environment can be created *virtually* at the location of the user or operator. In the former case, a user's sensory channels can be linked to remote sensors. For example, a user's vision can be linked to remote cameras, providing an exocentric or egocentric frame of reference. The user's actions in the proximal locality drive the movements of the remote actuators. In the latter case, the task environment surrounding the operator in which local actions are transformed into distant actions." G. Robinson and R. Lauria, "Legal Rights and Accountability of

continue to exist as "alter-egos" of humankind for the reasonably foreseeable future simply because they are and will continue to be programmed and controlled by humankind, transhumans, and possibly even post humans. At some point, however, telepresent and teleported entities would likely transition into the category of post-humans, no longer a taxonomic component of the human or transhuman species/subspecies. At that point such individuals would transition technologically into independent functioning and perhaps to self-replicating entities who will be subject to their own versions of Natural Law and resulting legal "philosophies" and implementing positive laws.

At a minimum, the present characteristics of cyberlaw[28] may provide hints of what and when distinctively unique post human space jurisprudence may come into existence, and how it will interface with traditional Natural Law Theory and the resulting variety of jurisprudential concepts formulated and practiced by what is quaintly referred to as "modern man," i.e., by Homo sapiens sapiens.

Another issue to consider is the applicability of space law dealing with the issue of "who owns space." Beyond the traditional tomes that have been written and published premised on the four primary treaties, this field is a bit unsettled to say the least. Technological advances have led to serious consideration of the use of space resources by both governments and entrepreneurs, but who in fact will have the right to assert ownership or

---

Cyberpresence: A Void in Space Law/Astrolaw Jurisprudence," in Annals of Air and Space Law, Vol. XXVIII 311 at p. 313 (2003). *See*, also, M. McGreevy, "The Presence of Field Geologists in Mars-like Terrain (1992) 1 *Presence*: Teleoperators and Virtual Environments 375 at 377; and by T.B. Sheridan, "Musings on Telepresence and Virtual Present," 3 *Presence*: Teleoperators and Virtual Environments, 120 at p. 120.

[27] Less relevant to the instant discussion at present is the concept of "teleportation," which can be defined as the hypothetical method of transportation in which matter or information (energy as "organized information") moves a solid object or person by psychokenesis (by kinetic information of the mind). It is a hypothetical method of transportation in which matter or information is dematerialized, usually instantaneously, and then re-materialized in a different physical location. *See*, therefore, "teleportation" in Webster's Ninth New Collegiate Dictionary, p. 1212 (1991).

[28] "Cyberlaw" refers to the exponentially developing area both of civil and criminal law applying to computers and their uses and activities conducted over the internet and other similar networks. The principal areas of cyberlaw practice relate to the protection of intellectual property rights, and also to freedom of speech and public access to information transmitted by electronic communication.
http://www.yourdictionary.com/cyberlaw and
http://www.businessdictionary.com/definitions/cyberlaw.html

control over such space resources?

As a matter of evolving law, in other words, who or what will be permitted to own portions of off-Earth space and/or its components? While the Outer Space Treaty emphasizes the "common heritage" principle, growing efforts by the international community to amend the treaty in certain respects to make it more reflective of current geopolitical realities and technological advancements in space include reconsidering the likely necessity of sovereign and private ownership of certain space resources.

The underlying agreements establishing the International Space Station governance and operating rights, both public and classified, serve as the ownership resolution positions of the participants. But what will be the relevance of these operating documents when cyberpersona and/or post humans function permanently and independently in such an outer space environment? Indeed, the law as it is presently constructed does not reflect the realities of nor the implications of advancing technology, and is therefore becoming irrelevant. For example, we can readily foresee issues arising in the context of nation states versus transhumans and post humans who intend to, or actually do, permanently occupy portions of interstitial space or various of its celestial bodies.

VIII. Conclusion

As discussed, there are many basic questions to pose and address when assessing the issues of Natural Law and jurisprudence as they pertain to cyberspace and cyberpersona, as well as to transhumans, post humans, sentient robots, and every entity that carries or embodies the human "essence." Furthermore, legal regimes will have to be devised to address every "one" in all of these categories who intends to or actually does permanently reside either physically off-Earth, or in cybernetic, virtual, non-physical, informational environments.

In parallel with these issues, we are increasingly aware of potential natural disasters that threaten the existence of modern humans, such as the possibility of catastrophic eruption of Yellowstone National Park, asteroids and threatening Near Earth Objects of which we are made aware by advanced telescopic and other technological sensing capabilities, and, of course, modern humans' own species-threatening technologies, such as nuclear warfare, biological warfare, the inability to identify and control disease bearing biological mutants, uncontrolled human population growth dynamics, not to mention catastrophic climate change, and the like.

A threat of a different sort is the possibility that Homo sapiens sapiens will be overtaken by its own technology, in the form of an artificial intelligence with independent decision-making capabilities. Humans now give up much of their life and death decision making options to both crude technology and highly advanced artificial intelligence. Law enforcement and military operations are excellent examples of this form of cyber-enhanced decision making. Technology is gradually usurping human reasoning capacity, accelerated by the willingness of humans to accept "whatever the computer says." The transition step leading to a new jurisprudence that addresses post human cognitive sentience may well be seen in the current legal questions and issues relating to the somewhat esoteric nature and functioning of humankind and cyberpersona, as reflected in the conflicting concepts of cyberorder versus cyberanarchy in cyberspace.

But how will that jurisprudence and the implementing laws emerge? Indeed, the shift from one legal regime to another can be fraught with difficulty and conflict, as we see today in the ongoing legal tug of war concerning issues such as gay marriage, and indeed civil and voting rights issues remain in dispute nearly one-half century after a set of basic legal principles was supposedly adopted in the US.

Given the nature of change, and the resistance that it often provokes, we can anticipate a prolonged dispute between those advocating new legal systems that deal with the manifold issues of cyber-sentience, post-humanity, off-Earth resource and territorial rights, and those who prefer to remain with the legal structures that they are already accustomed to, and a whole host of related and derivative issues that we can well anticipate lie ahead of us in the coming centuries as the space movement leads to extensive human presence throughout and perhaps beyond the Solar System.

The general theme of this book is international cooperation for the development of space, so what can our topic in this chapter tell us about that? Regardless of the specific issue or topic, one thing this discussion makes quite clear is that the development of extensive and permanent space habitats will lead to the formation of new nations or quasi nations, and thus an increasingly complex realm in which Natural Law will be evidently evolving, and in which the derivative jurisprudence must therefore evolve as well.

As the foregoing discussion hopefully has made clear, we can expect that some nations will lead in the development of these new capabilities, and in the design and implementation of the laws that reflect and support new capabilities, new needs, and indeed new forms of life à la the emergence of post-humanity, largely as a result of the space endeavor. Other nations will necessarily lag. Will the leaders and the laggards cooperate? Or will this become a matter of disagreement and dispute? Yes, and yes.

But ultimately what is at stake is not just disputes among species and subspecies, or disputes between a home world and maturing and independence-minded colonies. The underlying and enduring issue is human evolution and the survival of the human essence, whether that is in a form that we would recognize today as human, or in some other cyber, post human, or transhuman form of existence, one suited to micro-gravity or extra-gravity, one inhabiting orbiting or transiting craft. And five hundred or a thousand years from now, 100 generations in the future, will your descendants be sitting by the side of a lake, on Earth, at night, gazing up at the stars and planets, and wondering just how their great, great grandchildren are doing, voyaging way out there in the vast beyond, extending the reach of human evolution far beyond the reaches of our forgotten corner. What are they learning about the human essence as a result of that journey?

...

Dr. George S. Robinson

Dr. Robinson has been in the private and public practice of law since 1963, and has taught or lectured in space law at numerous universities around the world. After serving at NASA and the Smithsonian Institution in Washington, DC, for thirty years, he retired into private law practice and concentrates primarily on space matters. Dr. Robinson has served on hospital boards of directors, as well as various science and space-related boards of trustees and governmental and private advisory committees. He earned his AB degree from Bowdoin College, an LL.B. degree from the University of Virginia School of Law, an LL.M. degree from the McGill University Graduate Law Faculty, Institute of Air and Space Law, and the first Doctor of Civil Laws degree in space law awarded by that Institute. He presently is in private practice with his two sons and daughter-in-law. He is also author of Chapter 4 in this volume.

CHAPTER 21

Conclusion: Prospects and Ambitions for Cooperation

Langdon Morris
PwC

And

Kenneth J. Cox, Ph.D.
ATWG Founder

© Langdon Morris and Kenneth J. Cox, Ph.D., 2012. All Rights Reserved.

Big Ideas

So what insights may we take away from the previous 460 pages?

The great virtue of an anthology such as this is the opportunity it presents to juxtapose different voices, different themes, and different viewpoints in search of the deeper underlying patterns of analysis and previews of the future.

Here, then, is perhaps the briefest possible synopsis of each chapter, not as a substitute for actually reading what these erudite and thoughtful authors have graciously shared with us, but as a step toward finding a useful synthesis of their many insights and observations.

MICHAEL SIMPSON

There are several patterns of cooperation that describe the various ways that nations and other entities participate in the space adventure, from technology specialization, to economic return, to enabling private enterprise. the net result is that the base of participation is indeed becoming broader, which can only be a good thing. "Ultimately by broadening the base we will have provided the entire space sector, from the first participants to the most recent, with the foundation we need to make the future of space activity exceed even the wildest dreams of its greatest visionaries."

WALTER PEETERS

The development of humanity's capabilities in space began in a competitive environment of the Cold War, but has evolved to reflect widespread cooperation. Cooperation itself often proceeds through four stages, from relatively independent coordination to augmenting one another's capabilities, to interdependence, to full integration of effort. This progression can be taken as a model that in turn can be applied to global politics: "We can only hope that visionary politicians will grasp this unique opportunity to support the creation of a better society."

GEORGE ROBINSON

The law of space as it is written today is certainly not adequate for the needs of tomorrow, and thus it will evolve inevitably to reflect the requirements of nations (*individually and united?*), and inasmuch as space commerce will be an ever-increasing aspect of the space endeavor, it will also reflect the needs of entrepreneurs, and eventually of new, off-Earth communities. In addition, George has offered us a detailed list of 15 specific areas in which the law will necessarily evolve. "Space law in the twenty-first century must and will embrace and respond to the requirements of, and for change in, the private and public sectors – their work forces, military applications, private commercial ventures, transglobal trade agreements, and the like, all while recognizing the independent trading nature of long-duration and permanent space communities."

STEWART SANDERS AND CHRISTOPHER STOTT

The Space Data Association is a new model for international cooperation in space, one carefully organized to provide benefits to all industry participants through an open and collaborative organizational approach that addresses significant commercial, operational, and legal requirements. "Ultimately the entire space sector is benefitting from the creation of the Space Data Association and this proactive, voluntary move by the founding Satellite Operators not only in terms of enhanced and now formalized orbital situational awareness and point source interference identification and resolution, reducing costs and increasing safety and efficient use of resources throughout the space sector, but also in providing a viable model for international cooperation."

THOMAS DIEGELMAN

What are the essential commercial and technological issues that must be resolved as we move outward from Earth and into the cosmos? Tom offers a thoughtful review of key themes, among them transportation systems, speed, efficiency, modularization, and spaceport operations. "Hundreds of years from now, surely someone will write a book about the parallels of the journey of Columbus, the first transcontinental railroad across the United States, the Trans-Siberian railroad, the Suez and Panama Canals, and other great voyages and transportation firsts, and recognize commerce as the very common human thread uniting all of these endeavors. Yes, indeed: Houston, we have a profit ..."

CHRISTOPHER RILEY AND CHRISTOPHER WELCH

Yuri Gagarin's flight in 1961 was a transformative event in human history, and forty years later a small group of social entrepreneurs created a global phenomenon, Yuri's Night, to remember and celebrate that first human space flight. Ten years later, an estimated 100,000 people participated in the 2011 Yuri's Night celebrations around the world. Another celebration of Gagarin's flight is the film *First Orbit*, created specifically to commemorate the accomplishment, recapture the experience, and rekindle enthusiasm for space adventure. "The first decade of Yuri's Night celebrations since its inception in 2001 has shown just how strong the story of humanity's first spaceflight still appears to be. Fifty years after Gagarin's pioneering mission, the courage and spirit of adventure which human spaceflight epitomizes still appeal widely to the people who live today on the planet he first orbited."

Rita Lauria

Virtual reality simulations and virtual worlds offer the possibility of accelerating and enhancing astronaut training before leaving the ground, and permit participants to explore cultural, behavioral, interpersonal, and technical dimensions of their off-Earth work to better prepare for their impending adventures and challenges. As individuals from more cultures and countries join the astronaut corps, virtual reality training may become a necessity, as it is one of the few ways to prepare for cultural differences that will inevitably emerge in confined living spaces, and this will be a key tool to enable and enhance international cooperation. "By innovating together to build virtual worlds for the training of our envoy astronauts, including technical and mission training as well as training in the soft skills like communication, cross-cultural understanding, leadership, conflict management, teamwork, situation awareness, problem solving and decision making, international cooperation in the development of outer space will be facilitated."

Brett Biddington

Australia's unique position in the space community is assured by its location on the globe, its largely empty interior, and its political affinity with the Western democracies. Over the decades of the space age, Australia has played a critical support role in the space endeavor, and is now recognizing that a more prominent effort is in the interests of the nation, namely economic and security. "The [Australian] Government has now taken steps to create a coherent space narrative and to develop a critical mass of space qualified individuals within and beyond the public sector who can provide a necessary and sufficient mass of talent and expertize capable of lifting the profile of space in Australia and of Australia's profile in space internationally. ... In securing its own national interests, Australia is well-placed to ensure that the space-disadvantaged nations of the world can also enjoy the benefits of secure and assured access to services from space for national, regional, and global benefit."

Le Wang

China's progress as a spacefaring nation over the past few decades has led to significant accomplishments and a position as a leader among nations in its capabilities. Through alliances and partnerships with many nations, China has shared its expertise widely. With considerable plans for the future, including an expected presence on the Moon, China's ambitions are substantial. "The long term roadmap put forward by strategic study calls for a higher priority to be put on China's space program to provide much needed support to China's economic, social, scientific, and technological development. With the acceleration of social and economic development in China, its space program is anticipated to expand and improve at an ever faster speed, and international cooperation has become and will continue as a vital part of China's space program."

KAZUTO SUZUKI

Japan's approach to space was explicitly pacifist from the origin of the space age until 1998, when a perceived threat from North Korea aroused ample fears. Over the following decade, Japanese leaders grappled with the problems of self-defense and indeed the full scope of Japan's space ambitions. This brings the relationship between Japan's and China's motivations in space into sharp focus, as both countries strive for leadership in the Asia-Pacific region and globally. "The rivalry between APRSAF and APSCO is stimulating both Japan and China to use these organizations as vehicles for exercising their leadership. The unintended consequence of this leadership is the enhancement or empowerment of other Asian countries, as they are the beneficiaries of space-based services and technology transfer. Countries which have high ambitions for developing space capabilities now have easier access to high technology and possible international cooperation with experienced partners, and thus it can be said that this rivalry shall not be regarded as a space race, but as a healthy competition for providing public goods for the region."

CHRISTINA GIANNOPAPA

While the nations of Africa remain the generally most underdeveloped in the world, the potential for space-based technologies to contribute to African development is significant. There are a great many organizations across the continent that are becoming engaged in developing technologies and technical capabilities, and likewise many European organizations are beginning to partner across Africa to enhance and accelerate the process. "Africa is a continent on the rise, and while it will be still some time in the future before Africa will see full-fledged space powers (either nations or regions), it will steadily increase the use of space applications for economic and societal development."

VERONICA LA REGINA

European satellite communications policy has and will continue to have significant impact across Africa as a tool for development support and also in the case of crises and emergencies because the geographical proximity of the two continents means that many of the satellites that cover

Europe also can serve parts or all of Africa. In addition, due to the tectonic arrangement of the Africa and Eurasian plates, through which the African plate is being forced beneath the Eurasian, we can anticipate continuing geological disturbances across the plate boundary which runs more or less beneath the Mediterranean Sea, resulting in earthquakes, climatic variations, and disruptive volcanic activity on both the European and African sides. Because these incidents almost inevitably disrupt terrestrial communications systems, satellite-based systems provide an essential tool for disaster relief and management. Hence, policy decisions regarding satellite deployment and use will be a significant factor in African development throughout the coming century. "A coherent and appropriate satellite communications system for broadband, delivering services ... for the EU and for Africa is an opportunity to give substance to the fundamental principles of international space law that the exploration and use of outer space shall be carried out for the benefit and interest of all countries."

OLGA ZHDANOVICH

From the beginning of the USSR's space program at the height of the Cold War, the Soviet space organizations achieved many significant scientific and technological breakthroughs that are of course well known and appreciated. With the collapse of the USSR and the emergence of the Russian Federation, the approach has changed significantly from a purely independent pathway to one that relies more and more on collaboration and cooperation with various international partners. The use of space as a political tool for détente is well recognized through the Apollo-Soyuz program of the mid-1970s, which led to an unprecedented space docking of American and Soviet craft (which was also discussed in Chapter 3). "In the era marked by the beginning of private space tourism and broad commercialization of space and quite severe competition in Earth observation, telecommunications and other space integrated applications areas, there are still many opportunities left for international cooperation in space. ... People from different nations and cultures share many of the same goals and aspirations, and can effectively meld their technical expertise to accomplish demanding mission goals while developing mutual respect, friendship, and tremendous new learnings for the benefit of all."

CHRISTOPHER STONE AND BRENT ZIARNICK

This chapter questions the benefit of cooperation for cooperation's sake by recommending a strict metric be applied to determine if a proposed cooperation indeed provides a measurable yield. "This chapter argues for the 'Capability Criterion,' a method for individual actors (state or private) to decide to engage in international cooperation on a particular project or not based on a logical framework. The Capability Criterion is meant to be a benchmark or framework that any individual agent, whether organization, company, or nation, can use to decide whether or not they should engage in a particular international cooperation endeavor. The logic of the framework is that if the criterion is not met, the agent should presumably not cooperate. Conversely, if the criterion is met, then the engagement in the proposed cooperation makes sense. ... The Capability Criterion is this: 'An actor should engage in international cooperation if and only if completion of the cooperative effort's objective will result in the actor achieving a higher level of enhanced space capability than it would be able to achieve by acting alone while exerting a reasonable but strenuous level of political will and resources.'"

JAMES RENDLEMAN AND J. WALTER FAULCONER

There is a strong case to be made for each nation to consider cooperation with others in their space endeavors, as it enables each to access increased resources and may improve efficiency while reducing risk, expanding access to useful technologies, and enhancing prestige. There also may be benefits from a political perspective, as well as in workforce development and stability. A potential down side may exist, however, in the form of increased costs and complexity related to coordination, the risks of unwanted technology transfer. For nations with a strong nationalistic sensibility, cooperation may also be unwelcome. Motivated by clear self interest, the authors conclude that "the case for cooperation is strong and powerful. ... Differences among partners must be understood and respected. Leaders help by attempting to foster conditions that allow for reciprocity and transparency when they can occur, although these conditions will vary. Rapport, knowledge of partners, and patience among the assembled team members will also be essential. Successful international partnerships tend to have unifying attributes—fundamental trust with aligned goals."

THOMAS DIEGELMAN AND THOMAS DUNCAVAGE

The longevity of any nation depends on its capacity to sustain three critical factors: its sovereignty, its national security, and its economy. A shortcoming in any of the three leaves it vulnerable to decline and ultimately to collapse. In the modern world, scientific and technological capabilities are critical to both security and economic development, and capabilities in space are in turn crucial to sustaining science and technology on the leading edge of progress. NASA, however, has largely failed to engage the American public in a deep understanding of the importance of investing in space as a means of sustaining America's standing among the nations of the world. "NASA should be leading a national discussion on the role of space as it relates to issues including the national deficit, economic stability and growth, and national defense. From such a discussion can come a solid policy platform, from which it then and only then becomes possible to engage with global partners in the pursuit of new goals and ambitions in space."

JEFFREY NOSANOV AND MICHAEL POTTER

"The 100 Year Starship Study (100YSS) is an unprecedented, multi-disciplinary initiative that challenges humanity to identify commercially sustainable ways to design, develop, and build a manned vehicle for interstellar space travel within 100 years, by the year 2111. Aside from the obvious technical challenge, there are numerous organizational challenges to mounting a multi-generational endeavor to launch a human Starship mission. First and foremost is the need to develop a clear vision and mission statement for such an ambitious undertaking; the scope of such a mission is truly a global undertaking. ... The 100YSS challenge offers humanity the opportunity to choose a clear and compelling vision and a path to the stars, and just as importantly, to leave behind the current political muddling and chaos that holds back our planet's shared potential."

GEORGE ROBINSON

While Diegelman and Duncavage argue that nations are transitory, Robinson makes the same point with regard to species. "The humankind species is now shaping its ongoing survival by changing the life-support capabilities of Earth to the point where off-Earth migration may not be optional, but rather necessary for the continued evolution of the species, and the continuation of its odyssey to survive and inch toward understanding its own 'essence.'" And this evolutionary thrust may lead not only to new human capabilities, but also to new humanoid forms, post-humans capable of living beyond Earth in different gravitational conditions, or in various forms of human-cyber amalgams. Is our current law capable of addressing these emerging situations? "We can expect that some nations will lead in the development of these new capabilities, and in the design and implementation of the laws that reflect and support new capabilities, new needs, and indeed new forms of life à la the emergence of post-humanity, largely as a result of the space endeavor. Other nations will necessarily lag. Will the leaders and the laggards cooperate? Or will this become a matter of disagreement and dispute? Yes, and yes.

But ultimately what is at stake is not just disputes among species and subspecies, or disputes between a home world and maturing and independence-minded colonies. The underlying and enduring issue is human evolution and the survival of the human essence, whether that is in a form that we would recognize today as human, or in some other cyber, posthuman, or trans-human form of existence, one suited to micro-gravity or extra-gravity, one inhabiting orbiting or transiting craft."

•••

So there you have it, a summary in a handful of pages of the previous 460. Some of the key points ... cooperation is essential to humanity's endeavors in space, and it takes many different forms. It requires a lot of work, but it is the basis of future success, and offers significant benefits to all humans of all nations. It involves public and private sector efforts, engages all aspects of science from physics to psychology and everything in between, it is driven by courageous and clear-thinking leadership, and accomplished by equally courageous and clear thinking scientists, engineers, entrepreneurs, and astro-cosmo-taikonauts.

We realize, of course, that these authors had a lot more to say than just the few highlights we have evoked here, and we do not mean that these brief excerpts should be taken as a complete summary of their far richer thoughts and insights. Nevertheless, viewed together they do give a rather clear picture of both the situation and the possibilities.

A bit more detailed view reveals a picture like this: To date, nations have been the drivers of progress in the exploration and development of space, and they did so for a combination of reasons, including rivalry, threat, pride, curiosity, science, and economics. In specific, the Cold War rivalry between the USSR and the USA instigated the Space Race, and both sides feared that the militarization of space by the other would leave it in a disastrously inferior position in the event of conflict. Fortunately, that scenario never emerged, and the two former adversaries gradually moved toward awkward partnership and then to fully committed collaboration, as we see today in their joint participation in the ISS. Both nations, like a great many others subsequently, were also motivated to develop their capabilities in space as a matter of national pride. Likewise, the desire to know what's "out there," and to know that we can actually get there, has been a compelling impetus. The scientific dimension has also turned inward, back toward our home planet, and we have discovered that a presence in space enables us to achieve magnificent and now important insights about Earth – climatology, geology, agriculture, forestry, oceanography, etc.

And the economic benefits have likewise become unmistakable; the advanced technological capabilities resulting from the space programs have provided incalculable benefits to the Earth-bound economies of all participating nations, and in cases related to satellite communications, the economic benefits of being in space have also been significant.

All of these factors remain pertinent when we consider the reasons to invest in the space endeavor today, while additional reasons have also emerged. Many, including some of the authors of this book, argue that the journey to space is an evolutionary imperative, and tell us that due to our current state of knowledge we are irrevocably obligated to create alternative homes for the ongoing human experiment, that our species must spread beyond Earth as a matter of species destiny and self-preservation.

There is also today, certainly, a commercial imperative. The ascent, literally and figuratively, of private space ventures, brings a new form of economic possibility to the commercial dimension. The early entrepreneurial pioneers are establishing their high-risk ventures and hoping to achieve celestial returns; in so doing they are redefining the commercial landscape. If their expectations are correct, then history will remark on the folly of those who did not participate in the commercial development of space; we are not quite at that point today (today being defined as roughly the current decade, 2010 – 2020), but a decade hence may bring a considerably clearer picture. And should the commercial dimension emerge to the extent that it might, on the upper end of the projections, then the impact on the Earth-bound economy will be as significant as any of the recent technological breakthroughs that we celebrate today – the internet, mobile telephony, alternative energy, etc. It should be noted that the previous book in this series, fittingly titled *Space Commerce*, argues for and anticipates just such an impact, so we come to this question with a strong point of view and an equally strong expectation for the significant future of the commercial space endeavor. We agree that the definitive proof is not yet in hand, and at the same time the progress even in the last few years continues to build evidence for a huge future market, and not even so far in the future.

And tomorrow? What of beyond 2020?

We can expect, first of all, that the terrestrial economy will continue to be under duress, and this will lead to increasing stress on political and social systems. The modern world is a tremendous jumble of threats, traumas, and possibilities, with change coming ever faster and bringing breakthroughs and heartbreak side by side, moment by moment. Scientific discoveries, new technologies, and wonderful social progress occur along with wars, catastrophes, suffering, and mass murder.

It is in fact difficult to foresee anything but enduring economic challenges, compounded by demographic issues, ethnic, ideological and religious disputes, resource shortages, significant health concerns, and widespread psychological difficulties (Future Shock on a mass scale), all overlain by the potential for climate change catastrophe. It's hard to imagine any rosy scenarios, and quite easy to see the many shortcuts to widespread disaster.

How, then, should we think about the space endeavor in such a context? How do we reconcile our optimism for space commerce and our impetus to sustain our species by venturing outward, with what seems like entirely justified pessimism concerning our prospects here on Earth?

We struggle to pay for and provide food, water, and heath care for seven billion inhabitants; what justification can we make for an investment of billions of dollars to go elsewhere?

Certainly the space endeavor provides much knowledge that will help mitigate the challenges of Earth, and this has been one of the arguments from the very beginning for its large investment. Though the ROI is sketchy, there are clear indications of advantage. The increasingly sophisticated Earth observation from space may illuminate and help devise response strategies for the climate challenge, and may help address resource issues, as well as to optimize agriculture, aquaculture, oceanography and other sciences that will be crucial to the response to future environmental, health, and climate change hardships.

As we saw also in Chapters 5, 12, and 13 (The Space Data Association, The European-African Partnership, and European SatCom Policy) satellite communications offer significant potential to increase efficiencies in all forms of economic and social activity.

Resources of commercial importance may be forthcoming from other bodies such as the Moon or asteroids, and from orbiting plants and factories.

And profits from commercial space ventures may fund economic growth and development on Earth, the ever-welcome promise of those elusive "high paying jobs," which in other terms is the development of the workforce, another holy grail of economics.

Technology spin-offs from space science may well prove critical to future economies.

And there also may be political benefits to be gained should we succeed in engaging many nations in cooperative space ventures, helping to set a pattern of cooperation rather than confrontation. These of course are the main topics that have motivated the creation of this volume, and certainly our authors have shown many reasons why these are reasonable expectations. Every nation has reason to go to space, and when the benefits of cooperation far outweigh the costs (which we may assess according to the Capability Criterion, *à la* Chapter 15) collaborative efforts emerge entirely naturally.

Short of total economic collapse or total war, this is where we are headed. But what about war? Does increasing competition in space not set the stage for more conflict?

THE TRANSGLOBAL/GLOBAL ECONOMY

The character of military conflict, war, has also changed in our time, and this bears a brief exploration. For centuries, and perhaps millennia, wars were fought to pursue the imperial ambitions of the aggressors. Napoleon's successes, while temporary, epitomized the imperial mentality, and his war-fighting innovations soon became standard techniques among his adversaries. In fact, he stood on a hill at Waterloo and commented admiringly on British tactics. "Finally they're learning!" he announced to his aide de camp. Too well, it seems, as soon thereafter he was sailing off to St. Helena.[1]

Imperial ambition is rarely pursued in our world. Today's wars are not territorial, they are ideological, religious, and ethnic conflicts, and struggles for self-determination. Instead, the battleground between nations has firmly shifted to commerce, where the competition is very real, although the stakes and the modes of conflict are of course entirely different.

Hence, the likelihood of armed conflict between the US and China, which sixty years ago seemed nearly inevitable, now seems entirely unreasonable, even accounting for the uncertain status of Taiwan as a game

---

[1] Cronin, Vincent. *Napoleon.* Harpercollins, 1995.

piece trapped exactly between the two great superpowers. And why should this be? Because the two nations are so deeply intertwined economically that the conflict would be not only be ruinous in terms of the sheer destruction that today's massive weapons would cause on both sides, but because the commercial consequences would be perhaps even more devastating. China possesses trillions of US dollars and is rapidly acquiring more through its persistent trade surplus. That capital is in turn being used to remake the Chinese nation, largely paying for the infrastructure and institutions of its transformation into a modern society.

China's trade with the US and the rest of the world is creating the new jobs that China desperately needs, as it transitions from a rural, agricultural nation to an urban, industrialized one. It is estimated that 25 million Chinese are moving from the countryside to the cities ... each year. That means 25 million new jobs must be created, and 25 million new housing opportunities, and all of the electricity, water, sewage, and transportation infrastructure that 25 million (more) people require.

As a matter of scale, this is roughly equivalent to creating three Parises, or three Chicagos, or three Sydneys each year. If you have wondered at the profusion of construction cranes blanketing the skylines of Shanghai, Guangzhou, Shenzen, Bejing, and dozens of other cities, this is why.

The force of Chinese urbanization seems to be inevitable, unstoppable. To a great extent, it seems even that the longevity of the Chinese government is dependent on its ability to sustain, support, and enable this tidal wave of human migration; should the country's leadership fail to deliver the jobs and the houses, it is widely thought that public support for the regime would diminish precariously. Hence, the present government's tenure is linked most closely to its capacity to manage and sustain a rapidly expanding economy. Success, in other words, is being measured in terms of trade and economics as it occurs between nations, as the growth and acquisition of commercial assets, rather than domination or control of new territories or a larger population.

This phenomenon is widely referred to as "globalization," and the idea of a "flat world," as described by contemporary authors including Thomas Friedman is now widely accepted.[2] Yes, the current domain of competition is the global marketplace.

The qualities and characteristics that determine success in this

---

[2] Friedman, Thomas. *The World Is Flat: A Brief History of the 21ˢᵗ Century*. Farrar, Straus and Giroux, 2005.

marketplace are also quite clear. They are factors including science, technology, the capacity to manage complexity, and critical urban infrastructure, including transportation and utilities. Underlying these are more intangible factors that are also essential, including the information and communications infrastructures, education systems, learning accomplishments, knowledge, and the ever-elusive attribute, innovativeness. We recognize the items on this list as defining characteristics of civilization, and it seems that they will continue to be just as essential in the future.

So what, you may be wondering, does any of this have to do with space, or with international cooperation for its development? The answers are two-fold. First, it is not an exaggeration to say that commerce as it is now practiced is dependent on space. Satellite-based communication, information, observation, and location systems enable global commerce, and have contributed to the expansion of its scale, speed, and scope. Indeed, increasing the speed of commerce is critical to enduring success, as Tom Diegelman points out in Chapter 6.

And as we have seen repeatedly in the preceding chapters, every nation has or wants access to such space-based systems because they are incontrovertable: if you want to play in the global economy, and for good reason everyone does, then you simply have to have them, or least have access to them.

Hence, science, technology, education, competence, knowledge, and innovation are the driving forces underlying success in a game that continues to change due to abundant innovation and expanding technology, all driven by scientific insights and breakthroughs. And space is central to all of this.

The second reason that space is a critical part of this dialog is that, as we have explored in all of the books in this series, the development of the critical capabilities needed for space flight and space habitation turn out to be equally central to the innovative breakthroughs that drive terrestrial commerce; what we learn in space, and in the drive to get there, we apply to our lives on Earth, and to our businesses.

The transfer rate is accelerating, and as more and more nations develop their capabilities in space, the innovation rate is accelerating also. Hence, the ironic truth is that rate of change of the global, Earth-bound economy is significantly influenced by the quest to venture beyond Earth. In this environment, the danger for any given nation is that falling behind implies the risk of falling farther behind, becoming less competitive, less capable, and thus risking in effect future economic growth. Such a downward spiral is every economist's worst nightmare, and in their briefings to heads of state the dangers of this spiral are reiterated and reinforced: we *must* invest in education, in science, in technology, or we are simply doomed.

This brings also an additional dimension of urgency to the capability criterion suggested by Christopher Stone and Brent Ziarnick in Chapter 15. As military officers, perhaps their first thought concerning this simple but powerful assessment tool was related to national defense, but in fact a successful economy is one critical element of any nation's defenses, which Tom Diegelman and Tom Duncavage explore so cogently in Chapter 17.

So we can take away from a discussion of globalization the awareness that everyone wants to participate in the new economy beyond globalization, the Transglobal/Global Economy, but only because everyone needs to. (In Diegelman and Duncavage's words, "terrestrial going celestial..." – see p. 393.)

The costs, however, are intimidating. Hence, the virtue of partnership has two dimensions: sharing competence accelerates learning, which is critical to survival, and sharing cost reduces the burden on any one partner, which is critical to balancing the need to invest in space capability to assure long term success while still meeting immediate financial obligations for food, energy, health care, etc. The five partnering patterns that Michael Simpson identifies in Chapter 2 explain a great deal about the various strategies that are now being employed to get in the game. Similarly, in each of the chapters in Part II we read a great many specific instances of partnering for exactly the same reasons.

International cooperation, in other words, is now indispensible to the economic process, and economic success is central to the survival and success of every nation.

Consider, for example, the cooperative institutions that have grown up around global financial system. The list of global bodies that are actively engaged in setting standards, enforcing regulations, and providing technical assistance to nations and institutions includes the International Monetary Fund, the World Bank and regional development banks, the Bank for International Settlements, the Basel Committee on Banking Supervision, the Financial Stability Forum, the International Organization of Securities Commissions, the International Association of Insurance Supervisors, the International Accounting Standards Board.[3]

None of these institutions would be at all necessary if commerce

---

[3] Wolf, Martin. *Fixing Global Finance.* The Johns Hopkins University Press, 2008. P 184.

were strictly a national affair, but of course it's exactly the opposite. And as sustaining each of these institutions is a costly proposition, you can be sure that all of them do serve an important purpose or their funding would have already been cut off.

And what, we might ask, about the nations that are sitting slightly outside of the global community, the ones sometimes referred to as "rogue," such as North Korea and Iran? Bruce McCandless commented on this recently as follows:

> *I wonder if the "pull" of recognition and status derived from participation in international endeavors might not be a powerful tool. I recognize our national concern regarding technology transfer, but all nations are headed into space (eventually) whether we like it or not. While there are areas of commonality between weapons delivery and civil space programs there are also significant differences. Regarding reentry, the civil application strives to decelerate a "capsule" with moderate, limited, g-levels so as to transition to a parachute / soft-landing device over water or a generally clear area. The weapons delivery system tries to maintain maximum velocity right down to a detonation point at low altitude over a populated area. Civil launch vehicles using cryogenic propellants are not really the sort of thing that weapons system operators are looking for. Recall the old saw: "Keep your friends close, and your enemies even closer!"*[4]

The significant pull that Bruce refers to may also include joining the marketplace of nations to achieve fuller participation in global economic life. North Korean leader Kim Jong Un has recently made a point of emphasizing the importance of economic development to his nation, apparently shifting the focus of his government beyond the military issues that characterized his father's rule.[5] Similarly, the Western response to Iranian aggression has included significant economic and commercial sanctions, intended to make life in that country more difficult, and thereby to exert pressure on Iranian leaders to change their policy. It may or may not, however, be having an impact. Associated Press reports that, "Iran said August 4, 2012 that it had successfully test-fired an upgraded version

---

[4] McCandless II, Bruce. Private correspondence with the authors.
[5] Lee, Jean H., "North Korea's new leader makes diplomatic debut." *Bloomberg BusinessWeek*, August 3, 2012.

of a short-range ballistic missile with improved accuracy, increasing its capacity to strike land and naval targets."[6]

The following day came another report noting that, "Israel has upgraded its top-tier Arrow II missile defense, a Defense Ministry official confirmed on August 5, 2012, as the country girds for possible attacks from Iran and Syria. Sensors, command and control equipment and radar have been enhanced to improve reach and accuracy, the official confirmed without elaborating. Israel has developed a network of air defense systems to parry various threats it sees from its enemies, including the Arrow, a joint project with the Boeing Co. in the US that is designed to shoot down incoming missiles launched as far away as Iran."[7] We are not yet free of armed conflict, and space inevitably plays a role in space defense, and probably offensive capabilities, among many nations. Indeed it is widely reported that US military spending on space has exceeded NASA's budget in every year since 1982.[8]

Hence, the military will accompany, and often lead into space, but the role for civilian and commercial life is likewise unavoidable.

So let us now recap the discussion of the last few pages:

- International commerce is vital to every nation, and such commerce is essentially a form of competition that is played out within a highly collaborative framework.
- It is a competition driven by knowledge, and characterized by increasing speed and massive complexity.
- In this realm space plays a critical role, and a role that will only become more important as times passes.
- Further, the factors that lead to success in commerce are the same as those that lead to success in cooperative space endeavors, including rigor, mutual benefit, trust, and innovativeness.

And yet as we look at the space activities of many nations, we get a sense of fragmentation. As James Rendleman and Walter Faulconer point out in Chapter 16, due to various forms of international conflict and constraint, some countries are replicating knowledge that was already

---

[6] Associated Press. "Iran Says Test-Firing of Missile a Success." *The New York Times*, August 4, 2012.
[7] Associated Press. "Israel upgrades Arrow missile defense system." *Washington Post*, August 5, 2012.
[8] *The Economist* Magazine, "Spooks in orbit." July 2, 2011. p 68.

created by others, and needlessly wasting human capital in the process. So what we need, perhaps, is an overriding goal or ambition that can engage the best of human thinking and ingenuity, and which can inspire the faith and hope in the human capacity to transcend our current difficulties and challenges not as one nation's project, but a multi-national, all-nations one. By using space as a forum in which cooperation may lead to the development of new knowledge, skills, capabilities, and opportunities not for a nation, but for all nations and all peoples, we look forward, then, to defining a cooperative endeavor that may transcend nationalist politics to define a fitting next target for human aspiration in space.

THE MISSING PIECE

We would describe such a goal as a cohesive, driving force, an aspirational target that provides transcendent and entirely obvious benefits to humanity, an ambition or initiative that can engage the imaginations of vast numbers of people from every nation – millions of us – and that can inspire youth in vast numbers to choose the pursuit of space as their career. Such a goal would transcend nationalism and regionalism and speak instead to our deepest shared human goals and aspirations individually and collectively as a species. A goal that evokes friendliness, and mutual respect among the billions of neighbors on this small planet who understand that their fate and your fate and her fate and everyone's fate is intimately linked, that the bell tolls for each of us when it tolls for any of us. That we're in this together, and it can be for better; it does not have to be for worse.

That transcendent, unifying force is as yet absent from the dialog. Where may we find it?

Perhaps it is the 100 Year Starship Study, realized as an actual starship. Or if the Chinese space program sustains its current rate of progress, then the next moon landing with people aboard will likely be a Chinese crew, surely a magnificent accomplishment. Will that provoke a chorus of admiration, or a wave inferiority to sweep over the US and Europe, as the Sputnik satellite and Gagarin orbital flight did in the US? We cannot predict, nor is it clear what degree of prestige will accrue to China as a result; certainly there will be significant praise among people worldwide, but how much? Or conversely, will the accomplishment be viewed as the rerun of a show we've already seen 50 years previously? Again, it's impossible to predict.

Based on the uncertainty of the psychological and social impact, we put a manned Moon landing aside as the next great motivator; likewise an asteroid landing, or even a Mars landing. It is genuinely strange to be considering such monumental achievements in science, technology and courage, and yet to wonder if the broader public will notice and appreciate them on a mass scale; has history so deeply skewed our collective expectations? It is possible.

There is, however, another possibility that we believe holds interesting promise as a catalyst through which a deeper appreciation, a profound appreciation for the meaning and benefits of the development of space may be widely, universally shared. This is *a great world university in orbit.*

WORLD UNIVERSITY IN ORBIT

While it is not really implied by any of the authors herein, consider nonetheless the following scenario: Suppose that there was a university in Low Earth Orbit, a first-class research university, undergraduate and graduate, with thousands of students and faculty working diligently to address the greatest challenges of humanity from the unique vantage point of 300 – 500 miles above the Earth's surface. Suppose also that such a university were adjacent to a first class industrial park, an innovation park, a place where new space ventures are established as naturally as they are adjacent to every terrestrial world-class university. Connecting the two might be the space equivalent of Sand Hill Road, Silicon Valley's renowned venture capital boulevard, where the funds for a many start-ups is raised by those with the potential to achieve brilliant returns, for the great and perhaps transformative technologies and business models of the future.

Perhaps this is fantastical, but in the spirit of exploring this vision let's layer on another implausible dimension. Suppose that this university is not the property of, project of, or affiliate of any *one* nation; suppose instead that it exists as a consortium of *all* nations, or all that choose to participate, and that each nation sends its proportionate share of students and faculty to an orbiting facility, a college town, really, a global village that's not on the globe.

Would such an institution inspire hope among the beleaguered mass of the Earth-bound? Could such an institution repay its cost in the form of social, technical, and economic benefit to its many investor-nations? Could it be the jumping-off point for further off-world exploration and development?

It certainly could all achieve of these things.

The idea of a university in space isn't particularly original. In our previous volume, *Space Commerce*, both Frank White and Michael Simpson examined some of the possibilities for a university in space, and their observations remain pertinent now. Michael wrote,

> "If orbital manufacturing, asteroid mining, space tourism, lunar habitation and other such visions have any chance of commercial success they, too, will soon need a very specially prepared group of people to attend to the on-site tasks needed to achieve success. Beyond this classic vision of a need for an experienced, private space cadre, a combination of cosmic merchant marine and space-proven technical representatives, if you will, there is an even more pressing need. It emerges from the natural consequences of any or all of these early commercial visions for activity beyond earth being truly successful.
>
> "The implication of success is that human settlement will follow. It has always happened that way in the past, and there is no reason to believe that it will be otherwise this time. In the face of that expected reality we need a lot more than just ship drivers and equipment experts. We will need space-proven expertise in every field essential to the infrastructure of a human community. Hence, we will need education in space because after years of sending the occasional human to orbit, we are on the threshold of sending human society to space. Groups of humans can go anywhere without educational institutions, but human society exists only where there is also education. At ISU we see this as rooted in our mission."[9]

What we are here suggesting may be a bit beyond ISU's charter, but perhaps not. In any case our point is not whether this should be ISU, but that a global university in orbit could represent something significant, and it could achieve the extraordinary. With effective links and partnerships with the great universities of each nation and each continent, with joint research on the entire range of topics which a space community may legitimately address, which happens to be any and every topic of interest to science, such a project would represent an investment by humanity in the future of humanity, an investment in an intelligent and forward-looking way, one designed to achieve returns across many dimensions.

---

[9] Simpson, Michael. "To Plan for a Century: ISU's Vision of of Education in Space." *Space Commerce: The Inside Story by the People Who Are Making it Happen.* Aerospace Technology Working Group, Morris & Cox, Editors, 2010. p 367.

The ISS is a model, and one of its great accomplishments has been to establish a broader base of tangible cooperation among the nations that have contributed time, talent and and/or treasure to it. The proposed university is then the next step, a way to prepare entire new generations of space professionals in the fertile environment of cooperation, and not only in theoretical or simulated environments, but on a real orbiting platform/campus.

Getting such a campus established in space is no small undertaking, orders of magnitude larger than the ISS, and the cost and complexity cannot be underestimated; this would be a cathedral-like project, a construction site for generations. Nevertheless, the construction of such a university might not be as massive an undertaking as was the ISS. Inflatable module technologies have advanced considerably, and such techniques could create many of the physical facilities needed for building an educational institution in space. And perhaps the university would be located within "shuttle" distance of the ISS, enabling it to make use of ISS science facilities from time to time, while also providing docking and rendezvous training.

So let's take our vision a step further then, and imagine that a world university in orbit is up and operational. You, then, happen to be a student, bright and engaged, ready to go off to college yourself. Would you be interested in spending a year or two in orbit? Perhaps this is a tuition-free university, the cost paid by your home country, an investment in your future and its own future. Would such an educational opportunity be of interest to you now? It seems that it would indeed be a very attractive option for a lot of young people, and the potential spin-offs, social, commercial, economic, and scientific, seem unlimited.

If you were the parent of a qualified student, how would you feel then? Would you want your daughter or son to spend some months in orbit, learning about the latest technology, conducting cutting edge research, living and working and learning side by side with people from every continent, every culture, every nation? Fears aside, it seems hard to imagine a better learning situation, or a better opportunity to assure the foundation of your child's future success and contribution.

Perhaps there could be other, equally or more compelling space-based projects for humanity to engage in, but in any case this one does seem to be a worthwhile endeavor, a powerful means to harness the willingness of people and their nations to search for and engage in cooperative endeavors that hold vast promise for improving life on Earth, and extending human presence far beyond Earth as a matter of survival, and as a means of thriving as a species and a civilization.

One of the attractions that the adventure into space has always offered is the respite it gives from the daily crisis. It puts instead our attention on our greatest accomplishments and aspirations, on the long term future, beyond the bounds of Earth and the vicissitudes our troubled times.

We need such a respite now perhaps more than ever. It is time for a unified human endeavor to achieve a permanent presence, in space in the form of a noble on-world and off-world university, open to students of all nations, which aspires to address the grandest challenges that humanity faces, to use the collaborative acquisition of knowledge as the best way forward. The student body, multicultural generations of new leaders trained in space and nurtured by the overview effect[10] could be a wonderful contribution to human development. Certainly it would be a fitting monument to cooperation.

KNOWLEDGE, CAPABILITY, AND WILLINGNESS

Humanity's progress in space is a function of our *knowledge*, which must then be applied as *capability*, and our *willingness* to pay for such knowledge and its application. A world university in orbit represents an intentional transformation of the pattern underlying all of human civilization, the progressive acquisition and application of knowledge, into an intentional path forward for the benefit of not an elite few of a single nation or region or investor class, but inclusively and for the enduring benefit of all who live today and who will live tomorrow and beyond. It is an investment in the future of humanity.

International cooperation for the development of space should indeed aspire to such an accomplishment; no lesser goal is worthy of the challenge, the cost, or the effort. And no lesser goal is likely to achieve a result as enduring or significant to the future of the human species, or to the fate of one and all, or of our beloved home planet.

•••

---

[10] White, Frank. *The Overview Effect: Space Exploration and Human Evolution*, Houghton Mifflin, Boston, 1987 (1st edition). Second edition published by AIAA in 1998.

LANGDON MORRIS

Langdon Morris is a Director of PwC, and a leader in its Growth & Innovation practice. Formerly a partner of InnovationLabs (www.innovationlabs.com), he is recognized globally as a leader in the field of innovation. His recent clients include organizations such as NASA, GE, Gemalto, Total Oil, the Federal Reserve Bank of the US, Johnson & Johnson, Tata Group, France Telecom, Stanford University, Wipro, L'Oréal, Accor Hotels, and many others.

He is a member of the leadership team of ATWG, and is formerly Senior Practice Scholar of the Ackoff Center at the University of Pennsylvania. He is a Senior Fellow of the Economic Opportunities Program of the Aspen Institute, Associate Editor of the International Journal of Innovation Science, and a member of the Scientific Committee of Business Digest, Paris. He has taught MBA courses in strategy at the Ecole Nationale des Ponts et Chaussées in Paris and Universidad de Belgrano in Buenos Aires.

He is the author and co-author of numerous white papers and five highly acclaimed business books, with editions in Japanese, Chinese, Korean, and French. He is highly sought after as a speaker and workshop leader who participates frequently at conferences and workshops worldwide.

KENNETH J. COX, PH.D.

Dr. Kenneth J. Cox earned his bachelor's degree in 1953 and his master's degree in 1956 in electrical engineering from the University of Texas/Austin. He earned his PhD at Rice University in 1966.

From 1960-1962 Ken led the first evaluation of digital flight system applications for the Martin Company. In 1963 he joined NASA to develop the flight control system for the Little Joe II Booster Vehicle. Later, Dr. Cox became the Technical Manager for the Apollo Digital Control Systems, which included the Lunar Module, the Command Module and the Command/Service Module, the first spacecraft to fly with a digital flight control system. He later was Chief Technologist, a member of the NASA Institute of Advanced Concepts, Technical Manager of the Apollo-Soyuz Androgynous Docking Demonstration, Space Shuttle Technical Manager for Guidance, Navigation, and Control, and Division Chief of the Avonics System Division.

His awards include the AIAA Mechanics and Control Flight Award in 1971, the NASA Medal for Exceptional Engineering and Achievement in 1981, the AIAA Digital Avionics Award in 1986, and the NSS Space Pioneer Award in 2007.

Ken is coeditor of this book series, and coauthor of Chapter 1. He has been the leader of ATWG since it was established in 1990.

INDEX

1

100 Stories about Docking, 324
100 Year Starship Study, XIII, 425, 468

2

2001: A Space Odyssey, 4
2009 Defence White Paper, Australia, 184, 202

A

Abrams, Nancy Ellen, 426
Abuja Treaty, 270
Aerospace Technology Working Group, I, V, 498
Afghanistan, 323
Africa, II, 6, 88, 127, 138, 161, 162, 194, 204, 261, 262, 263, 264, 265, 266, 267, 268, 269, 270, 271, 272, 273, 274, 275, 276, 277, 278, 279, 280, 281, 283, 284, 285, 286, 287, 288, 292, 293, 294, 295, 296, 297, 298, 299, 300, 301, 302, 303, 304, 305, 306, 307, 308, 309, 310, 353, 400, 465
AfricaConnect, 299
Africa-EU Summit, 262
African Development Bank (ADB), 263, 267
African Monitoring of the Environment for Sustainable Development (AMESD), 264
African Union Commission, 268
Agency for Aerial Navigation Safety in Africa and Madagascar (ASECNA), 272
Agency for International Development, 9
agriculture, 4
Al-Saoud, Sultan bin Salman, IX
Alaska, 107
Albania, 320
Alcântara Launch Center, Brazil, 414
Alcatel Alenia Space, 367
Algeria, 270, 273
Alice Springs, Australia, 167
American Interplanetary Society, 346
Analytical Graphics Corporation, 37
Andoya Rocket Range, 36
Aouda.X Mars Space Suit Simulator, 10
Apollo Program, XV, 346, 396, 483
Apollo Raw and Uncut, 127
Apollo-Soyuz Test Project, 47, 314, 325
Arabsat, 307
Argentina, 415
Ariane, 36, 39
Ariane V, 378
Arrow II, 477
Association of Space Explorers (ASE), VIII, 49

Asia Pacific Space Centre (APSC), 178
Asia-Pacific Multilateral Cooperation in Space Technology and Applications, 248
Asia-Pacific Regional Space Agency Forum, 249
Asia-Pacific Space Cooperation Organization, 248
Aster Global Digital Elevation Map, 8
asteroid mining, 13, 150
asthma, 7
Asthmapolis, 7
Atlantis Space Shuttle, 8
Atoms for Peace, 102
Aerospace Technology Working Group (ATWG), V, 14
AUSMIN, 202
AUSSAT, 178, 185
Australia, II, VI, XII, 6, 10, 20, 22, 23, 24, 38, 54, 69, 137, 161, 165, 166, 167, 168, 169, 170, 171, 172, 173, 174, 175, 176, 177, 178, 179, 180, 181, 182, 183, 184, 185, 186, 187, 188, 189, 190, 191, 192, 194, 195, 196, 197, 198, 199, 200, 201, 202, 203, 205, 206, 207, 208, 249, 315, 341, 464
Australian Government Space Engagement: Policy Framework and Overview, 180
Australian Space Activities Act, No 123 of 1998, 178
Austria, 13, 314
Austrian Space Forum, 10
Automated Transfer Vehicle (ATV), 153

B

Backhauling, 291
BAIE (Barcelona Aeronàutica i de l'Espai [Barcelona Aeronautics and Space Association]), 34
Baikonur Cosmodrome, 319, 332
Bangladesh, 254
Bank for International Settlements, 475
Barcelona, 34
Basel Committee on Banking Supervision, 475

Basic Law for Space Activities, Japan, 247
Beazley, Kim, 174
Belakovskiy, Dr. Mark S., 313, 334, 342
Beyond Earth, VI
Biddington, Brett, II, XII, XIII, 165, 207, 464
Netanyahu, Binyamin, 420
Bigelow, Robert, 114
Bion satellite, 313, 326
Bismarck, Wilhelm, 403
Blagonravov-Dryden negotiations, 324
Block, Fred, 10
BlueStream space applications, 30
Boeing Space and Intelligence Systems, 9
Bolden, Charles, XIII, 365
Brand, Vance, 325, 340
Branson, Richard, 11, 114
Brazil, III, 6, 137, 229, 335, 413
British Interplanetary Society, 139
British National Space Centre (BNSC), 273
Buffet, Warren, 114
Bulgaria, 320
Burning Chrome, 146
Bush, US President George W., 326
business model, 289, 308, 479

C

C Band satellites, 307
Canada, 20, 137, 180, 334, 430
Canadarm, 21
Canadian Space Agency, 21
Capability Criterion, II, XI, 345, 348, 349, 350, 351, 353, 354, 355, 356, 358, 359, 360, 361, 467, 472, 475
Cape Verde, 269
Cape York, Australia, 177
CARDIOME, 336
Cassini-Huygens spacecraft, 5, 10, 19, 386
Coleman, Catherine, 130
Centario, 335
Central Committee of the Soviet Communist Party, 315
Chandrayaan-1 spacecraft, 19, 416
Chapman, Senator Grant, 181
Chesson, Bob, 129

China, II, XII, XIII, 6, 8, 11, 13, 23, 24, 39, 50, 51, 52, 55, 69, 145, 161, 184, 187, 195, 199, 201, 204, 205, 206, 209, 210, 211, 212, 213, 214, 215, 216, 217, 218, 219, 220, 221, 222, 223, 224, 225, 226, 227, 228, 229, 230, 231, 232, 233, 234, 235, 236, 237, 238, 239, 240, 241, 243, 244, 248, 249, 250, 251, 252, 253, 254, 255, 256, 257, 258, 297, 320, 334, 353, 355, 356, 358, 371, 372, 374, 377, 381, 383, 384, 385, 388, 396, 414, 417, 464, 465, 472, 473, 478
China Aerospace Science and Technology Corporation, 251
China National Space Agency (CNSA), 251
China's National Space Administration (CNSA), 211
China-Brazil Earth Resources Satellite 2 (CBERS-2), 254
Chinese Academy of Sciences (CAS), 233
Chinitz, Benjamin, 108
Christmas Island, 22, 178
Churchill, Winston, 401
Clancy, Paul, 29
Clarke, Arthur C., 339
ClearSpaceOne, 9
climate change, 4, 263, 288, 471
Climate Information for Development in Africa Programme (CLIMDEV Africa), 264
Clinton, US President William, 47
Clinton, US Secretary of State Hillary, 51
Cold War, 45, 50, 55, 168, 171, 244, 245, 314, 323, 347, 395, 431, 466, 469
Colorado Space Coalition, 32
Columbia Space Shuttle, 387, 419
Columbus module of ISS, 338
Commerce Department, US, XII, 358, 381
Committee on Science and Technology Industry for National Defence (COSTIND), 251
Conference on the Peaceful Uses of Outer Space (COPUOS), 171, 203, 226, 315
connectivity, 291

Constellation Exploration Program, 93, 120, 404, 411
containerization, 106
convergence, 289, 308
Cook, Captain James, 194
Corps of Engineers, US, 99
Cosmonautics Day, 122
COSPAR, 53
Council of Scientific and Industrial Research (CSIR), 273
Cowan-Sharp, Jessy, 145
Cox, Donald, 347, 357, 381
Cox, Kenneth, III, VI, I, III, 15, 90, 98, 461, 480, 483
Crick, Francis, 443
CTAE (Centre de Tecnologia Aeroespacial), Spain, 34
Cuba, 320
Cuban Missile Crisis, 50
Curiosity spacecraft, 12
Cutty Sark, 91
cyberlaw, 453
cyberpersona, 438
cyberspace, 144, 159
Czech Republic, 25
Czechoslovakia, 320

D

Dachstein, Austria, 10
Dahl, Eric J., 403
DARPA, XIII, 426
Deep Space Tracking Station, 175
De Gaulle, President Charles, 319
Denmark, 29, 107
deoxyribonucleic acid structure (DNA), 443, 445
Department of Defense, US, 9, 32, 419
Department of Industry, Innovation, Science, and Tertiary Education (DIISTE), Australia, 199
Department of State, US, 251
Deutsche Bank, 400
Development Cooperation Instrument (DCI), 263
Dibb, Paul, 173
Diegelman, Thomas, II, 89, 120, 393, 411, 463, 467, 468
Digital Agenda, 287, 290, 292, 301, 309
Digital Asia Research Center, 249

digital divide, 299
Digital Earth Scientific Platform, 234
Disaster Monitoring Constellation (DMC), 24
Dordain, Jean-Jacques, 367, 388
Dragon capsule, 10
Duncavage, Thomas, III, 393, 411, 467, 468
Dutch East India Company, 430

E

Earth, IV, V, IX, X, XII, XIII, 4, 5, 7, 8, 9, 10, 11, 12, 13, 21, 22, 24, 26, 30, 31, 32, 44, 50, 53, 54, 57, 58, 59, 61, 68, 70, 72, 74, 75, 77, 78, 86, 90, 110, 112, 113, 116, 117, 118, 126, 128, 129, 130, 132, 133, 137, 138, 140, 142, 144, 148, 150, 164, 166, 169, 181, 182, 183, 186, 187, 188, 189, 191, 192, 196, 197, 198, 200, 204, 205, 207, 208, 215, 216, 217, 219, 223, 224, 225, 227, 228, 231, 234, 249, 252, 254, 265, 272, 273, 293, 319, 320, 322, 323, 325, 340, 341, 346, 347, 358, 360, 365, 371, 385, 391, 396, 397, 404, 408, 414, 416, 418, 419, 421, 426, 430, 434, 437, 439, 440, 441, 442, 444, 445, 446, 447, 449, 450, 451, 452, 453, 454, 456, 457, 458, 462, 463, 464, 466, 468, 470, 471, 472, 474, 479, 481, 482
EASSy, 299
East India Company, 74
École Polytechnique Fédérale de Lausanne, 9
education, V, XV, 480
Egypt, 181, 270
Eisenhower, US President Dwight, 102, 323
Emma Maersk cargo ship, 91
Engdahl, F. William, 400
European Space Agency (ESA), IX, 5, 19, 21, 24, 25, 29, 30, 31, 34, 35, 36, 39, 45, 47, 53, 56, 86, 124, 129, 130, 131, 132, 134, 139, 142, 150, 161, 163, 170, 175, 195, 197, 198, 223, 224, 225, 226, 227, 228, 229, 233, 239, 245, 249, 271, 272, 280, 284, 286, 299, 313, 326, 327, 328, 329, 330, 332, 333, 334, 336, 338, 340, 341, 342, 343, 350, 366, 367, 368, 374, 378, 379, 382, 386, 387, 388, 415
Estonia, 29
EUMETSAT, 29, 264, 271, 280
EUROLAB, 336
Europe, II, IX, XII, 6, 29, 31, 35, 41, 50, 54, 88, 137, 152, 166, 183, 187, 201, 216, 233, 239, 244, 262, 267, 271, 273, 287, 288, 290, 292, 293, 294, 295, 296, 299, 300, 301, 302, 303, 304, 305, 308, 322, 327, 333, 334, 335, 338, 339, 345, 349, 351, 352, 353, 354, 355, 357, 365, 366, 367, 374, 378, 379, 381, 388, 466, 478
Europe 2020, 299
European Astronaut Centre, 153
European Development Fund (EDF), 263
European Launcher Development Organisation (ELDO), 170
European Neighbourhood Policy Instrument (ENPI), 263
European Organization for the Exploitation of Meteorological Satellites (EUMETSAT), 29, 264, 271, 280
European Physical Society Journal, 8
European Space Council, 288
European Union, 263, 351
Eutelsat, 307
Everest (Mount), 8
Ewald, Reinhold, I, IX
exceptionalism, 383
exoplanet HD 189733b, 12
Extravehicular Activity (EVA), 154

F

Facebook, 138
Falcon rocket, 9, 10
Farrands, John, 172
Faulconer, J. Walter, III, XIII, 363, 391, 467, 477
Federal Aviation Administration, US (FAA), V, 32
Ferguson, Chris, 7
Financial Stability Forum, 475
Finland, 29

First Action Plan 2008 – 2010, 298
First Orbit, 121, 122, 463
Fisher, Admiral Lord, 402
Five-Year Program for National Economic and Social Development, 230
forestry, 4
Foton satellite, 313, 326
Fractional Orbital Bombardment System (FOBS, 70
France, VI, 15, 56, 227, 273, 314, 318, 334, 383, 421, 483
French Guinea, 170
Friedman, Lou, 359, 429
Futron Corporation, 31, 164
Future Shock, 471

G

Gabon, 267
Gagarin, Yuri, 11, 50, 121, 323, 463
Gagarin Cosmonaut Training Centre (GCTC), 338
Galileo spacecraft, 264, 271, 385, 387
game theory, 394
Gass, Volker, 9
GEANT, 299
Gemini spacecraft, 15, 483
Geoscience Australia, 191
Geospace Double Star Exploration Program (DSP), 223
Geostationary Navigation Overlay Service (EGNOS), 264
Geosynchronous Satellite Launch Vehicle (GSLV), 416
Germany, 29, 334, 383, 400
Giannopapa, Christina, II, 261, 286, 465
Gibson, William, 146
Gifford, Congresswoman Gabriella, 131
Gini coefficient, 295
Glavkosmos, 317
Glenn, Senator John, 323
Global Exploration Strategy, 347
Global Exploration Strategy and Roadmap, 361
Global Monitoring for Environment and Security (GMES), 264, 271
Global Navigation Satellite System (GNSS), 187

Global Positioning System (GPS), 7, 186, 352, 428
globalization, 308
Google, 8, 13, 140
Gorbachev, Soviet President Mikhail, 326, 331, 327
Goss Gilroy Inc., 21
Great Wall Corporation, 24
Great Wall of China, 8
Gross Domestic Product (GDP), 94
Gubarev, A., 322
Gullish, Jay, 31

H

Haavelmo, Trygve, 291
Haipeng, Jing, 11
Hamilton, Paul, 306, 307
Hayabusa spacecraft, 19
Heavenly Ambitions: America's Quest to Dominate Space, 354
heliosphere, 11
Hershkowitz, Daniel, 419
homo sapiens, 443
homo sapiens sapiens, 455
House of Representatives, US, 10
Hu Jintao, Chinese President, 214, 371
Hudson's Bay Company, 74, 430
Human Behavior and Performance (HBP), 151
human evolution, 4, 119, 440, 441, 443, 444, 445, 447, 453, 458, 469
human genome project, 444
Human Research Roadmap, 150
Hungary, 320
Hurley, Doug, 8
hydrology, 4

I

Icarus Pathfinder spacecraft, 428
India, III, 6, 12, 137, 250, 369, 371, 383
Indian Science Congress, 416
Indian Space Research Organization, 12
Indonesia, 250
Information Gathering Satellite (IGS), 246
Infrastructure Trust Fund, 299
Ingold, Oliver, 55, 57

innovation, 15, 483
Institute of Medical and Biological Problems (IMBP), 313, 317, 334
Institute of Space Biology and Medicine, 317
Institute of Space research, 317
Intelsat, 307
Intercosmos-1, 322
International Accounting Standards Board, 475
International Association of Insurance Supervisors, 475
International Institute of Space Commerce (IISC), VII, 14, 43, 56, 498
International Monetary Fund, 261, 475
International Organization of Securities Commissions, 475
International Space Advisory Group (ISAG), 179
International Space Station (ISS), II, VIII, 5, 7, 10, 19, 21, 43, 44, 45, 46, 52, 55, 75, 106, 109, 115, 125, 126, 129, 139, 142, 144, 145, 150, 151, 152, 153, 157, 179, 198, 314, 326, 334, 336, 337, 340, 341, 347, 352, 360, 369, 372, 376, 384, 413, 414, 420, 431, 443, 456
International Space University, III, VI, VII, XI, 14, 22, 23, 24, 28, 41, 43, 54, 56, 88, 121, 123, 141, 142, 238, 311, 339, 341, 435, 498
International Standards Organization, ISO, 114
International Symposium on Personal and Commercial Spaceflight, 33
International Telecommunications Satellite Organization (INTELSAT), 365
Interregional Traffic Flow, 2012, 303
Interstate Commerce Commission (ICC), 107
Iran, 254, 476
Ireland, 30
Isle of Man, VII, 37, 81
International Organization for Standardization (ISO), 107
Israel, III, 6, 12, 383, 415, 418, 477
Israel Defense Forces, 421
Israeli Space Agency, 419
Italy, 273, 334

International Traffic in Arms Regulations (ITAR), XII, 74, 251, 254, 355, 356, 357, 358, 359, 379, 380, 381, 435

J

James Webb Space Telescope (JWST), 378
Japan, 6, 8, 15, 243, 334, 368, 383, 465, 483
Japanese Self-Defence Force (SDF), 245
JAXA, 195, 233, 249, 415
Jemison Foundation, 434
Jindalee Over the horizon Radar Network (JORN)., 169
Joel R. Primack, 426
Johannesburg World Summit on Sustainable Development (WSSD), 272
Johnson-Freese, Joan, 354
Johnson Space Center, 118, 120, 326, 411
Joint Defence Facility, 167
Joint World Space Observatory for Ultraviolet (WSO-UV), 229
Jupiter, 114

K

Kaspi, Yohai, 12
Kennedy Space Center, 33
Kennedy, US President John, 323, 358
Kenya, 267, 273
Khruschev, USSR President Nikita, 323
Kilimanjaro, 8
Kiruna, Spaceport Sweden, 35
Kondratyev, Dimitri, 130
Kongsberg Defense and Aerospace Company, 36
Korea, 23
Kosygin, USSR Premier Alexei, 320, 324
Kourou, French Guiana, 24, 36
Kozlov, Dmitry, 326
Ku Band satellites, 306

L

La Regina, Veronica, II, 287, 311, 465
Lagrange point, 116
Lasser, David, 346
Latvia, 25
Lauria, Rita, II, 143, 160, 464
Le Wang, II, XIII, 209, 238, 464
leadership, 15, 483
Leonov, Alexei, 48, 325
Liberal Democratic Party (LDP), 247
LIDAR, 21
Lithuania, 25, 27
Liu Wang, 11
Liu Yang, 11
Lockheed Martin, XV, 382
Long March rocket, 217, 254, 357
Loral, 357
Lux Research Corporation, 95

M

Madigan, Russel, 176
Magnus, Sandra, 8
Malaysia, 250, 253, 334
Malaccamax, 108
Mann, Adrian, 428
Maputo Declaration, 272
Mars, 12, 13, 54, 57, 103, 150
Mars Convention, I, 57
Matson, John, 419
Mawson Lakes, Australia, 23
McCandless, Bruce II, III, I, XI, XIV, XV, 476
Mediterranean Sea, 294
Menzies, Robert, 168
Mexico, 181
Mid-Atlantic Regional Spaceport, 33
Middle East, 309
military, 163
Millennium Development Goals, 266, 288, 297, 308, 310
Milton Bradley company, 395
Minister for Space and Space Development Strategy Headquarters, Japan, 247
Ministry of Industry and Information Technology (MIIT), China, 211
Ministry of Science & Technology (MOST), China, 211
Mir, 45, 326, 334
Mir Station, VIII
Mir-Shuttle Program, 314, 326, 334, 372
Mirra, Dr. Carlo, 313, 337
Mitchell, Edgar, I
Mongolia, 254, 320
Moon, 13, 39, 114, 132, 244, 478
Moon Treaty, 171
Moonwalk One – The Director's Cut, 127
Moore's Law, 438, 449
Morocco, 270
Morris, Langdon, I, 3, 15, 461, 483
MOSLAB, 328
Mozambique, 269
Musk, Elon, 114
Mwencha, Erastus Jarnalese Onkundi, 267

N

nano satellites, 421
nanotechnology, 416
Napoleon, 472
NASA, III, IV, V, XI, XV, 120, 287, 393, 411, 483
NASA Ames Research Center, 145
NASA Astronaut Office, 156
National Aeronautics and Space Act of 1958, 407
National Aerospace Plane (NASP), 93
National Broadband Network (NBN) initiative, 185
National Centre for Space Studies (CNES), France, 273
National Geographic Society, 431
National Meteorological Satellite Centre, Australia, 232
National Oceanographic and Atmospheric Administration (NOAA), US, 32, 86
Natural Law Theory, 439, 451
Navy, US, III, XI, XV, 411
Netherlands, 430
Neuromancer, 146
New Mexico Spaceport, 12, 33
New Universe and the Human Future: How a Shared Cosmology Could Transform the World, The, 426
New Zealand, 22, 134
NIGCOMSAT-1R, 24

Nigeria, 24, 229, 255, 270
NigeriaSat2, 24
NigeriaSat-X, 24
Nimmo, Francis, 11
Nixon, US President Richard, 324
Noordwijk, The Netherlands, 130
North American Air Defense Command (NORAD), 84, 180
NordicBaltSat, 26
North Korea, 246, 320, 476
Norway, 29, 35
Norwegian Ministry of Trade and Industry, 36
Nosanov, Jeffrey, III, 425, 435, 468
Nurrungar, Ground Station, Australia, 174

O

Obama, US President Barak, 10, 359, 371
oceanography, 4
Oil and the Origins of World War I, 400
O'Keefe, Sean, 382
Open Source Starship Alliance, 433
Optical, Geospatial, Radar and Elevation (OGRE), 196
Orbiting Mars Control Center for Exploration (MCCex), 117
Order of the Soviet Government N1388-618, 316
Organization of African Unity (OAU), 267
Orlon, 111
Orphans of Apollo, 435
Outer Space Treaty, XII, 54, 55, 57, 68, 69, 71, 73, 77, 78, 84, 166, 200, 201, 248, 370, 371, 375, 454, 456
overview effect, 482
Oxford Companion to American Military History, The, 428

P

Pacific Ocean, 127
Pakistan, 8, 253, 254
Pan-African Telecommunications Network, PANAFTEL, 303
Panama Canal, 98, 118
Panama Canal Authority, 99

Pearl Harbor, 399
Peeters, Walter, I, 43, 56, 462
Peres, Israeli President Shimon, 420
Peru, 254
Pettit, Donald R., 10
Plantary Resources Inc., 13
Planetary Society, The, 429
plate tectonics, 292
platinum, 13
Poland, 25, 320
Polar Satellite Launch Vehicle, 12, 416
Polvani, Lorenzo, 12
Pontes, Marcos Cesar, 335, 414
Port of Houston, 120, 411
Port of Houston Authority, 98, 118
Port of San Francisco, 108
posthuman, 453, 458, 469
Post-Panamax cargo ship, 94
Potter, Michael, III, 425, 435, 468

R

Ramon, Ilan, 419
Rascom, 307
REACH2020: Tele-reach for the Global South, 23
Reagan, US President Ronald, 47, 93
Remek, V. 322
Rendleman, James D., III, XIII, 363, 391, 467, 477
Riga Technical University, 28
Riley, Christopher, II, 121, 141, 463
Robinson, George, I, III, 67, 69, 72, 73, 74, 77, 79, 144, 437, 441, 444, 454, 459, 462, 468
Robonauts, 453
Rogers, Buck, 406
Romania, 25, 320
Roscosmos, 126, 227, 315, 331, 338
Rudd, Australian Prime Minister Kevin, 182, 203
Russia, II, XII, 6, 24, 45, 46, 47, 50, 69, 128, 134, 135, 137, 145, 161, 187, 204, 205, 225, 226, 227, 229, 233, 244, 249, 257, 293, 313, 314, 317, 318, 326, 327, 328, 329, 330, 331, 333, 334, 337, 338, 339, 341, 343, 350, 353, 354, 372, 373, 374, 377, 382, 383, 387, 388, 414, 415
Russian Federation, 313

Rwanda, 269

S

Sagan, Carl, 450
Salyut, 334
Samara Aerospace State University, Russia, 328, 332
Sand Hill Road, 479
Sanders, Stewart, I, 81, 87, 463
SatCom Policy, 287
Satellite Technology for the Asia-Pacific Region, 249
Saturn, 5, 10, 13
Savannah, NS, 102
Scandinavia, 292
Schaffer, A.M., 45
Schreiber, Michael, 8
Science and Technology Agency (STA), Japan, 245
Science without Borders, 415
Scientific American, 419
scientific divide, 299
Scully-Power, Dr. Paul, 179
SES company, 307
Shadow of the Moon, 127
Shenzhou 9, 11
Sheppard, Philip, 126
shipbuilding, 400
Showman, Adam, 12
Shukor, Sheikh Muszaphar, 336
Shuttle-Mir Program, 314, 326, 334, 372
Silanna company, Australia, 185
Simpson, Jim, 9
Simpson, Michael, I, XI, 19, 462, 475, 480
Singapore, 56
Singh, Indian President Manmohan, 371, 417
Singularity Principle, 438
SKYLAB, XV
Snow Crash, 146
Solà, Josep Comas, I, 35
Sonny Carter Training Facility, 113, 120, 411
South Africa, 270
South American Space Agency, 415
South Korea, 250, 334, 372, 383
Southern Cross Project, 415
Soviet Academy of Sciences, 315

Soyuz, 19, 46, 130, 325, 338
Space Act Agreements, 120, 145, 411
Space Commerce, III, XI, 4, 28, 38, 81, 88, 90, 93, 96, 98, 99, 101, 104, 109, 115, 120, 287, 311, 393, 411, 435, 470, 480
Space Data Association (SDA), 36, 39, 81, 463, 471
Space Debris Action Plan, 256
Space Development for Exclusively Peaceful Purposes, 245
Space Exploration Technologies Corporation (SpaceX), 10, 145
Space Florida, 32
Space Generation Forum 2000, 122
Space Industry Innovation Council, 192
Space Policy, 363
Space Policy Advisory Group (SPAG), 181
Space Science & Technology in China: A Roadmap to 2050, 233
Space Shuttle, IV, XV, 6, 7, 8, 15, 21, 33, 46, 105, 109, 125, 141, 326, 327, 335, 339, 369, 372, 373, 376, 377, 378, 382, 386, 387, 392, 483
Space Review, The, 359, 414
space tourism, 56, 480
spaceport, 89, 463
Spaceport America, 32
Spacepower: What it Means to You, 347
Spain, 34
Spiroscout, 7
Sputnik, 346
Square Kilometer Array, 22
Sriharikota Launch Site, India, 12
Stafford, Thomas, 48, 325
Star City, 49
Star Trek, 4
Star Wars, 4
State Department, US, XII, 358
Stephenson, Neal, 146
Stone, Captain Christopher M., II, XI, XII, 53, 345, 362, 417, 466, 475
Stott, Christopher, I, 81, 88, 463
Strategic Plan of the African Union, 262
Suez Canal, 118
supercomputing, 416
Surrey Satellite Technology Ltd. (SSTL), 24

Suzuki, Kazuto, II, XII, 259, 465
Sweden, 29, 314
Swiss Space Center, 9
Switzerland, 9, 13
Syromyatnikov, Vladimir, 324

T

Taepodong, 246
Tange, Arthur, 172
Tanzania, 269
Tel Aviv University Ultra Violet Experiment, 418
tele-education, 308
telegeography, 303
telemedicine, 308
terraforming, 54, 58
Thailand, 250, 253, 254
Thor-Agena rocket, 20
Three George Dam Project, 220
Tiangong, 11
Tiangong-1, 217, 252
Titan, 10
transhuman, 442, 458, 469
Trans-Siberian railroad, 118
Treacy, Noel, 31
Tripoli Declaration, 299
Tsiolkovsky, Konstantin, 340
Turkey, 254
twentieth century, X, 43
twenty-first century, X, 201, 462
Twitter, 137

U

Ukraine, 414
United Kingdom, 21, 24, 134, 168, 273
United Nations, 71, 280
United Nations Economic Commission for Africa (UNECA), 264, 267
United Nations Environment Programme (UNEP), 265
United Nations Food and Agriculture Organization (FAO, 269
United Nations General Assembly, 316
United Nations Outer Space Treaty of 1967, 57
United Nations Platform for Space-based Information for Disaster Management and Emergency Response (UN-SPIDER), 269
United States National Space Policy, 364
University of Bremen, 28
University of California at Santa Cruz, 11
University of California-Davis, 10
University of Latvia, 28
urbanization, 473
USSTRATCOM, 83

V

Vandenburg Air Force Base, 34
Variable Objects Monitor (SVOM), 229
Variable Specific Impulse Magnetoplasma Rocket (VASIMR), 103
Venezuela, 181, 229, 255
Venta-1, 28
Venus, 194
Verga, Antonio, 313, 342
Vietnam, 250, 321, 323
Viking Mars Lander, 35
Virgin Galactic, 11, 35
Virginia Commercial Spaceflight Authority, 32
Virginia Company, 74
virtual environments (VEs), 147
virtual reality (VR), 146, 464
virtual worlds (VWs), 145, 464
Volkov, Sergey Alexandrovish, 8
Von Braun, Wehrner, 92
Vostok-1, 127
Voyager spacecraft, 11

W

Walheim, Rex, 8
War & Peace in the Nuclear Age, 435
Watson, James, 443
weaponization, 257
Weizmann Institute of Science, 12
Welch, Christopher, II, 121, 141, 463
West Africa Cable System (WACS), 304
Wheelock, Douglas, 130
White, Frank, 480
Whitlam, Gough, 172

Whitson, Peggy, 156
Woomera Luanch Site, Australia, 22, 170
World Bank, 475
World Meteorological Organisation (WMO), 265, 269
World University in Orbit, 479

Y

Yellowstone National Park, 456

YouTube, 126
Yugoslavia, 321
Yuri's Night, II, 6, 121, 122, 123, 124, 125, 126, 135, 139, 140, 463
Yuriy, Alekseyev, 414

Z

Zhdanovich, Olga, II, 313, 466
Ziarnick, Captain Brent D., II, XI, XII, 345, 362, 466, 475

INTERNATIONAL COOPERATION FOR THE DEVELOPMENT OF SPACE

An Aerospace Technology Working Group Book

In Partnership with

The International Space University

and

The International Institute of Space Commerce

www.ingramcontent.com/pod-product-compliance
Lightning Source LLC
Chambersburg PA
CBHW020719180526
45163CB00001B/30